AF787524

Recent Advances in Antennas and Millimeter-Wave Applications for Mobile Communication Systems

Recent Advances in Antennas and Millimeter-Wave Applications for Mobile Communication Systems

Editors

Syed Muzahir Abbas
Yang Yang

MDPI

Basel • Beijing • Wuhan • Barcelona • Belgrade • Novi Sad • Cluj • Manchester

Editors
Syed Muzahir Abbas
School of Engineering
Macquarie University
North Ryde
Australia

Yang Yang
School of Artificial Intelligence
Beijing University of Posts and Telecommunications
Beijing
China

Editorial Office
MDPI
St. Alban-Anlage 66
4052 Basel, Switzerland

This is a reprint of articles from the Special Issue published online in the open access journal *Electronics* (ISSN 2079-9292) (available at: https://www.mdpi.com/journal/electronics/special_issues/18XA3F8OOB).

For citation purposes, cite each article independently as indicated on the article page online and as indicated below:

Lastname, A.A.; Lastname, B.B. Article Title. *Journal Name* **Year**, *Volume Number*, Page Range.

ISBN 978-3-7258-0840-3 (Hbk)
ISBN 978-3-7258-0839-7 (PDF)
doi.org/10.3390/books978-3-7258-0839-7

© 2024 by the authors. Articles in this book are Open Access and distributed under the Creative Commons Attribution (CC BY) license. The book as a whole is distributed by MDPI under the terms and conditions of the Creative Commons Attribution-NonCommercial-NoDerivs (CC BY-NC-ND) license.

Contents

About the Editors

Syed Muzahir Abbas

Syed Muzahir Abbas received his B.Sc. degree in electrical (telecommunication) engineering from the COMSATS Institute of Information Technology (CIIT), Islamabad, Pakistan, in 2006; his M.Sc. degree in computer engineering from the Center for Advanced Studies in Engineering (CASE), Islamabad, in 2009; and his Ph.D. in electronic engineering from Macquarie University, North Ryde, NSW, Australia, in 2016. He has been a Transmission Engineer with Alcatel-Lucent, Pakistan; an RF Engineer with CommScope, Australia; and a Senior Antenna Design Engineer and a Senior Principal Engineer with Benelec Technologies, Australia. He has taught various courses with CIIT in Islamabad and Western Sydney University, Macquarie University, and The University of Sydney in Australia. Currently, he is a Lead Antenna Design Engineer with GME, Australia. He has been a Visiting Researcher with the ElectroScience Laboratory, Ohio State University, USA, and Queen Mary University of London, UK. His research interests include base station antennas, 5G antennas, mmWave antennas, 3D-printed technology, metamaterials and metasurfaces, high impedance surfaces (HIS), frequency selective surfaces (FSS), electromagnetic bandgap structures (EBG), artificial magnetic conductor (AMC), beam steering, UWB, multiband antennas, flexible/embroidered antennas, CNT yarns, CNT/graphene-based antennas, reconfigurable antennas/electronics, and the development of antennas for UWB, VHF, UHF, and WBAN applications.

Yang Yang

Yang Yang received his Master's degree and Ph.D. from the School of Telecommunications Engineering, Beijing University of Posts and Telecommunications (BUPT), Beijing, China, in 2012 and 2015, respectively. From 2015 to 2017, he was a Post-Doctoral Researcher with the Tsinghua National Laboratory for Information Science and Technology (TNList), Department of Electronic Engineering, Tsinghua University, Beijing. He has been a visiting professor at Yonsei University, Seoul, South Korea, and a visiting scholar at Sungkyunkwan University, Suwon, South Korea. He has published more than 60 papers in IEEE journals and conferences. He received the exemplary reviewer award from IEEE TCOM. Currently, he is an Associate Professor at the School of Artificial Intelligence, BUPT. His research interests include AI-based wireless networks, machine learning, and millimeter-wave massive MIMO technology.

Preface

With the spectrum scarcity of sub-6 GHz bands and the increasing demands for extremely high data rates, millimeter-wave (mmWave) communications are regarded as a critical technology for future mobile communication systems. Millimeter-wave frequency bands, which offer abundant underutilized spectral resources, have been explored and exploited in the past several years to meet the requirements of emerging wireless services, including high data rates, ultra-reliability, and ultra-low delivery latency. Despite the benefits of high data rates and ultra-low latency, the unique characteristics of mmWave (e.g., continuous wide bandwidth, large path loss, and penetration loss), along with hardware constraints, introduce great challenges in designing efficient and robust mmWave communication systems. This requires a great deal of fundamental research on the design of all aspects of the antenna and mmWave application, including but not limited to antenna design, channel measurement, signal processing, and interference management.

Moreover, by incorporating high multiplexing gains achievable with massive antenna arrays, mmWave massive multiple-input–multiple-output (massive MIMO) systems show great potential to significantly raise user throughput, enhance spectral and energy efficiency, and increase mobile network capacity. Although the potential of mmWave massive MIMO is exciting, all aspects of system designs require innovative techniques to fulfill the full potential of mmWave massive MIMO, including the transceiver architecture, the design of precoding and antenna arrays, channel estimation and feedback, multiple access schemes, and resource allocation.

This reprint aims to show some of the latest works covering new research on topics related to mmWave communications and multiple-antenna technology, which provide effective ideas and insights for peers in the above-related fields.

Syed Muzahir Abbas and Yang Yang
Editors

electronics

MDPI

rticle

⁻rom 5G to beyond 5G: A Comprehensive Survey of Wireless Network Evolution, Challenges, and Promising Technologies

Ali Sufyan [1,*], Khan Bahadar Khan [1], Osama A. Khashan [2,*], Talha Mir [3] and Usama Mir [4]

1 Department of Information and Communication Engineering, The Islamia University of Bahawalpur (IUB), Bahawalpur 63100, Pakistan
2 Research and Innovation Centers, Rabdan Academy, Abu Dhabi 114646, United Arab Emirates
3 Department of Electronic Engineering, Baluchistan University of IT, Engineering and Management Sciences (BUITEMS), Quetta 87300, Pakistan
4 Department of Computer Science, University of Windsor, Windsor, ON N9B 3P4, Canada
* Correspondence: ali.sufyan@iub.edu.pk (A.S.); okhashan@ra.ac.ae (O.A.K.)

Abstract: The histrionic growth of mobile subscribers, disruptive ecosystems such as IoT-based applications, and astounding channel capacity requirements to connect trillions of devices are massive challenges of the earlier mobile generations, 5G turned up the key solution. The prime objective of the 5G network is not only to maintain a 1000-fold capacity gain and 10 Giga Bits per second delivered to a single user, but it also assured quality-of-service, higher spectral efficiency, the ultra-reliable and improved battery lifetime of devices and massive machine-type communication (mMTC). The huge traffic load and high amount of resource consumption in 5G applications, augmented reality and virtual reality for magnificent virtual experience, and wireless body area networks will seriously affect the channel capacity of cellular cells and interrupt the admission and service of other users which makes compulsory new means of channel capacity and spectral efficiency enhancement techniques. In this research, we review several key emerging wireless technologies to increase channel capacity and spectral efficiency that will not only lead to improve network performance but also meets the ever-increasing user demands. We investigate various benefits and current research challenges of using these technologies. We analyze massive multi-input multi-output technology (mMIMO) an efficient technique and promising solution for the 5G and Beyond 5G (B5G) networks with several benefits and features. Moreover, this paper will be of vast help to the researchers who will involve advance investigation and also to the wireless network operator industry that is in the search for smooth development of state-of-the-art 5G and B5G networks.

Keywords: mMIMO; 5G; augmented reality; virtual reality; wireless body area networks; IoT applications, beyond 5G

Citation: Sufyan, A.; Khan, K.B.; Khashan, O.A.; Mir, T.; Mir, U. From 5G to beyond 5G: A Comprehensive Survey of Wireless Network Evolution, Challenges, and Promising Technologies. *Electronics* **2023**, *12*, 2200. https://doi.org/10.3390/electronics12102200

Academic Editors: Syed Muzahir Abbas and Yang Yang

Received: 2 February 2023
Revised: 13 April 2023
Accepted: 17 April 2023
Published: 12 May 2023

Copyright: © 2023 by the authors. Licensee MDPI, Basel, Switzerland. This article is an open access article distributed under the terms and conditions of the Creative Commons Attribution (CC BY) license (https://creativecommons.org/licenses/by/4.0/).

1. Introduction

The deployment of fourth-generation long-term evolution network (4G-LTE), and its extension long-term evolution- Advanced (LTE-A) in various countries have not only accomplished the International Mobile Telecommunications Advanced (IMT-A) constraint by utilizing IP envisioned for all services but also maintained up and around 1 Gb/s less mobility and 100 Mb/s data rate for large mobility. The histrionic growth of more mobile data subscribers in recent years is the result of people's craving for faster internet while on the go. The Wireless World Research Forum (WWRF) forecaste around 7 trillion mobile products will provide 7 billion individuals in 2017; which is about 1000 times the world's population [1]. While as per Ericsson's technical mobility report published in 2017, almost 29 billion devices are forecast by 2020 including 18 billion IoT related [2]. Furthermore, Rangan et al. [3] predicted 50 billion devices by 2020 and will keep up growing exponentially by around 5 zetta-bytes per month in 2030 [4]. IoT applications demand such as smart homes/cities/grids etc., sensors networks, explosive big data, and wearable artificial intelligent devices are increasing exponentially [5,6]. That has raised substantial attention to form new mobile standards in the telecommunication market.

Against these requirements, the massive data; peta-bytes (10,005 bytes), internet speed in Giga-bits per second (GB/s), and connection to trillions of devices definitely need next-generation wireless communication systems. To encounter these huge challenges design and establish new standards for the next 5G communication, the academia, the standardization agencies, and the telecom industry are working in accordance [7,8]. Also drastic enhancements and new innovations are essential to be made during network design both in physical and upper layers [9]. Emerging technologies such as disruptive ecosystems like IoT named as connected community and machine-to-machine (M2K) communications are also in consideration to be an important part and known as tactile Internet, a newly invented term [10]. METIS [11] and 5GNOW [12] are two main European ventures to address 5G networks. To reduce the firm orthogonality and synchronization criteria in present systems, particularly, 5GNOW explored new physical layer patterns by using non-orthogonal waveforms.

The key objective of the 5G network is not only to maintain a 1000-fold capacity gain and 10 Giga Bits per second delivered to a single user but also to assure quality-of-service higher spectral efficiency (SE), the ultra-reliable and improved battery life of devices, less expensive and massive machine-type communication as dissipated in Figure 1. Only 5G networks can tackle these challenges and it will contribute a true universal boundless mobile experience through upgraded terminals, low latency, and ultra-reliable connectivity The major challenge of the 5G network is tremendous mobile traffic demand.

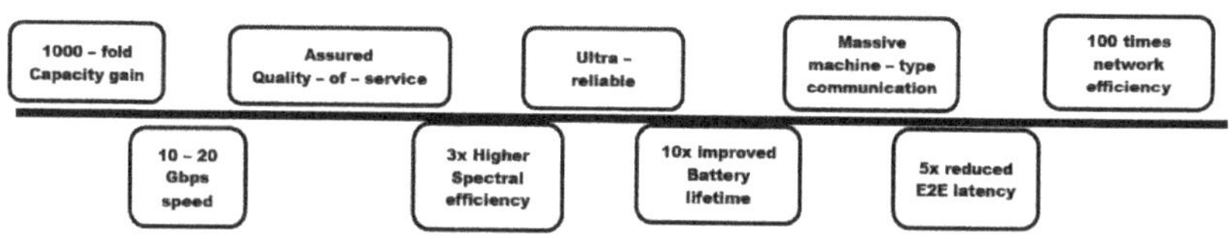

Figure 1. The prime objective of 5G.

This would certainly place a huge amount of traffic load at the edge and a huge amount of data will be processed in the cloud and consume a large number of resources which will not only affect the inadequate capacity of traditional cellular cells but also interrupt the admission and service of other subscribers. Another challenge for organizations is to process multiple repositories from multiple users using multiple applications in multiple environments simultaneously, producing a large amount of digital garbage and useless information. For example, data received from smart homes and health care are processed at the edge [13], processed at the fog to produce helpful information [14], and visualized in subscriber devices. Therefore, the edge, the fog, the cloud, and even the subscriber devices play an important role in the life process of the management of this data (i.e., smart cars, virtual reality, and health care). Therefore, it is a need to design, develop, and implement architectural models to produce on-demand and edge-fog-cloud processing systems to continuously handle big data.

The key problem with the ongoing development of mobile wireless systems is that it is thoroughly dependent upon network densifying to the cells or enhancing spectrum to attain the required throughput. The saturation point of these rare resources is almost reached. Moreover, cell densification and bandwidth increase also increases network latency and pay the cost of expensive hardware. Therefore, the throughput can effectively be increased by an untouched factor of SE without increasing cell densification and bandwidth to meet the current needs of wireless networks. Considering all these challenges, it also makes compulsory new means of channel capacity enhancements and SE techniques. Massive MIMO is an important part of the 5G key enabling technologies and is considered to be the solution to the above-discussed challenges.

The data rate that can be communicated by a certain bandwidth in a particular communication network is referred to as SE, measured in bits/Sec/HZ. Spectrum efficiency, spectral efficiency, or bandwidth efficiency is a technical quantity described as the data rate that can be communicated over a specified bandwidth and used to measure a frequency band. It can be enhanced through increasing modulation order.

We outline various capacity enhancements and SE improvement techniques in this research that will not only lead to enhanced network performance but also meets the ever-increasing user demands. We thoroughly discuss the benefits and current research challenges of using these technologies and presented subsequent contributions in this article.

- A brief summary of the 5G and beyond 5G, is represented.
- A significant review of key enabling technologies, in terms of enhancing channel capacity and spectral efficiency is described.
- Massive MIMO (mMIMO): a perfect candidate for achieving high data rate, channel capacity, and energy efficiency.
- Extensive description of the strengths and shortcomings of research efforts in implementing mMIMO.
- Comprehensive analyses of these challenges and open research problems as well as state-of-the-art solutions including various figures and tables are presented.

We also present previous 3GPP wireless technologies standardization and the current deployment of 5G in the world. Moreover, we identify various research challenges and open research issues that need to be addressed in future mMIMO systems for 5G and B5G. Considering these challenges, this study will surely be helpful for the researchers working on mMIMO for 5G. The remaining article is categorized into various sections. In Section 2, the evolution of the 5G network summarizes, in Section 3 related work and SE improvement techniques have been discussed while in Section 4, we discuss the features, challenges, and benefits of key enabling techniques. Section 5 briefly describes a framework for supporting mMIMO technology and system-level performance features as a solution for the present challenges particularly to the enhancement of channel capacity and SE, with state-of-the-art proposed solutions. Finally, the article is concluded in Section 6.

2. 5G Evolution

The 5G cellular system offers an extremely expandable and flexible network scheme to connect everything and everybody, everywhere. Several industries, e.g., DOCOMO, Huawei, ZTE, Ericsson, Qualcomm, Samsung, Vodafone, and Nokia Siemens Network have paid countless enthusiasm to develop 5G networks so far. Broadly, 5G is categorized into three domains:

- Ultra-Reliable Low Latency Communication (URLLC): fast and highly reliable with 100% coverage and uptime, applications to unmanned vehicles and smart factories
- enhanced Mobile Broadband (eMBB): whose goal to provide large data applications, massive device and user capacity for wireless broadband services
- Massive Machine Type Communication (mMTC): which permits a massive number of wireless devices connection density, energy efficiency, and reduced cost per device [15–17].

The early 5G spectrum in various countries is below 6 GHz but an additional wireless spectrum above 6 GHz and beyond is also proposed for enhanced capacity and SE [18]. The 3rd Generation Partnership Project was developed in 1998. It consists of seven regional/country telecommunications standards developing organizations whose aim is to regulate general policies and specifications, and produce reports that define 3GPP technologies. The 3rd Generation Partnership Project (3GPP) specifications are organized as releases that consist of several technical reports and specifications, each one of which may have concluded after various revisions [19]. A new release offers fresh radio access technology and/or improvements to an existing one and denoting to the achievement of certain milestones [20].

In Rel-13, mMTC and Narrow Band IoT (NB-IoT) were previously established by 3GPP to enable an extensive range of cellular devices, particularly designed for machine-to-machine, IoT applications and deployment scenarios [21]. In 2019, 3GPP has already initiated commercial deployment of Release 15 (Rel-15) focusing on eMBB and URLLC The 5G networks are aimed to work with existing 4G networks by using a range of cells; macro cells for wide area coverage and small cells for in-building, homes, hospitals, schools, and smart forms as shown in [22]. During 5G connection formation, the user device will connect to both 4G for control signaling and 5G for fast data connection. The next phase of 5G is named Release 16 (Rel-16). Phase 2 of the 5G network will be presented in 2023 decided by the ITU World Radio-communication Conference in 2019 (WRC-19) to secure an additional mobile spectrum to meet consumers and business needs [21]. Rel-16 truly emphasizes industrial internet-of-thing (IIoT) related enhancements for Industry 4.0 as well as enhanced URLLC, the Time-Sensitive Communication (TSC), a platform for Non-Public Networks (NPN) wireless and wire-line convergence and complete system resiliency [19].

Rel-15 not only offers extraordinary performance for 5G standards but also provides extensive backward compatibility for new releases in coming years with additional features for ultra-high reliable communication, enhanced data rate, low latency, and improved security characteristics in Rel-16 and in discussion Rel-17 [23]. Release 17 (Rel-17) is an evolution to 5G-advanced systems and connects the community and provides an improved platform for multi-access edge computing, functioning in frequency bands beyond 52 GHz, aiding for reduced capability (RedCap) user equipment(UE), and proximity services, IIoT framework [24], virtual reality [25], smart homes automation [26], multi and broadcast architecture, Non-Terrestrial Networks (NTN), autonomous vehicles [27], and unmanned aerial systems (drones) [28–31] as shown in Figure 2. Furthermore, a physical uplink shared channel (PUSCH) and physical uplink control channel (PUCCH) will be used in Rel-17 for uplink. Release 18 (Rel-18) is officially the evolution of 5G-Advanced. It will bring improvements in the extended reality and field of artificial intelligence by employing machine-learning-based techniques at multiple network levels, that will permit highly intelligent network solutions. This artificial intelligence (AI) based on machine learning (ML) solutions will use solve multi-dimensional optimization issues and intelligent network management with regard to non-real-time and real-time operations. Moreover, cyclic-prefix orthogonal frequency-division multiplexing (CP-OFDM) and discrete Fourier transform (DFT) spread OFDM (DFT-S-OFDM) in the uplink will be investigated. Ookla is a famous speed test provider to test the performance and internet speed of an internet connection [32]. As per Ookla 5G interactive map tracks, 5G roll-outs commercially in more than 132,031 locations across the globe by 216 operators including 220 pre-Release [33]. The world's first commercialization of 5G sub-6 GHz spectrum C-band aggregation by Qualcomm Technologies, Inc. and NTT DOCOMO, INC enabled in Japan [34]. Table 1, briefly discusses new performance requirements and targets of 5G as compared to B5G [35].

Table 1. Comparison of 5G and Beyond.

Attribute	5G	Beyond 5G
Types of application	- URLLC - enhanced Mobile Broadband - massive Machine Type Communication	- Hybrid emBB and URLLC - mMTC - URLLC - Reliable eMBB
Types of Device	- Tablets and Smartphones. - Drones - Sensors	- Tablets and Smartphones. - Drones - Sensors - Wearable appliances

Table 1. *Cont.*

Attribute	5G	Beyond 5G
Energy and SE	10× (bps/Hz/m^2/Joule)	100× (bps/Hz/m^2/Joule)
Data Rate	1 Giga bits/sond	100 Giga bits/sond
End-to-end Delay	5 (ms)	1 (ms)
Processing Delay	100 (ns)	10 (ns)
Spectrum	- MmWave - Sub-6 GHz	- Mm Wave - Sub-6 GHz

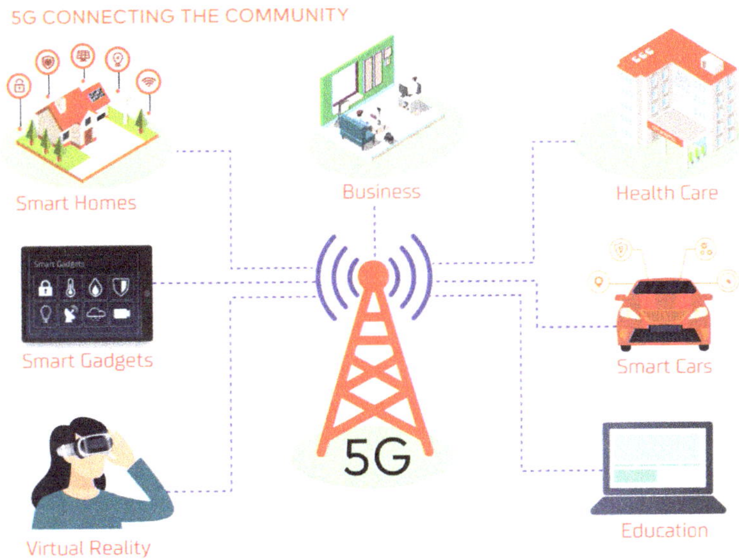

Figure 2. 5G, connecting the community.

3. Related Works

There is no comprehensive review of both channel capacity and SE enhancement techniques together in 5G networks and Beyond. However, some papers explore channel capacity enhancement techniques, while some SE improvement techniques are in subtopics. To deliver ultra-fast data speed by using increased spectral efficiency in 5G networks various authors have proposed different techniques including NOMA [36–38], OFDM [39,40], D2D communications [41,42], mmWave-SCN [23], MIMO [43], by enhancing modulation. As in [2], the authors mentioned that integration of NOMA with mmWave technology can also be a solution by the multiplexing of NOMA but they discussed it conceptually while the algorithmic designs to realize NOMA and the implementation methods are still a myth. Similarly, the authors in [44] declared mMIMO an efficient method for improving SE but were unable to explain how it will mitigate the challenges. Narcis et al. [45] also discussed key enabling techniques for 5G and considered mMIMO can significantly increase the SE but the results will not fully achieve the predicted requirements of 5G-PPP and IMT-2020, especially higher data rate of 10 Gb/s along with linked densities of 100 k–1 *M* devices/km^2. They also provide the solution to this problem by using super high and extremely high-frequency bands which will obvious drawback of giant propagation loss. A brief comparative overview of the existing surveys on 5G key enabling techniques has been investigated in Table 2.

Table 2. A comparative overview of existing surveys on 5G key enabling techniques.

Authors & Ref.	mMIMO	Beamforming	D2D	Small Cell	mmWave
Al-Falahy et al. [46]	✓	✓	✗	✓	✓
Shafique et al. [47]	✓	✗	✓	✗	✗
Sudhamani et al. [48]	✓	✗	✓	✓	✗
Sharma et al. [49]	✗	✗	✗	✗	✓
Hossain et al. [50]	✗	✗	✓	✗	✗
Akyildiz et al. [51]	✓	✗	✓	✗	✓
Adedoyin et al. [42]	✓	✗	✓	✗	✓
Nguyen et al. [36]	✗	✗	✗	✗	✓
Salah et al. [52]	✓	✓	✗	✓	✓
Saha et al. [53]	✗	✗	✗	✓	✓
Vaezi et al. [54]	✓	✗	✗	✗	✓
Ahmad et al. [55]	✓	✗	✗	✗	✓
Sufyan et al. [This Work]	✓	✓	✓	✓	✓

Spectral Efficiency Enhancement

Spectral efficiency refers to the amount of information that can be transmitted over a given amount of spectrum, while channel capacity refers to the maximum amount of information that can be transmitted over a communication channel. Increasing the SE of a system can lead to an increase in the channel capacity, as more information can be transmitted over the same amount of spectrum. Some common SE improvements techniques include:

- Multi-carrier modulation (MCM) techniques such as Orthogonal Frequency Division Multiplexing (OFDM), which divide the bandwidth into multiple sub-carriers and transmit data simultaneously on each sub-carrier.
- The mMIMO technology that uses multiple antennas to transmit and receive data simultaneously, increasing the data rate and improving the quality of the signal.
- Adaptive Modulation and Coding (AMC) that selects the best modulation scheme and coding rate based on the channel conditions to maximize the SE.
- Channel equalization which compensates for channel impairments, such as attenuation and distortion, to improve the signal quality and increase the data rate.
- Interference Management techniques such as power control, beamforming, and interference cancellation which reduce interference from other signals and improve the signal quality.
- Cognitive Radio which uses advanced signal processing techniques to dynamic allocation of spectrum resources to optimize SE while avoiding interference with other users.

Overall, these techniques are used in various communication systems, including wireless communication systems, to increase data rates, improve the quality of the signal, and make more efficient use of available bandwidth. Recently, an increasing number of research attention have been done on OFDM-IM (the combination of conventional OFDM with indexed modulation (IM) to provide a trade-off between SE and EE. In [56], a new concept of sparsely indexing modulation (SIM) is used to enhance SE. Another new combination of spacetime electromagnetic models of Tx/Rx antennas with OFDM, leading to the EM-OFDM, is proposed to enhance antenna SE by ratios up to 300% [57]. A promising and innovative digital modulation technology to provide tradeoff between SE and EE is Spatial modulation (SM), investigated in [58]. The primary goal behind SM is to transmit extra information using ON/OFF states of transmit antennas to reduce the implementation cost by minimizing radio-frequency chains. The authors in Refs. [46–48] have focused on massive-MIMO techniques—but not all in one. Similarly, the authors in Refs. [50,51] have focused on D2D communication as a suitable candidate for 5G networks. The authors in [51] have focused on mmWave, D2D communication, and mMIMO. In [36], the authors only discussed mmWave as a promising candidate for 5G key enabling techniques. The authors in Refs. [54,55] have focused on mMIMO and mmWav. Therefore, this comprehensive survey is motivated to yield broad and thorough information about all important 5G key enabling techniques for the improvement of channel capacity and SE.

Generally, new signal processing techniques for 5G communication networks can be characterized into the following four parts:

- New modulation, signal estimation, and coding techniques.
- Effective spectrum management and new schemes (licensed and unlicensed).
- Using new spatial processing schemes.
- New system-level enabling technologies.

4. Key Enabling Techniques

As Table 3 discusses various multiple access techniques for different mobile communication generation but there is no specific performance metric for 5G networks so far. Several new signal processing techniques are in the study by academia, industry, and telecom operators so the ultimate goal of the 5G system in terms of flexibility, enhanced channel capacity, SE, compatibility, power efficiency, reliability, peak service rates, etc. can be achieved successfully. Among these classes, various potential technologies have achieved the targeted setup for 5G networks, which are mentioned in Figure 3. Several of these technologies, when combined together can attain greater performance as compared to their individual performance. Each technology has exclusive potential advantages and challenges, described in detail.

Table 3. An overview of different generations, 3GPP Release's and multiple access techniques.

Technology	3GPP RELEASE	Access Techniques
2G system	Release 96 for 14.4 kbps user data Release 97 for GPRS Release 98 for EDGE	FDMA and TDMA
3G system	Release 99 for UMTS	CDMA
LTE (2 × 2 MIMO)	Release 9	OFDMA
4G LTE Advanced (4 × 4 MIMO)	Release 10	OFDMA and OFCDM
4G LTE-Advanced Pro	Release 13 and beyond	OFDMA and OFCDM
WiMAX	Release 9	OFDMA
5G systems	Release 15 for Phase 1 Release 16 for Phase 2	OMA for downlink, NOMA for uplink
5G-Advanced	Release 17 Release 18	CP-OFDM and DFT-S-OFDM TDD and FDD

4.1. Millimeter-Wave Communications (mmWave)

The majority of overwhelming communication networks are already operating in the microwave band, which makes it too scary.

The millimeter-wave (mmWave), becomes an important facilitator of 5G networks as a result of the availability of abundant spectrum resources at the millimeter-wave band from 30 to 300 GHz promising high data rate [59]. It is a key research area for wireless power transfer [31], solution for high bandwidth vehicular communication [60] and unmanned aerial vehicles (UAVs) [61,62], as it let huge antenna arrays to be arranged in lesser form-factor. Therefore, mmWave is considered to be the main candidate to achieve improved network capacity and data rate. According to the US Federal Communications Commission, various frequency spectrum's within the mmWave appear promising and suitable players for 5G networks and Beyond, as well as the local multi-point distribution service (LMDS) band ranging from 28 to 30 GHz, the non-licensed 60 GHz band, and 12.9 GHz located from 71–76 GHz, 81–86 GHz, and 92–95 GHz in the E-band [46]. Moreover, the mmWave is revolutionary because it has distinct propagation requirements such as hardware constraints and atmospheric absorption as compared to microwaves.

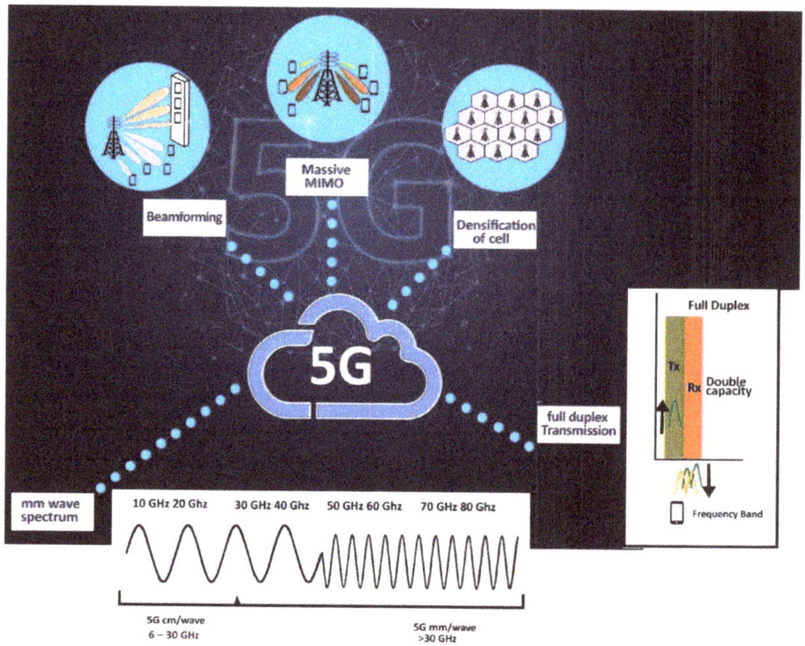

Figure 3. 5G key enabling technologies.

It is broadly acknowledged that the mmWave must be utilized together with a limited cell radius of less than 100 m to reduce high-rise path loss. High path loss in mmWave as compared to MW bands below 3 GHz is considered main challenge, give by:

$$L_{FreeSpace(\mathrm{dB})} = 32.4 + 20log_{10}f + 20log_{10}R \quad (1)$$

where L is free space path-loss measured in decibels (dB), f (GHz) the carrier frequency and the distance between transmitter and receiver is R, measured in meters. This can be understand clearly that an extra path loss of approx. 23 dB & 31 dB when moving the operational frequency from 2–28 GHz and 70 GHz, respectively. As the non-LOS (reflected) signal is very weak, the mmWave can use highly directional antennas for line-of-site (LOS) communication. Another serious challenge of mmWave is signal attenuation at high frequency spectrum. This is very problematic as it limits signal propagation because the energy of mmWave is absorbed by water vapor and oxygen. The third challenge of mmWave signal is its penetrates with high loss, makes it very sensitive for the buildings [46]. However, due to the limited number of simultaneous connections at these extremely high frequencies, the mmWave channels are sporadic nature (scattered) in the angle/spatial domains. Hence, by using either hybrid analog-to-digital beamforming with a large scale antennas of coupling mmWave with massive MIMO and NOMA can avoid this restriction and can yield very high SE gain [54].

To the bottleneck of existing wireless bandwidth, enhancement of channel capacity, and SE a solution is proposed in [63]. There are various challenges of mmWave communications do exist, such as beam training and tracking, device association, directivity, sensitivity to blockage, and high propagation loss which are under research and need to address [64]. Guang et al. [65] developed an adaptive low-latency approach that uses cooperative networking to overcome end-to-end latency and boost systems channel capacity. In [23], a new technique mmWave small cell network (mmWave-SCN) is proposed that is a combination of mmWave, mMIMO, and network densification to combine additional infrastructure nodes and spectrum. The resulting mmWave-SCN contains features from mmWave and SCN, such as flexible deployment and management, ultra-high data rate support, and ironic obtainable spectral resources. Peng. Yu et al. [66] proposed the mmWave Aerial Base Station (mAeBS) by the grouping of the aerial base station (AeBS) and mmWave. This technique provides efficient and strong maneuverability, adequate spectrum resources, and also overcomes the shortcoming of proneness to obstruction by flexible adjustment of mmWave position. In [67], smaller antennas are made by combining even smaller antenna arrays which demonstrate the effectiveness of beam steering and narrowing (BSN) techniques for enhanced channel capacity. Hamed et al. [68] presented a framework to incorporate the high data rate and growing user demand in mmWave cellular network in the 28 and 73 GHz band by dividing a dense mmWave hexagonal cellular network into various smaller cells with their own base stations. The analysis shows a better spectrum and energy efficiency.

4.2. Multiple Access Techniques

Various 5G multiple access techniques have been proposed so far by the telecom industry and academia including Interleave Division Multiple Access (IDMA) [69], Low Density Spreading Multiple Access (LDSMA) [70], Lattice Partition Multiple Access (LPMA) [71], Sparse Code Multiple Access (SCMA) [72,73], Pattern Division Multiple Access (PDMA) [74–77], Power-domain NOMA [78,79] and various others are briefly explained in Figure 4.

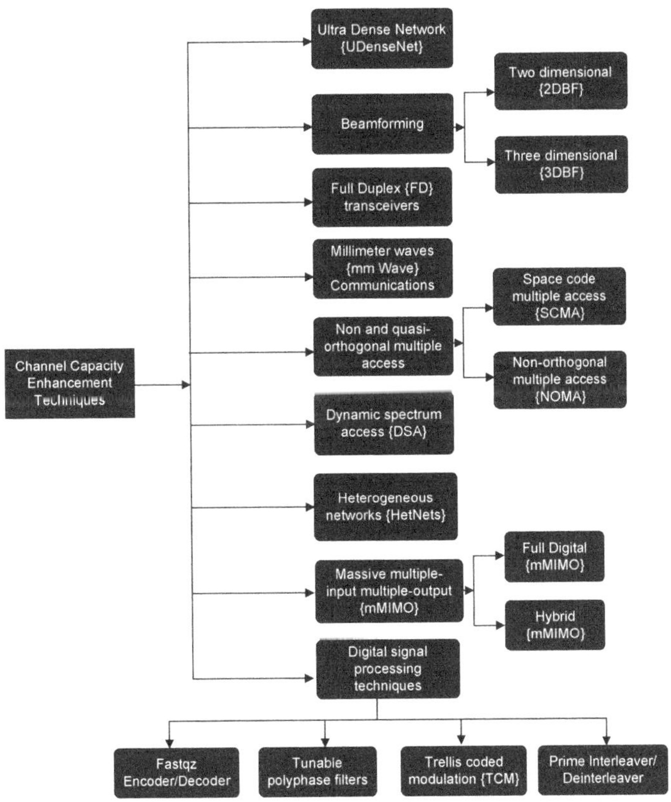

Figure 4. Summary of key enabling techniques for channel capacity and SE improvement.

4.2.1. Non-Orthogonal Multiple Access (NOMA)

NOMA is an effective and favorable technique to enhance user capacity, user fairness and SE for 5G and beyond wireless networks [80]. In NOMA, several users are allocated similar and single resource sections (time and frequency) in a specified time. The subscriber separation is attained by various interference cancellation schemes [81,82]. Shin et al. [83] categorized into single-cell NOMA and multi-cell NOMA while Mahmoud et al. [84] classified NOMA into two general types: code-domain and power-domain multiplexing. The Performance analysis of cooperative NOMA, practical implementation aspects, and research challenges such as hardware complexity, error propagation, carrier frequency offset and timing offset estimation are discussed in detail. Even though the implementation of NOMA in wireless communication is comparatively new but key ideas such as successive interference cancellation (SIC), superposition coding, and message passing algorithm (MPA) have previously been developed over and above two decades ago [85–88]. The key features of NOMA are a balanced trade-off between system throughput, user served fairly, and greater throughput yield than Orthogonal-Multiple access (OMA). The article [89] provides a systematic combination of dynamic spectrum access and MIMO, features, challenges and futuristics in the standardization activities regarding the execution of NOMA in 5G communications. In [54], an extensive research has been done on NOMA-enabled massive MIMO, and the spatial domain in designing subscriber pairings, pilot allocations, and relevant signal processing methods have solely been exploited.

4.2.2. Sparse Code Multiple Access (SCMA)

SCMA allows non-orthogonal communication of various user signals amongst code and power domains, which may well improve SE and provide stronger connectivity with limited resources. The authors in [90] conducted SCMA to attain an improved link-level performance, delivering additional multiple access capability through reasonable complexity and energy consumption hence, considered SCMA as an enhanced energy-efficient approach while Nikopour et al. [91] developed a technique for SCMA to increase the downlink throughput of a deeply loaded network. He found that SCMA can increase link adaptation as a result of reduced colored interference and enhanced SE of wireless networks. In [92], the authors derived and implemented SCMA-MIMO by combining MIMO and SCMA technologies to accommodate the ever-increasing number of users, reliability, and SE improvement for the next-generation 5G networks. Jienan et al. [93] proposed an efficient and affordable decoding algorithm based on a Bayesian program learning scheme and Monte Carlo Markov Chain (MCMC) which can reduce 60% computation complexity and only 0.5-dB performance loss related to a similar decoding algorithm.

4.3. *Ultra-Dense Network (UDense Net)*

Cell densification has become the major contributor to enhance capacity in previous generations of wireless communications, about 4–5 macro-cell base stations (MBSs) /km^2 in 3G, around 8–10 micro-cell BSs/km^2 in the 4G and expected increase of 40–50 small-cell BSs (SBSs)/km^2 in 5G [94]. The principal goal of cell densification is to address the coverage problem by spatial frequency-reusing and offloading the data traffic to the SCs. High splitting gain by dense small cells was a major challenge. Hence, an ultra-dense network (UDN or UDense Net), with one macro-cell BS that serves a huge variety of small-cell BS's, is widely considered a key player in achieving capacity. UDense Net is known as a promising technology to enhance SE, capacity, and significant system performance by utilizing spectrum opportunities locally for every single hertz. It also possesses the advantages of low transmission power, effective expansion of coverage, the throughput of the network, and flexible deployment [95,96]. However, several challenges including security, interference, mobility support, flexible back-haul connectivity, and energy consumption also exist. Similarly, In [59], two architectures: self-back-hauled small cells for UDense Net and direct access were investigated and results showed a 30% rate improvement. In [69], increased handover rates by using a trace methodical approach to secure the network by continuous monitoring. Wang et al. [97] suggested a network architecture for UDense Net by two effective localized mobility management techniques which have performance evaluation results of the average handover signaling costs, average packet delivery cost, average latency, higher handover latency and average signaling load to the core network. Li et al. [98] addressed the problem of anti-jamming communication in an unknown environment for UDense Net by a deep reinforcement learning base anti-jamming algorithm. This algorithm is represented in an actor-critic framework. For the selection of anti-jamming action, a convolution neural network (CNN) is used as an actor, and a deep neural network (DNN) as the critic for the value estimation.

4.4. Dynamic Spectrum Access (DSA)

As per 5G vision, spectrum bands have two categories: Below 6 GHz Band and above 6 GHz. The scarce and costly below 6 GHz band is exceptionally crowded as it is already been allocated to several communication systems. The challenging demand for data by a factor of 1000× can be addressed by exploiting additional bands, licensed and unlicensed. The cellular operators find unlicensed spectrum effective, economical, and less crowded supplement to enhance the capacity of existing soon-to-be overloaded wireless networks as a radar might be operating in the same band. An advanced dynamic channel access strategy to enhance the quality of experience has been investigated in [88]. The DSA is an access technique of sharing licensed-plus-unlicensed heterogeneous spectrum, as a means to feed emerging wireless network technologies. It has the advantages of low cost, enhanced coverage, channel capacity and bandwidth, high quality of service, and less interference [99–101]. In [102], extended dynamic spectrum access, a solution proposed for the unbalanced spectrum loads and perceived capacity bottleneck. The DSA has two broad classifications: cognitive-inspired radio access and co-operative-inspired radio access. A survey of advanced schemes for spectrum sharing to a significant increase of the spectrum efficiency in 5G networks which is extensively endorsed by both the industry and the academia [103].

4.5. Full Duplex (FD)

The name "duplex" in communication system denotes the ability of two systems are able to transmit and receive data. However, simultaneous data flow capability of the system is termed as Full-Duplex (FD). FD wireless has gained substantial research interest recently, its numerous benefits at higher layers, and doubling the network capacity by simultaneous transmission and reception on a single carrier without acquiring a new spectrum. Thus, data loss can be minimize and uninterrupted users transmission for channel sensing is achieved by using FD. Hence, FD increase the utilization of spectrum and in the process, improves network capacity but by paying the cost of increased hardware complexity and energy consumption. However, another challenge related to FD receivers is the transmitter's self-interference (SI) which is so powerful (a billion to a trillion times) and ranges from 90 to 20 dB. This high power at the transmitter side suppresses the desired signal at the receiver. So, to have full utilization of the FD network, it needs to be redesigned not only at the physical layer but also at the medium access control layer [104–108]. To reduce the SI on an acceptable level to the thermal noise researchers have developed several efficient techniques in the past such as a combination of passive and active cancellation methods. Columbia University conducted recent research on integrated full duplex radios by using non-magnetic complementary metal oxide semi-conductor circulators [109]. The use of a single antenna and recent evolution's in SI cancellation/suppression methods turned FD communications a promising candidate to increase spectrum efficiency and channel capacity for 5G and B5G networks.

4.6. Beamforming

The smart antenna technique which consists of a closed-loop pre-coding where we can steer the wireless beam flexibly toward the desired subscriber is called beamforming. It is proven technology for improved array gain, signal quality, network coverage, higher channel capacity, SE, and reduced inter-cell interference. It employs digital signal processing techniques that identify the directional antenna beam patterns by using several antennas at the transmitter side. Smart antennas equipped with beamforming abilities can be organized such that the transmitted signal can be directed to the desired location with exact precision and avoid unwanted signals from an undesired path simultaneously. It can be rotated electronically by phase shifting in 5G networks. The impact of suitable antenna configurations for directional multi-antenna beamforming have shown to improve the Ricean factor gain, increased SNR and low root-mean-squared (RMS) delay spread because of multi-path dispersion at the receiver. Beamforming consists of two-dimensional beamforming (2DBF) and three-dimensional beamforming (3DBF). By using narrow pencil beams, spatial degrees of freedom can be exploited by avoiding inter-cell interference [18,110,111]. Therefore, beamforming is considered to be key player for improving SE.

5. Massive MIMO

The Multi-input multi-output technology (MIMO) has been almost a decade, but equipping base stations (BS) with multiple antennas is a new concept in 5G and beyond 5G (B5G), called massive MIMO (mMIMO) [95,112–114]. The mMIMO has proven an efficient player for industrial Internet of things (IIoT) networks [115], mobile edge computing [116], virtual reality [117], 5G wireless communication networks [118], autonomous driving [119], augmented reality [120] and wireless sensor networks [121–123].

The advantage of mMIMO includes a massive shoot-up in SE, a minimized latency, and an expandable air interface arrangement. It is a multi-subscriber technology, in which every BS is equipped with a large number of active antennas M, arranged in an array and simultaneously communicates a set of single-element K customers on a similar frequency and time resource such that $M >> K$ making the signal pre-coding process simple and increase SE [124,125]. The difference between the 4G sector antenna and 5G massive-MIMO geometry is shown in Figure 5 which consists of >100 elements. The huge nature of mMIMO itself is the cause of reduced latency as it removes the channel frequency dependency which stops the effect of frequency selective fading on the signal strength. Therefore, every transmitted signal safely arrives at the receiver without suffering channel distortion with decreased latency.

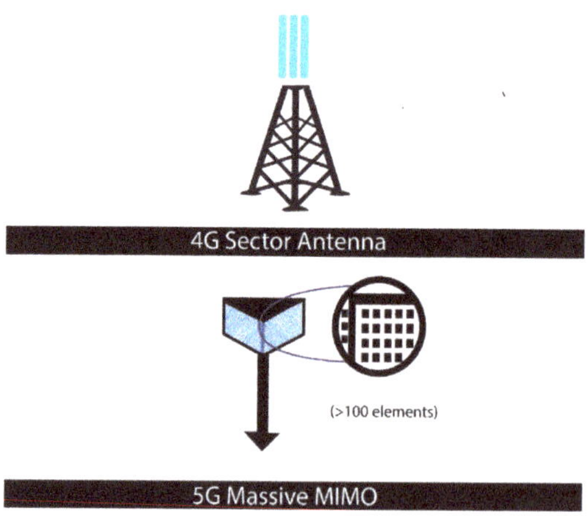

Figure 5. Difference between 4G and 5G massive-MIMO antenna.

The mMIMO can be implemented by two feasible approaches: full digital configuration and hybrid configuration. In fully digital mMIMO a wide-range digital signal processing unit is required along with a digital-to-analog converter for every antenna element [126,127]. The hybrid (analog-digital) mMIMO system that merges analog beamforming and digital MIMO signal which overcomes the amount of digital-to-analog converters [128–130]. The first implementation of mMIMO in real time and public implementation is, with $M = 100$ and $K = 10$ [131]. In mMIMO, an antenna array of the base station contains M dipole antennas with λ wavelength, each consisting of an effective size of $\lambda/2 \times \lambda/2$. This implies that an array of 1 m^2 can fit a hundred antennas at 1.5 GHz and 400 antennas at 3 GHz carrier frequency [132]. The array can be of any geometry; cylindrical, distributed arrays, linear or rectangular.

Table 4, describes the unique features of mMIMO which make it prominent from other key enabling technologies effectively. The SE of a specific group j is impacted by the model signaling completed in different cells. Assume, the pilot reuse factor f is an integer with the distribution of L sections into f split cell groups: $f = \tau_p / K$.

Table 4. Features of mMIMO on other key enabling techniques.

Sr. No.	Characteristics	Reference
1	Increased Throughput	[133]
2	Reduced Radiated Power	[134]
3	Unlimited Capacity	[135]
4	Hardware Efficiency	[136]
5	Energy Efficiency	[137]
6	Multi-user Gain	[138]
7	Enhanced Antenna Design	[139]
8	No Cost for Extra Site	[108]
9	Single-carrier Transmission	[140]
10	Enhanced Spectral Efficiency	[141]
11	Highly Secure	[142]
12	Low Latency	[16]
13	Anti-jamming	[143]
14	Robust	[144]
15	Enhanced QoS	[145]
16	Reliability	[146]
17	Omni-directional	[147]

The global pilot reuse $f = 1$ and non-global pilot reuse $f > 1$ are very well known. As the hexagonal cell topology contains six cells in each stage, the lowest pilot reuse factors which increase to symmetric pilot reuse patterns are $f = 1, 3, 4$. Consider an mMIMO scenario with base station antennas M = 200 and $\tau_c = 400$ symbols a coherence interval, where customers are homogeneously distributed in the cell except for the 10% cell center. These channels show uncorrelated Rayleigh fading by a distant dependent channel attenuation having a path loss exponent of 3.7. By taking $f = 1, 3, 4$ suppose an easy policy for power allocation.

$$\rho_{j,k} = \frac{\delta}{\beta^{j}_{j,k}} \tag{2}$$

where $j = 1, 2, \cdots, L$ and $k = 1, 2, \cdots, K$ Here $\delta > 0$ is a configuration framework that calculates the achieved SNR at every BS antenna and is known as statistical channel inversion power allocation:

$$\rho_{j,k} \times \frac{\beta^{j}_{j,k}}{\sigma^2_{BS}} = \frac{\delta}{\sigma^2_{BS}} \tag{3}$$

The average SE is an outcome of the different users with non-identical pilot reuse factors and processing schemes by considering two separate SNR levels $\delta/\sigma^2_{BS} = 0$ and 20 dB. The two different SNR levels yield essentially similar performance as evident in [132]. Therefore, SE can be increased by increasing the number of antenna elements because of increased data rates to the subscribers, relatively low cost to the network operators, coverage improvement, and increased service reliability [52].

As the array gain is not noise-limited and constructs the SE interference-limited, i demonstrates that Massive MIMO works equally fine at low and high SNR's. The seconc observation is that dissimilar pilot reuse factors are required at dissimilar customer load (i.e., No. of subscribers K). The pilot reuse of $f = 3$ is required at less load while $f = 1$ i desired to minimize the Prelog factor $(1- fK/\tau_c)$ when the value of K is very large. B choosing the appropriate f, mMIMO can offer a high SE above an extensive range of variou customers. The mMIMO delivers stable SE for any $K > 10$ and does not require comple scheduling as every active subscriber can essentially be served at the same time in eacl coherence interval or a minimum of τ_c /2 subscribers, resulting in half of the coherenc interval for data, which is normally greater than a hundred; the high amount of SE i then shared among all the subscribers. The significant design parameter in mMIMO i a pilot reuse factor which depends on the user load, total number of BS antennas, anc propagation environment. In 3GPP Release 18, the evolution of MIMO will continue During the commercial deployment, significant performance loss due to outdated channe state information (CSI) by UE has been investigated with moderate or high mobility ir multi-user MIMO (MU-MIMO) scenarios. To increase the performance of medium o high-mobility UE will be explored for potential CSI enhancements [32].

It is also observed that a suitable trade-off between SE and energy efficiency (EE) has ar important impact on the 5G network and Beyond 5G. Hence, a balance between SE and EE is urgently needed. Also, effective algorithms need to be proposed to achieve a satisfactory trade-off. The SE in mMIMO can significantly be increased by the following methods:

- By enhancing transmit power.
- By utilizing the uplink and downlink space division multiple access (SDMA) techniques.
- By acquiring an array gain.
- By obtaining channel state information (CSI).

Therefore, mMIMO is considered the ultimate solution for high data rate, channel capacity, and spectrum efficiency in 5G and B5G.

5.1. *Classification of Applications*

Massive MIMO is considered a key technology for the 5G and beyond networks because it can significantly enhance spectrum efficiency, reduce frequency spectrum constraints, and completely utilize existing space resources. The mMIMO can be classified into two applications categories smartphones and base stations as shown in Figure 6.

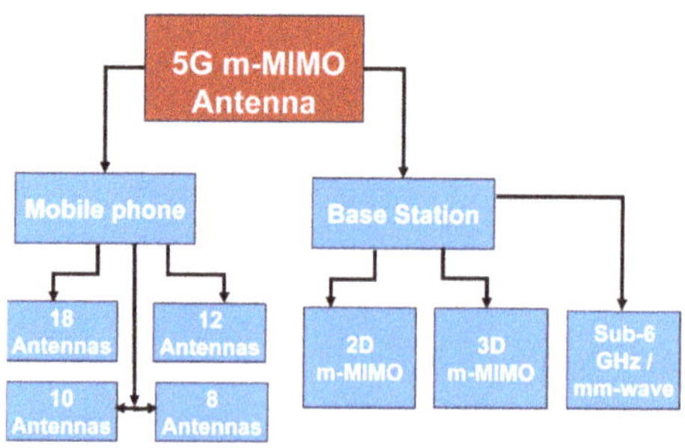

Figure 6. Classifications of applications for 5G massive-MIMO antennas.

5.1.1. Mobile Phone

Various mMIMO systems for mobile phone applications for 5G and beyond networks have been investigated that are categorized by a number of antenna elements such as 18, 12, 10, and 8. In [148], an 18-port 5G mMIMO antenna operating in the sub-6 GHz band is investigated. The 18-slot antennas were simply adjusted on the 6-inch board, performing as a radiator which is used for modern smartphones. The measured results of this mMIMO antenna have high isolation > 20 dB, and peak efficiency > 87% over the operating frequency. A 12-element MIMO antenna is presented in [149] for 45/5G mobile phones. As per the measured results, the ergodic channel capacity of 34 b/s/Hz and 26.5 b/s/Hz, were achieved in LTE bands 42/43 and 46 respectively. Although, this design achieved enhanced channel capacity, a trade-off between the isolation of <-12 dB, and total efficiency of >40% has been observed. Various 10-elements mMIMO antennas for the 5G smartphones have also been investigated [150–154]. All designs were proposed for 4G/5G smartphones operating in the sub-6 GHz band and prototypes were fabricated to evaluate isolation, efficiency, and MIMO diversity performances. It has been observed that a 10-element MIMO antenna array can achieve the maximum ergodic channel capacity of 51.4 bps/Hz.

Similarly, various 8-elements mMIMO antennas systems for the 5G smartphones have been presented operating in sub-6 GHz frequency band [155–158]. In order to consider the importance of backward compatibility, these designs are equally capable of working in two frequency bands including multi-and or wide-band structures. Although the antenna elements of mMIMO in the mobile phones must be in equal or greater number than eight antennas, various designs are not able to support the massive-MIMO system in the 5G mobile phones regardless of employing the same number of mMIMO antennas. Nevertheless, all such designs are observed to be suitable candidates for the mMIMO systems.

5.1.2. Base Station

For the design of 5G mMIMO base stations, the 2D model is the utmost simple planar mMIMO arrangement is an array of $N \times M$ configuration of the planar array. The various designs (cylindrical, planner, and circular) significantly affect the mMIMO system performance. As the beam can be adjusted only horizontally, usually the circular or planar array design shows a considerable reduction. Moreover, these designs also fail to increase channel capacity demands. Therefore, it is proposed that 3D massive arrays such as hexagons, cylinders, triangles, etc. arrangements should be adopted [159].

Many 5G base station antenna designs have been investigated in this study [160–163]. For an indoor MIMO base station, an ultra-wide-band portable multi-element antenna consisting of 121 physical antennas to construct a 484-port is investigated in [164]. This antenna is effectively working on a wide frequency range from 6 to 8.5 GHZ and uses 3D radiation patterns to measure total efficiency of around 70%. It has been observed that 64 QAM can simultaneously be transmitted for a single user with 8 antennas, and 5 Gbps is feasible at 200 MHz. Moreover, a 10 Gbps throughput for multiple users in an outdoor environment is also possible [165]. Another high isolation 32 elements mMIMO antenna with a measured bandwidth of 250 MHz For next generation 5G base stations is investigated in [166]. The decoupling of 32 dB among array elements with a reduced ECC of 0.0001 is observed by this proposed MTM-based method.

In [167], an mMIMO antenna system having 288-elements and 72 ports is presented for 5G base stations. A 3-layer configuration with 24-ports on every side of the antenna system. The single port gain of 9.41 dBi and 64% efficiency was measured. It has been observed that increasing the antenna elements significantly increases channel capacity in the mMIMO system. A significant variation in gain patterns also has been experimentally evident from various antenna elements in a finite array. The mutual coupling and edge effect are the main players in the beamforming which is directly dependent on the arrival angle. The key concept of mMIMO is to obtain all the functions of traditional MIMO but over a large scale. The mMIMO is analyzed as an efficient technique for B5G networks with several benefits and features, listed in Table 5 which briefly describes a comparison of various key enabling techniques with mMIMO [168]. The key problem with the deployment of mMIMO is a radio-frequency pattern that increases the difficulty of the symbol detectors can be addressed by combining analog and digital beamforming, called hybrid beamforming structure [169–172]. In [173], a detection algorithm for perfect mMIMO has been proposed to increase performance and decrease complexity. This paper briefly discusses various mMIMO detection algorithms in detail. Likewise, A practical test at Lund University was conducted utilizing around 100 BS antennas along with up and around 50 field programmable gate arrays. The obtained results show that mMIMO can simultaneously work for a number of users in a static inside/outside environment using a similar frequency and time band [174]. As conventional wireless communications systems are unable to provide the SE's that 5G applications require, the mMIMO is becoming a reality now because it can simultaneously provide in the downlink and uplink direction of B5G due to uplink-downlink duality. In [132], SE is expressed as an outcome of the number of base station antennas M. To achieve the highest SE, the functional subscribers are optimized for each M. As per IMT-Advanced, the performance baseline SE is in the range of 2–3 bit/s/Hz/cell, subject to simulation scenario. The results shown in [175], with $M = 100$ antennas an achievement of 52 bit/s/Hz/cell which is 17× to 26×, and with $M = 400$ antennas an unbelievable 38× to 57× enhancement over IMT-Advanced. Importantly, the number of active subscribers grows together with the SE. The SE for each user can obtain by dividing the top curve by the bottom curve and surprisingly the SE per user exist in the modest range of 1–2.5 bit/s/Hz [176]. Similarly, Yoshio et al. [177] also evaluated the transmission features of mMIMO by using asymptotic eigenvalue distribution of a Wishart matrix and determined about 20% enhancement in the channel capacity by using spatial correlation. Moreover, MmWave base station 256-element antenna arrays and mobile antenna of 32-element arrays are already commercially available [44]. Furthermore, results in [178], show that the Massive MIMO is important to not only enhance the SE but surely able to be the driving force to achieve increased area throughput in 5G networks. The mMIMO technology is proven to be true for providing 10-fold or even fifty-fold enhancement in SE over IMT-Advanced by serving several users at the same time.

Table 5. Comparison of mMIMO with other key enabling techniques.

Technology	Enhanced Data	Energy Efficiency	Increased Channel Capacity	Enhanced Spectral Efficiency	Substantial Device Support
mMIMO	✓	✓	✓	✓	✓
FD	✗	✗	✓	✓	✗
mmWave	✓	✓	✓	✓	✓
DSA	✓	✗	✓	✓	✗
UDense Net	✓	Partial	✓	✓	Partial
NOMA	✓	Partial	✓	✓	✓
SCMA	✓	✓	✓	✓	✓
Beamforming	✓	✓	✓	✓	✓

5.2. Open Challenges of mMIMO

The mMIMO is obviously high-caliber and remarkable to the conventional multiple antenna networks. Although, massive MIMO and 5G technology can be wonders for future wireless networks. However, several hardware issues such as the choice of the material, the limited size of the mobile phone chassis, overall cost, and characteristic parameters (bandwidth, mutual coupling, gain, efficiency, etc.) can be observed for both of the above-discussed applications. The unlimited variety of smartphones, operating in different frequency bands is another problem for antenna designers.

Although, the impact of mutual coupling and limited space in the 5G mMIMO system is challenging. However, the decoupling techniques and compact antenna size can be used to enhance isolation among the antenna elements. In this part, we report the most recent advancement of research on challenges in mMIMO frameworks with regard to the most serious problem of mutual coupling by targeting the recent literature published between 2018 to 2023.

To implement MIMO in 5G and beyond networks, there are various research challenges that need to be addressed such as mutual coupling, channel estimation, signal detection, energy efficiency, and pilot contamination. The narrow space in smartphones is a major challenge and the effect of coupling between adjacent antenna elements severely decreases the MIMO performance. Hence, the reduction of mutual coupling is highly desirable as it offers high isolation, efficiency, and MIMO diversity performance. Various decoupling techniques have been addressed in Table 6 to reduce mutual coupling.

Table 6. Futuristic suggested solutions for MIMO antenna to reduce mutual coupling.

Ref.	Year	Solution Suggested	Frequency GHz	Outcome Type
[179]	2018	Metal meandering strips	3.2–5	Simulated
[180]	2018	Pattern multiplicity	2.5–2.64	Simulated and tested
[181]	2019	Transmission- line decoupling	2.45	Simulated and tested
[182]	2019	Molecule-shaped structure	2.4–10.6	Simulated and tested
[183]	2019	Frequency-selective surface	3.5–4.9	Simulated and tested
[184]	2020	Decoupling network	3.3–4.5	Simulated and tested
[185]	2020	Decoupling super-strate	3.3–4.5	Simulated and tested
[186]	2021	Sub-Miniaturization	0.61–0.96 −1.7–5	Simulated and tested
[187]	2021	Split Ring Resonators	2.73–3.12 −4.33–4.68	Simulated and tested
[152]	2022	Defected Ground Structure	5.82–5.94	Simulated and tested
[188]	2022	SICL Feeding Structure	24.5–26.5	Simulated and tested
[189]	2023	Defective Ground Structure	2.2–2.64	Simulated and tested

Metal meandering consisting of metal-strip lines is an effective technique to overcome mutual coupling. This solution was proposed in [179] to achieve effective decoupling by meandering strip lines vertically and horizontally among 16 antenna elements. A six-element, three ports antenna is proposed in [180] that demonstrates pattern diversity by loading periodical inter-digital capacitors on the radiating elements. The proposed dual-polarized antenna was found to have an omni- directional radiation pattern and efficient mutual coupling of below −20 dB between any two ports. This design is considered an important milestone in the implementation of the mMIMO network for 5G and B5G due to the low coupling effect. The isolation can be improved by orthogonal geometry of the antenna elements using molecule fractal structure given in [182]. This technique shows improved isolation of higher than 20 dB. Another efficient solution reported in [183], is to reduce mutual coupling by inserting a frequency-selective surface scheme among the array elements. To improve isolation further, a frequency-selective surface decoupling structure in the substrate is used which blocks the propagation of the electromagnetic wave (EM) in the substrate. The effect of both the coupled magnetic and electric fields was removed by using the Frequency-selective surface (FSS) decoupling structure. An embedded decoupling scheme is introduced in [184] by combining parallel reversed C-shaped metal strips with two inverted U-shaped metal strips that produce additional coupling while placed near a pair of closely spaced dipole antennas. The reported result was a 10 dB mutual coupling reduction over 3.3–4.5 GHz and demonstrates the advantages of the wide-band frequency spectrum, embed-able, radiation pattern distortion alleviation, and dual-polarized capabilities for MIMO applications. Another, mutual coupling reduction for large antenna arrays is proposed in [185], using ϵ-negative meta-surface superstrate. This decoupling meta-surface structure is very useful to restore the radiation patterns, maximum mutual coupling reduction of 25 dB, broaden the bandwidth of the array, and minimize the active voltage standing wave ratio. The work published in [186] also represents low mutual coupling, good decoupling of the radiators, and impedance matching. In [187], a dual-band MIMO antenna is proposed over a frequency range of 2.7–3.1 GHz and 4.3–4.6 GHz showing improved isolation of higher than 21 dB and efficiency of 80% and good MIMO diversity performances. The mutual coupling was reduced by using a split ring resonator structure on the radiator. Another hybrid decoupling technique to reduce mutual coupling is presented in [152] using the defected ground structure (DGS) and circular ring parasites. The DGS reduced the effect of coupling between antenna elements whereas, the circular ring parasites reduced the mutual coupling between perpendicular antenna elements. A nove 64-element dual-polarized mMIMO antenna operating in 24.5–26.5 GHz frequency banc is presented in [188] and mutual coupling is reduced to −23 dB for the entire bandwidtl by using substrate-integrated coaxial line feeding structure. As the Substrate-integratec Coaxial Line Feeding (SICL) transmission contains a microstrip line, shielded by two roll of metallic vias on both sides and two ground planes above and below, aiming for ar extremely low coupling feature. In [189], L-shaped stubs and DGS are used to reduc mutual coupling effects. The L-shaped stubs are placed on one side of the two patche which generate a new coupling current for antenna excitation and create an additiona coupling path.

The huge amount of channel state information in beamforming will surely be ques tionable, particularly for the downlink. Therefore, mMIMO can only be operational in th time division duplexing because of feedback schemes, restrictive cost of channel estimatio and channel reciprocity. Thus, it will be unsuitable for frequency division duplexing bu pilot contamination in time division duplexing is also a huge challenge in mMIMO. So, nev means of channel estimation, feedback measures, and solutions need to be proposed for 5C and B5G. Moreover, if the transmission power is excessive in an amount that is typicall 3–5×, the mMIMO will experience thermal noise and pilot contamination from other cell which require passive cooling. Thirdly, The researcher will be unable to justify technique and algorithms due to insufficient channel models for mMIMO as to cope with the hug amount of data, tremendously fast algorithms will be needed. Furthermore, enhancing the antenna elements creates consequential challenges not only to equipment but also t operators and manufacturers.

6. Conclusions

As the need for high-speed internet increases substantially, the 5G system should have the ability to meet these requirements and provide aid for multi-fold improvement in channel capacity and network connectivity. This paper explains a comprehensive survey that carried on the basics of challenges involved in previous wireless networks, evolution and implementation of 5G wireless network that has been described in terms of enhanced data channel capacity, and spectral efficiency.

To deliver ultra-fast data speed, SE improvement is vital in 5G networks that can be improved through D2D communications, mMIMO, by enhancing modulation order, and acquiring new effective transmission waveforms. To encounter substantial traffic growth 5G wireless systems are expected to achieve higher channel capacity by utilizing mm-wave band, dense small cell deployment, beamforming and mMIMO technology. The critical review illustrates mMIMO technology is the solution to not only the increase in SE but also a motivation in relation to accomplishing higher orders of area throughput in 5G and B5G. However, it faces various critical challenges which need to be solved prior to the implementation of the next-generation wireless networks. Although the mMIMO is investigated as truly a key player amongst others due to its various features, still there are various secondary key emerging technologies such as FD, mmWave, UDense Net, and beamforming, their challenges and features with the focus on the enhancement of data rate, channel capacity, and SE. Moreover, this paper inspires researchers for the enhanced outcome of various problems, trends research gaps, and future directions in 5G and B5G networks.

Author Contributions: Formal analysis, U.M.; Investigation, O.A.K.; Project administration, T.M.; Writing—original draft, A.S.; Writing—review and editing, K.B.K. All authors have read and agreed to the published version of the manuscript.

Funding: This research received no external funding.

Data Availability Statement: Data analyzed in this article was a re-analysis of existing data, which is openly available at locations cited in the reference section.

Acknowledgments: The author is very grateful to all the co-authors for their contribution and support to this manuscript. The author would also like to thank the anonymous reviewers for their insightful suggestions that will help us to improve the quality of this paper. The author would like to gratitue special thanks to Muhammad Ali Qureshi, Zeshan Rasheed and Engr. Shahid Wasim for their help during write-up.

Conflicts of Interest: The authors declare no conflict of interest.

References

1. Wang, C.X.; Haider, F.; Gao, X.; You, X.H.; Yang, Y.; Yuan, D.; Aggoune, H.M.; Haas, H.; Fletcher, S.; Hepsaydir, E. Cellular architecture and key technologies for 5G wireless communication networks. *IEEE Commun. Mag.* **2014**, *52*, 122–130. [CrossRef]
2. Yadav, A.; Dobre, O.A. All technologies work together for good: A glance at future mobile networks. *IEEE Wirel. Commun.* **2018**, *25*, 10–16. [CrossRef]
3. Rangan, S.; Rappaport, T.S.; Erkip, E. Millimeter-wave cellular wireless networks: Potentials and challenges. *Proc. IEEE* **2014**, *102*, 366–385. [CrossRef]
4. Bariah, L.; Mohjazi, L.; Muhaidat, S.; Sofotasios, P.C.; Kurt, G.K.; Yanikomeroglu, H.; Dobre, O.A. A prospective look: Key enabling technologies, applications and open research topics in 6G networks. *IEEE Access* **2020**, *8*, 174792–174820. [CrossRef]
5. Series, E.E. 5G and EMF Explained 5G. 2019. Available online: http://www.emfexplained.info/site/misc/emf/downloads/5G&EMF%20Explained_AMTA_23Aug_2019_20.pdf (accessed on 14 February 2020).
6. Katti, R.; Prince, S. A survey on role of photonic technologies in 5G communication systems. *Photonic Netw. Commun.* **2019**, *38*, 185–205. [CrossRef]
7. Andrews, J.G.; Buzzi, S.; Choi, W.; Hanly, S.V.; Lozano, A.; Soong, A.C.; Zhang, J.C. What will 5G be? *IEEE J. Sel. Areas Commun.* **2014**, *32*, 1065–1082. [CrossRef]
8. Westberg, E.; Staudinger, J.; Annes, J.; Shilimkar, V. 5G Infrastructure RF Solutions: Challenges and Opportunities. *IEEE Microw. Mag.* **2019**, *20*, 51–58. [CrossRef]
9. Matinmikko-Blue, M.; Yrjölä, S.; Seppänen, V.; Ahokangas, P.; Hämmäinen, H.; Latva-Aho, M. Analysis of spectrum valuation elements for local 5G networks: Case study of 3.5-GHz band. *IEEE Trans. Cogn. Commun. Netw.* **2019**, *5*, 741–753. [CrossRef]
10. Fettweis, G.P. The tactile internet: Applications and challenges. *IEEE Veh. Technol. Mag.* **2014**, *9*, 64–70. [CrossRef]
11. Mobile. Wireless Communications Enablers for Twenty-Twenty (2020) Information Society. FP 7 European Project 317669. 2012. Available online: https://www.metis2020.com/ (accessed on 24 March 2022).
12. Wunder, G.; Jung, P.; Kasparick, M.; Wild, T.; Schaich, F.; Chen, Y.; Ten Brink, S.; Gaspar, I.; Michailow, N.; Festag, A.; et al. 5GNOW: non-orthogonal, asynchronous waveforms for future mobile applications. *IEEE Commun. Mag.* **2014**, *52*, 97–105. [CrossRef]
13. Sánchez-Gallegos, D.D.; Galaviz-Mosqueda, A.; Gonzalez-Compean, J.; Villarreal-Reyes, S.; Perez-Ramos, A.E.; Carrizales-Espinoza, D.; Carretero, J. On the continuous processing of health data in edge-fog-cloud computing by using micro/nanoservice composition. *IEEE Access* **2020**, *8*, 120255–120281. [CrossRef]
14. Goli-Malekabadi, Z.; Sargolzaei-Javan, M.; Akbari, M.K. An effective model for store and retrieve big health data in cloud computing. *Comput. Methods Programs Biomed.* **2016**, *132*, 75–82. [CrossRef] [PubMed]
15. Prasad, K.S.V.; Hossain, E.; Bhargava, V.K. Energy efficiency in massive MIMO-based 5G networks: Opportunities and challenges. *IEEE Wirel. Commun.* **2017**, *24*, 86–94. [CrossRef]
16. Pandey, B.C.; Mohammed, S.K. A Low-Latency Transmission Method for Massive MIMO Systems with Low Control Signaling Overhead. *IEEE Trans. Commun.* **2019**, *67*, 3292–3308. [CrossRef]
17. Liu, T.; Tong, J.; Guo, Q.; Xi, J.; Yu, Y.; Xiao, Z. Energy Efficiency of Massive MIMO Systems with Low-Resolution ADCs and Successive Interference Cancellation. *IEEE Trans. Wirel. Commun.* **2019**, *18*, 3987–4002. [CrossRef]
18. Ahmed, I.; Khammari, H.; Shahid, A.; Musa, A.; Kim, K.S.; De Poorter, E.; Moerman, I. A survey on hybrid beamforming techniques in 5G: Architecture and system model perspectives. *IEEE Commun. Surv. Tutor.* **2018**, *20*, 3060–3097. [CrossRef]
19. ETSI TS 123 501 V15.2.0 (2018-06). 5G: System Architecture for the 5G System. 2018. Available online: https://www.etsi.org/deliver/etsi_ts/123500_123599/123501/15.02.00_60/ts_123501v150200p.pdf (accessed on 19 April 2022).
20. Asif, S. *5G Mobile Communications: Concepts and Technologies*; CRC Press: Boca Raton, FL, USA, 2018.
21. GSMA Public Policy Position. 5G: System Architecture for the 5G System. 2019. Available online: https://www.gsma.com/spectrum/wp-content/uploads/2019/09/5G-Spectrum-Positions.pdf (accessed on 03 February 2020).
22. Fayek, H.H. 5G Poor and Rich Novel Control Scheme Based Load Frequency Regulation of a Two-Area System with 100% Renewables in Africa. *Fractal Fract.* **2021**, *5*, 2. [CrossRef]
23. Rong, B.; Dianati, M.; Zhou, L.; Karagiannidis, G.K.; Wang, C. 5G MmWave Small Cell Networks: Architecture, Self-Organization, and Management. *IEEE Wirel. Commun.* **2018**, *25*, 8–9. [CrossRef]
24. Paniagua, C.; Delsing, J. Industrial Frameworks for Internet of Things: A Survey. *IEEE Syst. J.* **2020**, *15*, 1149–1159. [CrossRef]
25. Lv, Z.; Chen, D.; Lou, R.; Song, H. Industrial security solution for virtual reality. *IEEE Internet Things J.* **2020**. [CrossRef]
26. Sarker, E.; Halder, P.; Seyedmahmoudian, M.; Jamei, E.; Horan, B.; Mekhilef, S.; Stojcevski, A. Progress on the demand side management in smart grid and optimization approaches. *Int. J. Energy Res.* **2021**, *45*, 36–64. [CrossRef]

27. Martinho, A.; Herber, N.; Kroesen, M.; Chorus, C. Ethical issues in focus by the autonomous vehicles industry. *Transp. Rev.* **2021**, *41*, 556–577. [CrossRef]
28. Zhang, Z.; Xiao, Y.; Ma, Z.; Xiao, M.; Ding, Z.; Lei, X.; Karagiannidis, G.K.; Fan, P. 6G wireless networks: Vision, requirements, architecture, and key technologies. *IEEE Veh. Technol. Mag.* **2019**, *14*, 28–41. [CrossRef]
29. Zhong, M.; Yang, Y.; Yao, H.; Fu, X.; Dobre, O.A.; Postolache, O. 5G and IoT: Towards a new era of communications and measurements. *IEEE Instrum. Meas. Mag.* **2019**, *22*, 18–26. [CrossRef]
30. Ghosh, A.; Maeder, A.; Baker, M.; Chandramouli, D. 5G evolution: A view on 5G cellular technology beyond 3GPP release 15 *IEEE Access* **2019**, *7*, 127639–127651. [CrossRef]
31. Khan, T.A.; Heath, R.W., Jr. Wireless Power Transfer in Millimeter Wave. *Wirel. Inf. Power Transf. Theory Pract.* **2018**, Book Chapter 8, 139–156.
32. Lin, X. An overview of 5G advanced evolution in 3GPP release 18. *IEEE Commun. Stand. Mag.* **2022**, *6*, 77–83. [CrossRef]
33. OOKLA. 5G MAP. 2020. Available online: https://www.speedtest.net/ookla-5g-map (accessed on 24 March 2023).
34. Qualcomm Technologies, I. Qualcomm and NTT DOCOMO Enable World's First. 2020. Available online: https://www.qualcomm.com/news/releases/2020/12/07/qualcomm-and-ntt-docomo-enable-worlds-first-commercialization-5g-sub-6-ghz (accessed on 15 September 2021).
35. Saad, W.; Bennis, M.; Chen, M. A vision of 6G wireless systems: Applications, trends, technologies, and open research problems. *IEEE Netw.* **2019**, *34*, 134–142. [CrossRef]
36. Nguyen, H.V.; Kim, H.M.; Kang, G.M.; Nguyen, K.H.; Bui, V.P.; Shin, O.S. A survey on non-orthogonal multiple access: From the perspective of spectral efficiency and energy efficiency. *Energies* **2020**, *13*, 4106. [CrossRef]
37. Dai, L.; Wang, B.; Yuan, Y.; Han, S.; Chih-Lin, I.; Wang, Z. Non-orthogonal multiple access for 5G: Solutions, challenges, opportunities, and future research trends. *IEEE Commun. Mag.* **2015**, *53*, 74–81. [CrossRef]
38. Basharat, M.; Ejaz, W.; Naeem, M.; Khattak, A.M.; Anpalagan, A. A survey and taxonomy on nonorthogonal multiple-access schemes for 5G networks. *Trans. Emerg. Telecommun. Technol.* **2018**, *29*, e3202. [CrossRef]
39. Ahmadinejad, H.; Falahati, A. Spectral efficiency in non-terrestrial heterogeneous networks with spectrum underlay access. *Phys. Commun.* **2021**, *46*, 101313. [CrossRef]
40. Banelli, P.; Buzzi, S.; Colavolpe, G.; Modenini, A.; Rusek, F.; Ugolini, A. Modulation formats and waveforms for 5G networks: Who will be the heir of OFDM? An overview of alternative modulation schemes for improved spectral efficiency. *IEEE Signal Process. Mag.* **2014**, *31*, 80–93. [CrossRef]
41. Gandotra, P.; Jha, R.K.; Jain, S. A survey on device-to-device (D2D) communication: Architecture and security issues. *J. Netw. Comput. Appl.* **2017**, *78*, 9–29. [CrossRef]
42. Adedoyin, M.A.; Falowo, O.E. Combination of ultra-dense networks and other 5G enabling technologies: A survey. *IEEE Access* **2020**, *8*, 22893–22932. [CrossRef]
43. Liu, P.; Luo, K.; Chen, D.; Jiang, T. Spectral efficiency analysis of cell-free massive MIMO systems with zero-forcing detector. *IEEE Trans. Wirel. Commun.* **2019**, *19*, 795–807. [CrossRef]
44. Kanhere, O.; Rappaport, T.S. Position location for futuristic cellular communications: 5G and beyond. *IEEE Commun. Mag.* **2021**, *59*, 70–75. [CrossRef]
45. Cardona, N.; Correia, L.M.; Calabuig, D. Key enabling technologies for 5G: Millimeter-wave and massive MIMO. *Int. J. Wirel. Inf. Netw.* **2017**, *24*, 201–203. [CrossRef]
46. Al-Falahy, N.; Alani, O.Y. Technologies for 5G networks: Challenges and opportunities. *It Prof.* **2017**, *19*, 12–20. [CrossRef]
47. Shafique, K.; Khawaja, B.A.; Sabir, F.; Qazi, S.; Mustaqim, M. Internet of things (IoT) for next-generation smart systems: A review of current challenges, future trends and prospects for emerging 5G-IoT scenarios. *IEEE Access* **2020**, *8*, 23022–23040. [CrossRef]
48. Sudhamani, C.; Roslee, M.; Tiang, J.J.; Rehman, A.U. A Survey on 5G Coverage Improvement Techniques: Issues and Future Challenges. *Sensors* **2023**, *23*, 2356. [CrossRef] [PubMed]
49. Sharma, T.; Chehri, A.; Fortier, P. Review of optical and wireless backhaul networks and emerging trends of next generation 5G and 6G technologies. *Trans. Emerg. Telecommun. Technol.* **2021**, *32*, e4155. [CrossRef]
50. Hossain, E.; Hasan, M. 5G cellular: Key enabling technologies and research challenges. *IEEE Instrum. Meas. Mag.* **2015**, *18*, 11–21. [CrossRef]
51. Akyildiz, I.F.; Nie, S.; Lin, S.C.; Chandrasekaran, M. 5G roadmap: 10 key enabling technologies. *Comput. Netw.* **2016**, *106*, 17–48. [CrossRef]
52. Salah, I.; Mabrook, M.M.; Hussein, A.I.; Rahouma, K.H. Comparative study of efficiency enhancement technologies in 5G networks—A survey. *Procedia Comput. Sci.* **2021**, *182*, 150–158. [CrossRef]
53. Saha, R.K. In-Building Capacity Enhancement using Small Cells in Mobile Networks: An Overview. *Eng. J.* **2022**, *26*, 53–71. [CrossRef]
54. Vaezi, M.; Baduge, G.A.A.; Liu, Y.; Arafa, A.; Fang, F.; Ding, Z. Interplay between NOMA and other emerging technologies: A survey. *IEEE Trans. Cogn. Commun. Netw.* **2019**, *5*, 900–919. [CrossRef]
55. Ahmad, W.S.H.M.W.; Radzi, N.A.M.; Samidi, F.; Ismail, A.; Abdullah, F.; Jamaludin, M.Z.; Zakaria, M. 5G technology: Towards dynamic spectrum sharing using cognitive radio networks. *IEEE Access* **2020**, *8*, 14460–14488. [CrossRef]
56. Salah, M.; Omer, O.A.; Mohammed, U.S. Spectral efficiency enhancement based on sparsely indexed modulation for green radio communication. *IEEE Access* **2019**, *7*, 31913–31925. [CrossRef]

7. Mikki, S.; Hanoon, A. Spectral efficiency enhancement using an antenna-based orthogonal frequency division multiaccess technique. *Int. J. Microw. Comput.-Aided Eng.* **2020**, *30*, e22404. [CrossRef]
8. Wen, M.; Zheng, B.; Kim, K.J.; Di Renzo, M.; Tsiftsis, T.A.; Chen, K.C.; Al-Dhahir, N. A survey on spatial modulation in emerging wireless systems: Research progresses and applications. *IEEE J. Sel. Areas Commun.* **2019**, *37*, 1949–1972. [CrossRef]
9. Dhariwal, A.; Mohammed, V.N. Metamorphosis of 5G Wireless Communication: A Review. *Telecommun. Radio Eng.* **2019**, *78*, 1567–1588. [CrossRef]
0. Shaham, S.; Ding, M.; Kokshoorn, M.; Lin, Z.; Dang, S.; Abbas, R. Fast channel estimation and beam tracking for millimeter wave vehicular communications. *IEEE Access* **2019**, *7*, 141104–141118. [CrossRef]
1. Zhang, C.; Zhang, W.; Wang, W.; Yang, L.; Zhang, W. Research challenges and opportunities of UAV millimeter-wave communications. *IEEE Wirel. Commun.* **2019**, *26*, 58–62. [CrossRef]
2. Zhang, L.; Zhao, H.; Hou, S.; Zhao, Z.; Xu, H.; Wu, X.; Wu, Q.; Zhang, R. A survey on 5G millimeter wave communications for UAV-assisted wireless networks. *IEEE Access* **2019**, *7*, 117460–117504. [CrossRef]
3. Niu, Y.; Li, Y.; Jin, D.; Su, L.; Vasilakos, A.V. A survey of millimeter wave communications (mmWave) for 5G: Opportunities and challenges. *Wirel. Netw.* **2015**, *21*, 2657–2676. [CrossRef]
4. Palacios, J.; Bielsa, G.; Casari, P.; Widmer, J. Communication-driven localization and mapping for millimeter wave networks. In Proceedings of the IEEE INFOCOM 2018-IEEE Conference on Computer Communications, Honolulu, HI, USA, 16–19 April 2018; pp. 2402–2410.
5. Yang, G.; Xiao, M.; Alam, M.; Huang, Y. Low-latency heterogeneous networks with millimeter-wave communications. *IEEE Commun. Mag.* **2018**, *56*, 124–129. [CrossRef]
6. Yu, P.; Li, W.; Zhou, F.; Feng, L.; Yin, M.; Guo, S.; Gao, Z.; Qiu, X. Capacity enhancement for 5G networks using mmWave aerial base stations: Self-organizing architecture and approach. *IEEE Wirel. Commun.* **2018**, *25*, 58–64. [CrossRef]
7. Akbar, M.N.; Atique, S.; Saquib, M.; Ali, M. Capacity Enhancement of Indoor 5G mmWave Communication by Beam Steering and Narrowing. In Proceedings of the 2018 10th International Conference on Electrical and Computer Engineering (ICECE), Dhaka, Bangladesh, 20–22 December 2018; pp. 85–88.
8. Hamed, A.M.; Rao, R.K. Spectral and energy efficiencies in mmwave cellular networks for optimal utilization. *Wirel. Commun. Mob. Comput.* **2018**, *2018*, 3097094. [CrossRef]
9. Chopra, G.; Jha, R.K.; Jain, S. Security issues in ultra dense network for 5G scenario. In Proceedings of the 2018 10th International Conference on Communication Systems & Networks (COMSNETS), Bengaluru, India, 3–7 January 2018; pp. 510–512.
0. Rahmani, Z. Implementation of New Multiple Access Technique Encoder for 5G Wireless Telecomunication Networks. Ph.D. Thesis, École Polytechnique de Montréal, Montréal, QC, Canada, 2017.
1. Fang, D.; Huang, Y.C.; Ding, Z.; Geraci, G.; Shieh, S.L.; Claussen, H. Lattice partition multiple access: A new method of downlink non-orthogonal multiuser transmissions. In Proceedings of the 2016 IEEE Global Communications Conference (GLOBECOM), Washington, DC, USA, 4–8 December 2016; pp. 1–6.
2. Duel-Hallen, A.; Holtzman, J.; Zvonar, Z. Multiuser detection for CDMA systems. *IEEE Pers. Commun.* **1995**, *2*, 46–58. [CrossRef]
3. Al-Imari, M.; Imran, M.A. Low Density Spreading Multiple Access. In *Multiple Access Techniques for 5G Wireless Networks and Beyond*; Springer: Berlin/Heidelberg, Germany, 2019; pp. 493–514.
4. Dai, X.; Zhang, Z.; Bai, B.; Chen, S.; Sun, S. Pattern division multiple access: A new multiple access technology for 5G. *IEEE Wirel. Commun.* **2018**, *25*, 54–60. [CrossRef]
5. Tayade, P.P.; Rohokale, V.M. Enhancement of spectral efficiency, coverage and channel capacity for wireless communication towards 5G. In Proceedings of the 2015 International Conference on Pervasive Computing (ICPC), Pune, India, 8–10 January 2015.
6. Vaezi, M., Ding, Z.; Poor, H.V. *Multiple Access Techniques for 5G Wireless Networks and Beyond*; Springer: Berlin/Heidelberg, Germany, 2019.
7. Maric, S.; Velimirovic, L.Z. Application of Quasi Orthogonal Short Sequence Families in Pattern Division Multiple Access—A Non Orthogonal Multiple Access Technique. In Proceedings of the 2018 IEEE 5G World Forum (5GWF), Silicon Valley, CA, USA, 9–11 July 2018; pp. 373–376.
8. Islam, S.R.; Avazov, N.; Dobre, O.A.; Kwak, K.S. Power-domain non-orthogonal multiple access (NOMA) in 5G systems: Potentials and challenges. *IEEE Commun. Surv. Tutor.* **2016**, *19*, 721–742. [CrossRef]
9. Liu, C.H.; Liang, D.C. Heterogeneous networks with power-domain NOMA: Coverage, throughput, and power allocation analysis. *IEEE Trans. Wirel. Commun.* **2018**, *17*, 3524–3539. [CrossRef]
30. Mathur, H.; Deepa, T. A Survey on Advanced Multiple Access Techniques for 5G and Beyond Wireless Communications. *Wirel. Pers. Commun.* **2021**, *118*, 1775–1792. [CrossRef]
31. Verdu, S. *Multiuser Detection*; Cambridge University Press: Cambridge, UK, 1998.
32. Wang, H.; Zhang, Z.; Chen, X. Resource allocation for downlink joint space-time and power domain non-orthogonal multiple access. In Proceedings of the 2017 9th International Conference on Wireless Communications and Signal Processing (WCSP), Nanjing, China, 11–13 October 2017; pp. 1–6.
33. Shin, W.; Vaezi, M.; Lee, B.; Love, D.J.; Lee, J.; Poor, H.V. Non-orthogonal multiple access in multi-cell networks: Theory, performance, and practical challenges. *IEEE Commun. Mag.* **2017**, *55*, 176–183. [CrossRef]
34. Aldababsa, M.; Toka, M.; Gökçeli, S.; Kurt, G.K.; Kucur, O. A tutorial on nonorthogonal multiple access for 5G and beyond. *Wirel. Commun. Mob. Comput.* **2018**, *2018*, 9713450. [CrossRef]

85. Wang, X.; Poor, H.V. *Wireless Communication Systems: Advanced Techniques for Signal Reception*; Prentice Hall Professional: Hoboken, NJ, USA, 2004.
86. Ding, Z.; Liu, Y.; Choi, J.; Sun, Q.; Elkashlan, M.; Chih-Lin, I.; Poor, H.V. Application of non-orthogonal multiple access in LTE and 5G networks. *IEEE Commun. Mag.* **2017**, *55*, 185–191. [CrossRef]
87. Wang, P.; Xiao, J.; Li, P. Comparison of orthogonal and non-orthogonal approaches to future wireless cellular systems. *IEEE Veh. Technol. Mag.* **2006**, *1*, 4–11. [CrossRef]
88. Lin, S.; Kong, L.; Gao, Q.; Khan, M.K.; Zhong, Z.; Jin, X.; Zeng, P. Advanced dynamic channel access strategy in spectrum sharing 5G systems. *IEEE Wirel. Commun.* **2017**, *24*, 74–80. [CrossRef]
89. Ouyang, F. Massive MIMO for dynamic spectrum access. In Proceedings of the 2017 IEEE International Conference on Consumer Electronics (ICCE), Las Vegas, NV, USA, 8–10 January 2017; pp. 9–12.
90. Zhang, S.; Xu, X.; Lu, L.; Wu, Y.; He, G.; Chen, Y. Sparse code multiple access: An energy efficient uplink approach for 5G wireless systems. In Proceedings of the 2014 IEEE Global Communications Conference, Austin, TX, USA, 8–12 December 2014; pp. 4782–4787.
91. Nikopour, H.; Yi, E.; Bayesteh, A.; Au, K.; Hawryluck, M.; Baligh, H.; Ma, J. SCMA for downlink multiple access of 5G wireless networks. In Proceedings of the 2014 IEEE Global Communications Conference, Austin, TX, USA, 8–12 December 2014; pp. 3940–3945.
92. Elkawafi, S.; Younis, A.; Mesleh, R. Performance Analysis of Sparse Code Multiple Access MIMO Systems. In Proceedings of the 2019 IEEE 30th Annual International Symposium on Personal, Indoor and Mobile Radio Communications (PIMRC), Istanbul, Turkey, 8–11 September 2019; pp. 1–6.
93. Chen, J.; Zhang, Z.; Fu, S.; Hu, J. A Joint Update Parallel MCMC-Method-Based Sparse Code Multiple Access Decoder. *IEEE Trans. Veh. Technol.* **2018**, *67*, 1280–1291. [CrossRef]
94. Ge, X.; Tu, S.; Mao, G.; Wang, C.; Han, T. 5G Ultra-Dense Cellular Networks. *IEEE Wirel. Commun.* **2016**, *23*, 72–79. [CrossRef]
95. Hoeher, P.A.; Doose, N. A massive MIMO terminal concept based on small-size multi-mode antennas. *Trans. Emerg. Telecommun. Technol.* **2017**, *28*, e2934. [CrossRef]
96. Hao, P.; Yan, X.; Yu-Ngok, R.; Yuan, Y. Ultra dense network: Challenges enabling technologies and new trends. *China Commun.* **2016**, *13*, 30–40.
97. Wang, H.; Chen, S.; Ai, M.; Xu, H. Localized Mobility Management for 5G Ultra Dense Network. *IEEE Trans. Veh. Technol.* **2017**, *66*, 8535–8552. [CrossRef]
98. Li, W.; Wang, J.; Li, L.; Zhang, G.; Dang, Z.; Li, S. Intelligent Anti-Jamming Communication with Continuous Action Decision for Ultra-Dense Network. In Proceedings of the ICC 2019–2019 IEEE International Conference on Communications (ICC), Shanghai, China, 20–24 May 2019.
99. Garcia-Rodriguez, A.; Geraci, G.; Giordano, L.G.; Bonfante, A.; Ding, M.; López-Pérez, D. Massive MIMO unlicensed: A new approach to dynamic spectrum access. *IEEE Commun. Mag.* **2017**, *56*, 186–192. [CrossRef]
100. Mumtaz, S.; Al-Dulaimi, A.; Frascolla, V.; Niyato, D.; Briggs, K. Dynamic Spectrum Management for 5G. *IEEE Wirel. Commun.* **2017**, *24*, 12–13. [CrossRef]
101. Wu, Q.; Li, G.Y.; Chen, W.; Ng, D.W.K.; Schober, R. An overview of sustainable green 5G networks. *IEEE Wirel. Commun.* **2017**, *24*, 72–80. [CrossRef]
102. Belikaidis, I.P.; Georgakopoulos, A.; Demestichas, P.; Miscopein, B.; Filo, M.; Vahid, S.; Okyere, B.; Fitch, M. Multi-RAT dynamic spectrum access for 5G heterogeneous networks: The SPEED-5G approach. *IEEE Wirel. Commun.* **2017**, *24*, 14–22. [CrossRef]
103. Zhang, L.; Xiao, M.; Wu, G.; Alam, M.; Liang, Y.C.; Li, S. A survey of advanced techniques for spectrum sharing in 5G networks. *IEEE Wirel. Commun.* **2017**, *24*, 44–51. [CrossRef]
104. Zhou, J.; Reiskarimian, N.; Diakonikolas, J.; Dinc, T.; Chen, T.; Zussman, G.; Krishnaswamy, H. Integrated full duplex radios. *IEEE Commun. Mag.* **2017**, *55*, 142–151. [CrossRef]
105. Li, R.; Chen, Y.; Li, G.Y.; Liu, G. Full-duplex cellular networks. *IEEE Commun. Mag.* **2017**, *55*, 184–191. [CrossRef]
106. Chen, T.; Baraani Dastjerdi, M.; Welles, J.; Zhou, J.; Krishnaswamy, H.; Zussman, G. Poster: Enabling Wideband Full-Duplex Wireless via Frequency-Domain Equalization. In Proceedings of the 25th Annual International Conference on Mobile Computing and Networking, Los Cabos, Mexico, 21–25 October 2019; pp. 1–3.
107. Luo, F.L.; Zhang, C.J. *Signal Processing for 5G: Algorithms and Implementations*; John Wiley & Sons: Hoboken, NJ, USA, 2016.
108. Zhang, Z.; Chai, X.; Long, K.; Vasilakos, A.V.; Hanzo, L. Full duplex techniques for 5G networks: self-interference cancellation, protocol design, and relay selection. *IEEE Commun. Mag.* **2015**, *53*, 128–137. [CrossRef]
109. Reiskarimian, N.; Dastjerdi, M.B.; Zhou, J.; Krishnaswamy, H. 18.2 highly-linear integrated magnetic-free circulator-receiver for full-duplex wireless. In Proceedings of the 2017 IEEE International Solid-State Circuits Conference (ISSCC), San Francisco, CA, USA, 5–9 February 2017; pp. 316–317.
110. Razavizadeh, S.M.; Ahn, M.; Lee, I. Three-dimensional beamforming: A new enabling technology for 5G wireless networks. *IEEE Signal Process. Mag.* **2014**, *31*, 94–101. [CrossRef]
111. Yang, B.; Yu, Z.; Lan, J.; Zhang, R.; Zhou, J.; Hong, W. Digital beamforming-based massive MIMO transceiver for 5G millimeter-wave communications. *IEEE Trans. Microw. Theory Tech.* **2018**, *66*, 3403–3418. [CrossRef]
112. Bian, J.; Wang, C.X.; Gao, X.; You, X.; Zhang, M. A general 3D non-stationary wireless channel model for 5G and beyond. *IEEE Trans. Wirel. Commun.* **2021**, *20*, 3211–3224. [CrossRef]

13. Bonfante, A.; Giordano, L.G.; López-Pérez, D.; Garcia-Rodriguez, A.; Geraci, G.; Baracca, P.; Butt, M.M.; Marchetti, N. 5G massive MIMO architectures: Self-backhauled small cells versus direct access. *IEEE Trans. Veh. Technol.* **2019**, *68*, 10003–10017. [CrossRef]
14. Benzaghta, M.; Rabie, K.M. Massive MIMO systems for 5G: A systematic mapping study on antenna design challenges and channel estimation open issues. *IET Commun.* **2021**, *15*, 1677–1690. [CrossRef]
15. Lee, B.M. Adaptive Switching Scheme for RS Overhead Reduction in Massive MIMO with Industrial Internet of Things. *IEEE Internet Things J.* **2020**, *8*, 2585–2602. [CrossRef]
16. Zeng, M.; Hao, W.; Dobre, O.A.; Ding, Z.; Poor, H.V. Massive MIMO-Assisted Mobile Edge Computing: Exciting Possibilities for Computation Offloading. *IEEE Veh. Technol. Mag.* **2020**, *15*, 31–38. [CrossRef]
17. Liu, Z.; Zhang, L.; Ding, Z. Overcoming the Channel Estimation Barrier in Massive MIMO Communication via Deep Learning. *IEEE Wirel. Commun.* **2020**, *27*, 104–111. [CrossRef]
18. Maruta, K.; Falcone, F. Massive MIMO Systems: Present and Future. *Electronics* **2020**, *9*, 385. [CrossRef]
19. Mchangama, A.; Ayadi, J.; Jiménez, V.P.G.; Consoli, A. MmWave massive MIMO small cells for 5G and beyond mobile networks: An overview. In Proceedings of the 2020 12th International Symposium on Communication Systems, Networks and Digital Signal Processing (CSNDSP), Porto, Portugal, 20–22 July 2020; pp. 1–6.
120. Quazi, A.; Rawat, P.; Gupta, S.K. Overview of Massive MIMO System and Capacity Evaluation for 5G Cellular Networks. In Proceedings of the Industry Interactive Innovations in Science, Engineering & Technology (I3SET2K19), Chennai, India, 16–17 May 2020.
121. Sur, S.N.; Bera, R.; Kumar Bhoi, A.; Shaik, M.; Marques, G. Capacity Analysis of Lattice Reduction Aided Equalizers for Massive MIMO Systems. *Information* **2020**, *11*, 301. [CrossRef]
122. Chataut, R.; Akl, R. Massive MIMO Systems for 5G and beyond Networks—Overview, Recent Trends, Challenges, and Future Research Direction. *Sensors* **2020**, *20*, 2753. [CrossRef] [PubMed]
123. Wei, G.; Zhang, B.; Ding, G.; Zhao, B.; Wei, Y.; Guo, D. Massive MIMO-Based Distributed Signal Detection in Multi-Antenna Wireless Sensor Networks. *Sensors* **2020**, *20*, 2005. [CrossRef]
124. Dicandia, F.A.; Genovesi, S. Exploitation of triangular lattice arrays for improved spectral efficiency in massive MIMO 5G systems. *IEEE Access* **2021**, *9*, 17530–17543. [CrossRef]
125. Patrudu, B.C.K.; Sridevi, P. Comparison of Uplink Spectral Efficiency in Massive MIMO Systems. In *Intelligent Computing in Control and Communication*; Springer: Berlin/Heidelberg, Germany, 2021; pp. 13–22.
126. Papageorgiou, G.K.; Sellathurai, M.; Ntougias, K.; Papadias, C.B. A Stochastic Optimization Approach to Hybrid Processing in Massive MIMO Systems. *IEEE Wirel. Commun. Lett.* **2020**, *9*, 770–773. [CrossRef]
127. Gao, P.; Sanada, Y. Effect of quantization range limit for low-resolution analog-to-digital converters in full-digital massive MIMO system. In Proceedings of the 2019 IEEE VTS Asia Pacific Wireless Communications Symposium (APWCS), Singapore, 28–30 August 2019; pp. 1–5.
128. Pfadler, A.; Ballesteros, C.; Romeu, J.; Jofre, L. Hybrid Massive MIMO for Urban V2I: Sub-6 GHz vs mmWave Performance Assessment. *IEEE Trans. Veh. Technol.* **2020**, *69*, 4652–4662. [CrossRef]
129. Ning, B.; Chen, Z.; Chen, W.; Du, Y.; Fang, J. Terahertz multi-user massive MIMO with intelligent reflecting surface: Beam training and hybrid beamforming. *IEEE Trans. Veh. Technol.* **2021**, *70*, 1376–1393. [CrossRef]
130. Zhai, X.; Chen, X.; Xu, J.; Ng, D.W.K. Hybrid beamforming for massive MIMO over-the-air computation. *IEEE Trans. Commun.* **2021**, *69*, 2737–2751. [CrossRef]
131. Vieira, J.; Malkowsky, S.; Nieman, K.; Miers, Z.; Kundargi, N.; Liu, L.; Wong, I.; Öwall, V.; Edfors, O.; Tufvesson, F. A flexible 100-antenna testbed for massive MIMO. In Proceedings of the 2014 IEEE Globecom Workshops (GC Wkshps), Austin, TX, USA, 8–12 December 2014; pp. 287–293.
132. Xiang, W.; Zheng, K.; Shen, X.S. *5G Mobile Communications*; Springer: Berlin/Heidelberg, Germany, 2016.
133. Busari, S.A.; Huq, K.M.S.; Mumtaz, S.; Rodriguez, J. Terahertz massive MIMO for beyond-5G wireless communication. In Proceedings of the ICC 2019–2019 IEEE International Conference on Communications (ICC), Shanghai, China, 20–24 May 2019.
134. Mollén, C.; Larsson, E.G.; Gustavsson, U.; Eriksson, T.; Heath, R.W. Out-of-band radiation from large antenna arrays. *IEEE Commun. Mag.* **2018**, *56*, 196–203. [CrossRef]
135. Björnson, E.; Hoydis, J.; Sanguinetti, L. Massive MIMO has unlimited capacity. *IEEE Trans. Wirel. Commun.* **2017**, *17*, 574–590. [CrossRef]
136. Björnson, E.; Hoydis, J.; Sanguinetti, L. Massive MIMO networks: Spectral, energy, and hardware efficiency. *Found. Trends Signal Process.* **2017**, *11*, 154–655. [CrossRef]
137. Mir, T.; Siddiqi, Z.; Mir, U.; MacKenzie, R.; Hao, M. One-Bit Hybrid Precoding for Wideband Millimeter-Wave Massive MIMO Systems. In Proceedings of the 2019 IEEE 89th Vehicular Technology Conference (VTC2019-Spring), Kuala Lumpur, Malaysia, 28 April–1 May 2019; pp. 1–6.
138. Okyere, B.; Musavian, L.; Mumtaz, R. Multi-User Massive MIMO and Physical Layer Network Coding. In Proceedings of the 2019 IEEE Globecom Workshops (GC Wkshps), Waikoloa, HI, USA, 9–13 December 2019; pp. 1–6.
139. Xu, C.; Hu, Y.; Liang, C.; Ma, J.; Ping, L. Massive MIMO, non-orthogonal multiple access and interleave division multiple access. *IEEE Access* **2017**, *5*, 14728–14748. [CrossRef]
140. Jedda, H.; Nossek, J.A. On PAPR in Single-Carrier Massive MIMO Systems. In Proceedings of the WSA 2019; 23rd International ITG Workshop on Smart Antennas, Vienna, Austria, 24–26 April 2019; pp. 1–5.

141. Shlezinger, N.; Eldar, Y.C. Spectral Efficiency of Noncooperative Uplink Massive MIMO Systems with Joint Decoding. In Proceedings of the ICASSP 2019–2019 IEEE International Conference on Acoustics, Speech and Signal Processing (ICASSP) Brighton, UK, 12–17 May 2019; pp. 4779–4783.
142. Al-nahari, A.; Sakran, H.; Su, W.; Tarbosh, S. Energy and spectral efficiency of secure massive MIMO downlink systems. *IET Commun.* **2019**, *13*, 1364–1372. [CrossRef]
143. Akhlaghpasand, H.; Björnson, E.; Razavizadeh, S.M. Jamming Suppression in Massive MIMO Systems. *IEEE Trans. Circuits Syst II Express Briefs* **2019**, *67*, 182–186. [CrossRef]
144. Hasan, W.B.; Harris, P.; Bromell, H.; Doufexi, A.; Beach, M. Adaptive User Grouping Based on EVM Prediction for Efficient & Robust Massive MIMO in TDD. *IEEE Access* **2019**, *7*, 162683–162696.
145. Azeem, H.; Du, L.; Ullah, A.; Mughal, M.A.; Aslam, M.M.; Ikram, M. Sub-array Based Antenna Selection Scheme for Massive MIMO in 5G. In *Cyberspace Data and Intelligence, and Cyber-Living, Syndrome, and Health*; Springer: Berlin/Heidelberg, Germany, 2019; pp. 38–50.
146. Albreem, M.A.; Juntti, M.; Shahabuddin, S. Massive MIMO detection techniques: a survey. *IEEE Commun. Surv. Tutor.* **2019**, *21*, 3109–3132. [CrossRef]
147. Lu, A.A.; Gao, X.; Meng, X.; Xia, X.G. Omnidirectional Precoding for 3D Massive MIMO with Uniform Planar Arrays. *IEEE Trans. Wirel. Commun.* **2020**, *19*, 2628–2642. [CrossRef]
148. Jaglan, N.; Gupta, S.D.; Sharawi, M.S. 18 element massive MIMO/diversity 5G smartphones antenna design for sub-6 GHz LTE bands 42/43 applications. *IEEE Open J. Antennas Propag.* **2021**, *2*, 533–545. [CrossRef]
149. Li, Y.; Sim, C.-Y.-D.; Luo, Y.; Yang, G. 12-port 5G massive MIMO antenna array in sub-6GHz mobile handset for LTE bands 42/43/46 applications. *IEEE Access* **2017**, *6*, 344–354. [CrossRef]
150. Deng, J.Y.; Yao, J.; Sun, D.Q.; Guo, L.X. Ten-element MIMO antenna for 5G terminals. *Microw. Opt. Technol. Lett.* **2018**, *60*, 3045–3049. [CrossRef]
151. Ullah, R.; Ullah, S.; Ullah, R.; Faisal, F.; Mabrouk, I.B.; Al Hasan, M.J. A 10-ports MIMO antenna system for 5G smart-phone applications. *IEEE Access* **2020**, *8*, 218477–218488. [CrossRef]
152. Sufian, M.A.; Hussain, N.; Abbas, A.; Lee, J.; Park, S.G.; Kim, N. Mutual coupling reduction of a circularly polarized MIMO antenna using parasitic elements and DGS for V2X communications. *IEEE Access* **2022**, *10*, 56388–56400. [CrossRef]
153. Jaglan, N.; Gupta, S.D.; Kanaujia, B.K.; Sharawi, M.S. 10 element sub-6-GHz multi-band double-T based MIMO antenna system for 5G smartphones. *IEEE Access* **2021**, *9*, 118662–118672. [CrossRef]
154. Li, Y.; Sim, C.-Y.-D.; Luo, Y.; Yang, G. Multiband 10-antenna array for sub-6 GHz MIMO applications in 5-G smartphones. *IEEE Access* **2018**, *6*, 28041–28053. [CrossRef]
155. Cui, L.; Guo, J.; Liu, Y.; Sim, C.-Y.-D. An 8-element dual-band MIMO antenna with decoupling stub for 5G smartphone applications. *IEEE Antennas Wirel. Propag. Lett.* **2019**, *18*, 2095–2099. [CrossRef]
156. Hei, Y.Q.; He, J.G.; Li, W.T. Wideband decoupled 8-element MIMO antenna for 5G mobile terminal applications. *IEEE Antennas Wirel. Propag. Lett.* **2021**, *20*, 1448–1452. [CrossRef]
157. Shi, X.; Zhang, M.; Xu, S.; Liu, D.; Wen, H.; Wang, J. Dual-band 8-element MIMO antenna with short neutral line for 5G mobile handset. In Proceedings of the 2017 11th European conference on antennas and propagation (EUCAP), Paris, France, 19–24 March 2017; pp. 3140–3142.
158. Serghiou, D.; Khalily, M.; Singh, V.; Araghi, A.; Tafazolli, R. Sub-6 GHz dual-band 8 × 8 MIMO antenna for 5G smartphones. *IEEE Antennas Wirel. Propag. Lett.* **2020**, *19*, 1546–1550. [CrossRef]
159. Ibrahim, S.K.; Singh, M.J.; Al-Bawri, S.S.; Ibrahim, H.H.; Islam, M.T.; Islam, M.S.; Alzamil, A.; Abdulkawi, W.M. Design, Challenges and Developments for 5G Massive MIMO Antenna Systems at Sub 6-GHz Band: A Review. *Nanomaterials* **2023**, *13*, 520. [CrossRef]
160. Wasim, M.; Khera, S.; Malik, P.K.; Kumari, S.V.; Das, S.; El-Shafai, W.; Aly, M.H. Base Station MIMO Antenna in 1 × 6 Array Configurations with Reflector Design for Sub-6 GHz 5G Applications. *Electronics* **2023**, *12*, 669. [CrossRef]
161. Bellary, A.; Kandasamy, K.; Rao, P.H. Analysis of wave propagation models with radio network planning using dual polarized MIMO antenna for 5G base station applications. *IEEE Access* **2022**, *10*, 29183–29193. [CrossRef]
162. Elsakka, A.; Farsaei, A.; van den Biggelaar, A.; Reniers, A.C.F.; Johansson, M.N.; Maaskant, R.; Johannsen, U.; Iupikov, O.A.; Smolders, A.B.; Ivashina, M.V. A mm-wave phased-array fed torus reflector antenna with ±30° scan range for massive-mimo base-station applications. *IEEE Trans. Antennas Propag.* **2022**, *70*, 3398–3410. [CrossRef]
163. Wali, S.Q.; Sali, A.; Allami, J.K.; Osman, A.F. RF-EMF exposure measurement for 5G over mm-wave base station with MIMO antenna. *IEEE Access* **2022**, *10*, 9048–9058. [CrossRef]
164. Manteuffel, D.; Martens, R. Compact multimode multielement antenna for indoor UWB massive MIMO. *IEEE Trans. Antennas Propag.* **2016**, *64*, 2689–2697. [CrossRef]
165. Hu, H.; Gao, H.; Li, Z.; Zhu, Y. A sub 6GHz massive MIMO system for 5G new radio. In Proceedings of the 2017 IEEE 85th Vehicular Technology Conference (VTC Spring), Sydney, NSW, Australia, 4–7 June 2017; pp. 1–5.
166. Al-Bawri, S.S.; Islam, M.T.; Islam, M.S.; Singh, M.J.; Alsaif, H. Massive metamaterial system-loaded MIMO antenna array for 5G base stations. *Sci. Rep.* **2022**, *12*, 14311. [CrossRef] [PubMed]
167. Al-Tarifi, M.A.; Sharawi, M.S.; Shamim, A. Massive MIMO antenna system for 5G base stations with directive ports and switched beamsteering capabilities. *IET Microwaves Antennas Propag.* **2018**, *12*, 1709–1718. [CrossRef]

68. Idowu-Bismark, O.; Okokpujie, K.O.; Ryan, H.; Adedokun, M.O. 5G Wireless Communication Network Architecture and Its Key Enabling Technologies. *Int. Rev. Aerosp. Eng.* **2019**, *12*, 70–82. [CrossRef]
69. Ioushua, S.S.; Eldar, Y.C. A family of hybrid analog–digital beamforming methods for massive MIMO systems. *IEEE Trans. Signal Process.* **2019**, *67*, 3243–3257. [CrossRef]
70. Ying, K.; Gao, Z.; Lyu, S.; Wu, Y.; Wang, H.; Alouini, M.S. GMD-Based Hybrid Beamforming for Large Reconfigurable Intelligent Surface Assisted Millimeter-Wave Massive MIMO. *arXiv* **2020**, arXiv:2001.05763.
71. Ardah, K.; Fodor, G.; Silva, Y.C.; Cruz, W.; Almeida, A. Hybrid Analog-Digital Beamforming Design for SE and EE Maximization in Massive MIMO Networks. *IEEE Trans. Veh. Technol.* **2019**, *69*, 377–389. [CrossRef]
72. Payami, S.; Sellathurai, M.; Nikitopoulos, K. Low-complexity hybrid beamforming for massive MIMO systems in frequency-selective channels. *IEEE Access* **2019**, *7*, 36195–36206. [CrossRef]
73. Alluhaibi, O.; Kampert, E.; Jennings, P.A.; Higgins, M.D. Impact of Overlapped AoAs on the Achievable Uplink Rate of Hybrid Beamforming for Massive MIMO mm-Wave Systems for Industrial Environments. *IEEE Access* **2019**, *7*, 101178–101194. [CrossRef]
74. Malkowsky, S.; Vieira, J.; Liu, L.; Harris, P.; Nieman, K.; Kundargi, N.; Wong, I.C.; Tufvesson, F.; Öwall, V.; Edfors, O. The world's first real-time testbed for massive MIMO: Design, implementation, and validation. *IEEE Access* **2017**, *5*, 9073–9088. [CrossRef]
75. Björnson, E.; Larsson, E.G.; Debbah, M. Massive MIMO for maximal spectral efficiency: How many users and pilots should be allocated? *IEEE Trans. Wirel. Commun.* **2015**, *15*, 1293–1308. [CrossRef]
76. Jungnickel, V.; Manolakis, K.; Zirwas, W.; Panzner, B.; Braun, V.; Lossow, M.; Sternad, M.; Apelfrojd, R.; Svensson, T. The role of small cells, coordinated multipoint, and massive MIMO in 5G. *IEEE Commun. Mag.* **2014**, *52*, 44–51. [CrossRef]
77. Karasawa, Y. Channel Capacity of Massive MIMO with Selected Multi-Stream Transmission in Spatially Correlated Fading Environments. *IEEE Trans. Veh. Technol.* **2020**, *69*, 5320–5330. [CrossRef]
78. Wang, Y.; Chen, X.; Cai, Y.; Hanzo, L. Stochastic Hybrid Combining Design for Quantized Massive MIMO Systems. *IEEE Trans. Veh. Technol.* **2020**, *69*, 16224–16229. [CrossRef]
79. Qin, Y.; Cui, Y.; Li, R. A New Decoupling Method for Massive MIMO Antennas. In Proceedings of the 2018 IEEE International Symposium on Antennas and Propagation & USNC/URSI National Radio Science Meeting, Boston, MA, USA, 8–13 July 2018; pp. 1577–1578.
80. Saurav, K.; Mallat, N.K.; Antar, Y.M. A three-port polarization and pattern diversity ring antenna. *IEEE Antennas Wirel. Propag. Lett.* **2018**, *17*, 1324–1328. [CrossRef]
81. Zhang, Y.M.; Zhang, S.; Li, J.L.; Pedersen, G.F. A transmission-line-based decoupling method for MIMO antenna arrays. *IEEE Trans. Antennas Propag.* **2019**, *67*, 3117–3131. [CrossRef]
82. Rajkumar, S.; Amala, A.A.; Selvan, K.T. Isolation improvement of UWB MIMO antenna utilising molecule fractal structure. *Electron. Lett.* **2019**, *55*, 576–579. [CrossRef]
83. Zhu, Y.; Chen, Y.; Yang, S. Decoupling and low-profile design of dual-band dual-polarized base station antennas using frequency-selective surface. *IEEE Trans. Antennas Propag.* **2019**, *67*, 5272–5281. [CrossRef]
84. Qin, Y.; Li, R.; Cui, Y. Embeddable Structure for Reducing Mutual Coupling in Massive MIMO Antennas. *IEEE Access* **2020**, *8*, 195102–195112. [CrossRef]
85. Tang, J.; Faraz, F.; Chen, X.; Zhang, Q.; Li, Q.; Li, Y.; Zhang, S. A metasurface superstrate for mutual coupling reduction of large antenna arrays. *IEEE Access* **2020**, *8*, 126859–126867. [CrossRef]
86. Toprak, Z.; Senega, S.; Lindenmeier, S. A novel automotive Ultra-Wideband 5G-MIMO-Antenna Array printed on a foil. In Proceedings of the 2020 50th European Microwave Conference (EuMC), Utrecht, The Netherlands, 12–14 January 2021; pp. 216–219.
87. Ishteyaq, I.; Masoodi, I.S.; Muzaffar, K. Eight-port double band printed MIMO antenna investigated for mutual-coupling and SAR effects for sub-6 GHz 5G mobile applications. *Prog. Electromagn. Res.* **2021**, *113*, 111–122. [CrossRef]
88. Filgueiras, H.R.D.; Sodré, A.C. A 64-element and dual-polarized sicl-based slot antenna array development applied to tdd massive mimo. *IEEE Antennas Wirel. Propag. Lett.* **2022**, *21*, 750–754. [CrossRef]
89. Li, Y.; Bian, L.A.; Liu, Y.; Wang, Y.; Chen, R.; Xie, S. Mutual coupling reduction for monopole MIMO antenna using L-shaped stubs, defective ground and chip resistors. *AEU-Int. J. Electron. Commun.* **2023**, *160*, 154524. [CrossRef]

Disclaimer/Publisher's Note: The statements, opinions and data contained in all publications are solely those of the individual author(s) and contributor(s) and not of MDPI and/or the editor(s). MDPI and/or the editor(s) disclaim responsibility for any injury to people or property resulting from any ideas, methods, instructions or products referred to in the content.

electronics

MDPI

Review

MIMO Dielectric Resonator Antennas for 5G Applications: A Review

Hamza Ahmad [1], Mohd Haizal Jamaluddin [1,*], Fauziahanim Che Seman [2] and Muhibur Rahman [3,*]

[1] Wireless Communication Center, Faculty of Electrical Engineering, Universiti Teknologi Malaysia, Johor Bahru 81310, Malaysia; hamzaahmad@graduate.utm.my
[2] Faculti Kejuruteraan Elektrik dan Elektronik, Univeristi Tun Hussein Onn Malaysia, Parit Raja 86400, Malaysia; fauziahs@uthm.edu.my
[3] Department of Electrical Engineering, Polytechnique Montreal, Montreal, QC H3T 1J4, Canada
* Correspondence: haizal@fke.utm.my (M.H.J.); muhibur.rahman@polymtl.ca (M.R.)

Abstract: This article presents a thorough literature review of published designs of multiple-input multiple-output (MIMO) dielectric resonator antennas (DRAs) specifically designed for 5G applications. The performance of these designs is discussed in detail, considering various parameters such as gain, isolation, size, bandwidth, profile, and radiation characteristics. The primary objective of this work is to appreciate the significant progress made in this vital area of research. This article also aims to identify any existing gaps in the literature and provide potential directions for future work.

Keywords: MIMO; dielectric resonator antenna; 5G

Citation: Ahmad, H.; Jamaluddin, M.H.; Seman, F.C.; Rahman, M. MIMO Dielectric Resonator Antennas for 5G Applications: A Review. *Electronics* **2023**, *12*, 3469. https://doi.org/10.3390/electronics12163469

Academic Editors: Syed Muzahir Abbas and Yang Yang

Received: 16 July 2023
Revised: 5 August 2023
Accepted: 14 August 2023
Published: 16 August 2023

Copyright: © 2023 by the authors. Licensee MDPI, Basel, Switzerland. This article is an open access article distributed under the terms and conditions of the Creative Commons Attribution (CC BY) license (https://creativecommons.org/licenses/by/4.0/).

1. Introduction

Modern wireless communication heavily relies on 5G technology, which offers significant improvements compared to its predecessor, LTE/4G technology. 5G provides higher speeds, lower latency, increased reliability, and greater capacity. The bands used in 5G can be broadly categorized into two groups: sub-6 GHz and mm-wave (above 24 GHz). The sub-6 GHz band offers extensive coverage and is suitable for outdoor communication, while the mm-wave band has a narrower coverage and finds applications in indoor communication.

To meet the requirements of 5G, multiple-input multiple-output (MIMO) antennas are employed. In a MIMO antenna system, multiple antennas are compactly mounted on a common ground plane to reduce multipath fading, increase channel capacity, and improve link reliability. However, the design of MIMO antennas poses challenges such as reducing the size of antenna elements and managing the mutual coupling between them, which can degrade MIMO performance [1]. Previous studies have reported various printed antennas with different decoupling mechanisms for MIMO configurations, each with its own advantages and disadvantages.

In recent years, dielectric resonator antennas (DRAs) have gained popularity due to their appealing features [2]. Unlike printed antennas, DRAs do not suffer from conductor loss and surface waves at higher frequencies. These antennas are made from dielectric materials, resulting in high radiation efficiency. They offer simple design relations, a small size, a wide bandwidth, straightforward excitation methods, and easy fabrication [3,4]. Given the numerous advantages of DRAs, there has been increasing attention paid towards the design of MIMO DRAs for various applications [5].

This article focuses on MIMO dielectric resonator antennas designed for different 5G bands as reported in the literature. The published designs will be thoroughly analyzed, examining their advantages and disadvantages, acknowledging the contributions made, and identifying potential research gaps for future work in this rapidly evolving field.

2. Categorization of MIMO DRAs for 5G Applications

The classification of the MIMO DRAs for 5G applications can be performed by grouping them in different categories based on their operating 5G frequency band, type of DRAs, decoupling mechanism, polarization, and so on. In this work, we will review the literature by grouping the designs on the basis of isolation techniques as follows:

2.1. Diversity Techniques

The mutual coupling between the DRA elements in a MIMO antenna can be improved by utilizing the diversity techniques of spatial, polarization, and pattern diversity, which requires the careful placing of the DRAs with respect to each other. These techniques do not require any extra decoupling structure as opposed to the methods in which complex additional structures are incorporated in the designs.

2.1.1. Circularly Polarized Designs

MIMO CP DRAs are advantageous in terms of channel capacity, data rate, and link reliability. In this regard, a circularly polarized sub-6 GHz 5G MIMO antenna was proposed using two offset-fed rectangular DRAs in [6]. One DRA was excited by a conformal feed, while the other was excited by a microstrip-slot feed as shown in Figure 1. It has been demonstrated that an offset conformal feed produces two orthogonal degenerate modes, which satisfies one condition for circular polarization; however, it lacks a phase difference of 90° which is also a necessary condition for circular polarization. To cater to this, an offset notch was removed from the DRAs to attain phase-quadrature. The position of the feed and the notch have been optimized by a parametric analysis. The impedance bandwidth for the proposed design is between 5.15–6.12 GHz, while the Axial Ratio bandwidth is between 5.30–5.87 GHz. Pattern diversity ensured isolation by placing the DRAs on opposite sides of the substrate, radiating in different directions. This arrangement offered a compact size but with an increased profile.

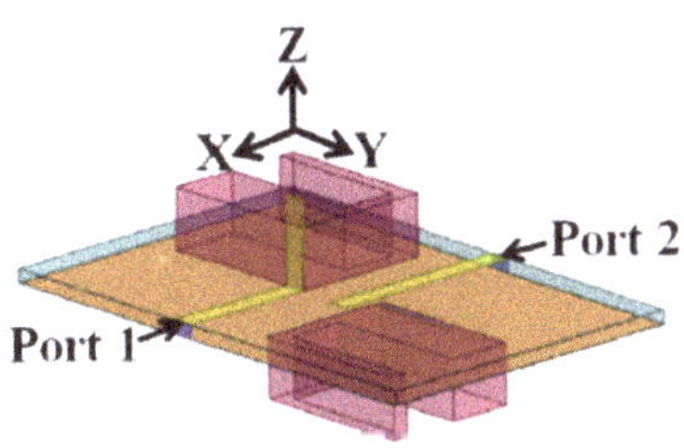

Figure 1. Offset-conformal-strip-fed MIMO DRA [6].

In [7], orthogonally placed graphene strips enabled circular polarization in a reconfigurable hemispherical MIMO DRA, but with added complexity. The feeding layer below the ground plane consists of a microstrip line that is surrounded by mushroom ridge-gap waveguide (RGW) unit cells. The RGW helps to stop the energy from the microstrip line from spreading and confines it right beneath the slot which eventually increases the gain to 10.3 dB. In [8], a two-element MIMO cylindrical DRA with circular polarization and spatial diversity for isolation improvement was presented with simulated results only.

In [9], circular polarization was achieved through modified annular ring slots and conformal feeding networks in a four-port cylindrical MIMO DRA. Polarization and pattern diversity were used for isolation. However, the design had an increased profile due to antennas on both sides of the substrate. In [10], circular polarization is achieved by a z-shaped slot in the ground plane which excites hybrid degenerate modes in the DRAs while the isolation between the DRAs is achieved through polarization diversity.

2.1.2. Multiband Designs

Multiband antennas are a useful tool for any professional who is on the road and connected to the critical communications community. These antennas can operate on multiple radio frequencies simultaneously while also providing coverage for other technologies. In this connection, a dual-band MIMO DRA with four ports for sub-6 GHz 5G applications was presented in [11]. It utilized rectangular DRAs excited by a slot in the patch radiator placed beneath the DRA on a substrate backed by the ground plane, as shown in Figure 2. The design achieved resonances at 3.3 GHz and 3.9 GHz, with impedance bandwidths ranging from 2.86–3.48 GHz and 3.67–4.25 GHz. Mutual coupling was reduced by orthogonal placement of adjacent DRAs. Triangular slots in the feed enabled dual-band operation in the rectangular MIMO DRA presented in [12]. The orthogonal placement of adjacent DRAs improved isolation, utilizing polarization diversity. However, the design has the drawback of limited impedance bandwidth.

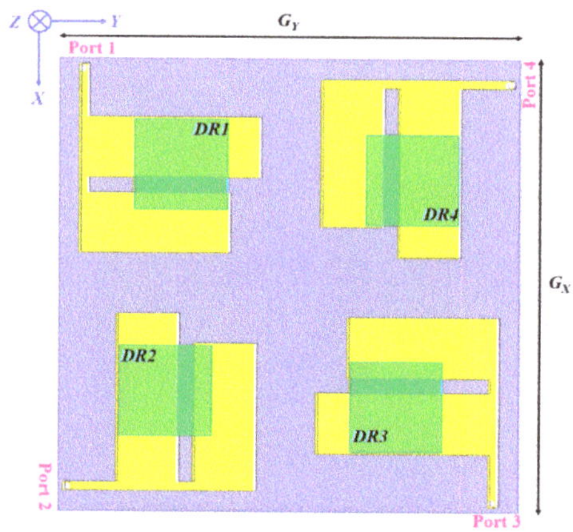

Figure 2. MIMO DRA with patch radiator [11].

A hybrid MIMO DRA with multiband characteristics was introduced in [13]. It used two rectangular DRAs on a substrate with a partial ground plane. The design achieved dual-band operation (3.4–4.32 GHz and 4.96–5.05 GHz) through a microstrip feeding network and parasitic patch. The partial ground plane widened the bandwidth and improved matching. The orthogonal placement of DRAs and the feeding network resulted in isolation exceeding 22 dB for both bands.

In [14], a compact MIMO DRA for 5G mm-wave applications was proposed. It utilized two rectangular DRAs on opposite sides of a substrate, fed by coplanar waveguide feeds with ground plane slots. Operating at 28 GHz and 38 GHz, the antenna achieved gains of 6.2 dBi and 7.57 dBi, respectively. The impedance bandwidth was approximately 5 GHz (18% and 13% at 28 GHz and 38 GHz). The design improved isolation by 27 dB without additional structures for mutual coupling reduction and with a reduced structural complexity, although the height was increased due to the proposed arrangement.

In [15], a dual-band MIMO DRA for vehicular applications was presented. It used rectangular microstrip lines with S-shaped apertures to excite higher order modes. Ring-shaped dielectric resonators provided wider bandwidth, and the asymmetric aperture generated multiple resonances and circular polarization. The design achieved LHCP and RHCP polarization diversity with high gain (8.1 dBi) and radiation efficiency (>90%) by using mirrored apertures and a conducting plate for isolation. A simulated cylindrical MIMO DRA utilizing spatial diversity for isolation was presented in [16], which employs microstrip feed with a circular patch for dual-band operation.

In [17], a dual-band MIMO dielectric resonator antenna (DRA) for 5G mm-wave applications was presented. Four rectangular DRAs were arranged in a cross-shaped structure in which each DRA was excited by a substrate integrated waveguide through a slot as shown in Figure 3. The impedance bandwidth obtained was between 24.5–27.5 GHz and 33–37 GHz. Polarization diversity isolated adjacent DRAs, while spatial diversity reduced coupling between DRAs in parallel. The design achieved an isolation greater than 22 dB and 17 dB (for the lower and upper bands, respectively) and high gain (9.9 dBi) with high efficiency (>96%) across the operating bandwidth.

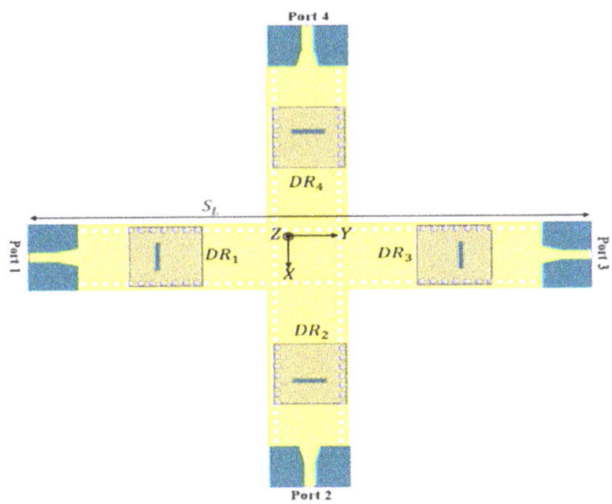

Figure 3. SIW-fed MIMO DRA [17].

In [18], a dual-band MIMO DRA for the sub-6 GHz 5G and other bands was presented. It utilized five rectangular DRAs arranged cubically with PEC layers, achieving resonances at 3.5 GHz and 5.9 GHz. An offset coaxial probe feed was used, and orthogonal placement reduced coupling for isolation exceeding 25 dB. The design offered greater capacity and a smaller footprint but occupied a large volume.

In [19], a two-port dual-band antenna was presented, consisting of a slot antenna and a DRA on separate substrates. The slot antenna operated at 5.2 GHz (5.15–5.25 GHz bandwidth), while the DRA operated at 24 GHz (23.87–24.36 GHz bandwidth). Maximum gains achieved were 3.93 dBi (lower band) and 6.32 dBi (higher band). Port isolation exceeded 35 dB through orthogonal feeding using polarization diversity. However, the design had limitations such as a narrow bandwidth and fabrication complexity.

2.1.3. High Gain Designs

High gain designs are advantageous in term of their high level of functionality and security. These antennas have a narrow radio beam, which increases the strength of the signal. A MIMO DRA for 5G mm-wave applications utilizing eight cylindrical DRAs on a substrate, excited by rectangular slots, was presented in [20]. Four DRAs formed two linear arrays in the proposed structure. Isolation between DRAs was achieved by tilting array beams in opposite directions for pattern diversity. A peak gain greater than 7 dBi is achieved by the proposed array with a radiation efficiency of 80%. Advantages include high gain, good isolation, and a compact size without additional structures for mutual coupling reduction, with the drawback of increased sidelobe levels. A high gain of 9.43 dBi was achieved by a 3 × 3 MIMO array in [21].

In [22], a compact MIMO DRA with pattern diversity has been presented. The proposed design consists of two epsilon-shaped DRAs placed on top of a substrate facing each other and two cylindrical DRAs between the epsilon-shaped DRAs. The distance between the epsilon shaped DRAs is less than a wavelength and so is the distance between the cylindrical DRAs. As a result, the entire structure acts as a single DRA. It achieved high gain (6.5 dBi), efficiency (97%), and a compact size with enhanced isolation using rectangular slots.

In [23], a wideband MIMO DRA for sub-6 GHz 5G applications was proposed. I featured A-shaped DRAs on a substrate, with conformal feeds. Orthogonal placemen reduced mutual coupling, achieving an isolation greater than 20 dB. The wide impedance bandwidth achieved by perturbing the triangular DRA into the proposed A-shape ranges between 3.2–6 GHz with a measured gain ranging between 6.03–7.45 dBi. In [24], a simulated design was presented, utilizing spatial diversity for decoupling, and improved gain by exciting DRA in a higher-order mode.

A simulated four-port, closely spaced MIMO DRA consisting of four arrays of sixteen cylindrical DRAs utilizing pattern diversity for 5G mm-wave applications was presented in [25]. With the proposed array, a gain of 10 dBi was achieved. A similar design with two ports consisting of two arrays each with four DRAs, achieving a gain of 9 dBi, was proposed in [26]. In [27], polarization diversity is utilized to reduce the mutual coupling in a four-port wideband MIMO antenna for 5G mm-wave applications where the impedance bandwidth ranges between 26.5 GHz to 43.7 GHz and the maximum gain is greater than 8 dBi.

2.1.4. High Capacity Designs

A sixteen-port eight-element MIMO DRA with a built-in decoupling mechanism was introduced in [28]. The design comprised four DRAs on the top side and four DRAs beneath, all placed on an FR4 substrate as shown in Figure 4. Each top DRA was excited by two orthogonally positioned CPW-fed conformal strip lines, while each bottom DRA was excited by two orthogonally positioned microstrip-fed conformal strip lines. This arrangement formed an eight-element MIMO, enabling a sixteen-port MIMO configuration to enhance the channel capacity. Isolation between the DRAs was achieved through all the three diversity techniques, i.e., polarization, pattern, and spatial diversity. The increased profile due to the placement of DRAs on both sides of the substrate is a major disadvantage

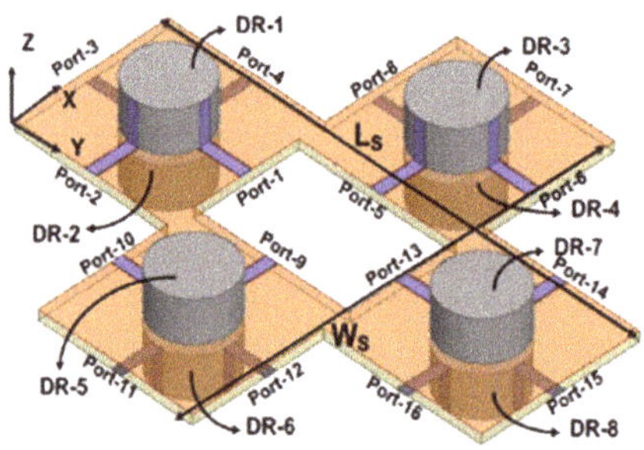

Figure 4. Sixteen-port eight-element MIMO DRA [28].

In [29], an eight-port MIMO antenna with a quad-directional pattern was presented. The design featured four cylindrical DRAs placed in a box configuration radiating in four directions, with each DRA being excited by probe feeding and conformal-strip feeding, achieving orthogonal feeding and improving isolation. In [30], an eight-port MIMO antenna exhibiting high channel capacity utilizes spatial diversity and polarization diversity to achieve an isolation greater than 12 dB within an operating frequency of 3.4–3.6 GHz. The design has the advantage of mutual coupling reduction without incorporating any structure but with an increased size.

2.1.5. High Isolation Designs

In [31], a quad-port MIMO DRA has been presented for 5G mm-wave applications. The proposed design consists of a rectangular DRA that is mounted on a substrate integrated waveguide and excited differentially through two narrow slots on the top surface of SIW.

The DRA resonates at 28 GHz with an impedance bandwidth of 27.5–28.4 GHz. The radiators in the MIMO configuration are arranged in such a manner that the DRAs that are adjacent to each other are isolated from each other due to orthogonal polarization, while the DRAs that are placed opposite to each other are isolated from each other due to the larger gap between them. Thus, the proposed design utilizes a combination of polarization and spatial diversity to attain a high isolation greater than 40 dB in the entire band of operation.

In [32], a four-port MIMO antenna achieves an isolation greater than 25 dB by employing polarization and spatial diversity to operate at the 35 GHz mm-wave band. A similar approach has been used in [33] for obtaining an isolation greater than 27 dB in an SIW-fed MIMO DRA operating in the mm-wave band of 26.64–28.55 GHz. A comparison of several designs reported so far utilizing diversity techniques for mutual coupling reduction has been given in Table 1.

Table 1. Comparison of reported designs utilizing diversity techniques.

Ref.	Gain (dBi)	Bandwidth (%)	Isolation (dB)	Polarization
[6]	4.7	17.2	17.1–27.6	Circular
[11]	5.8, 6.2	19.5, 14.6	>20, >26	Linear
[17]	9.1, 9.9	11.5, 12	>22, >17	Linear
[23]	6.03–7.45	60.8	>20	Linear
[25]	10	6.7	>15	Linear
[28]	5–6.5	8.8	>17	Linear
[31]	4.2	3.2	>40	Linear
[7]	5	30.2	>22	Circular

From the reported designs utilizing diversity techniques for mutual coupling reduction, it can be observed that high isolation can be achieved only if a combination of diversity techniques is employed. Moreover, diversity techniques will not improve the gain or bandwidth, or achieve circular polarization. Separate strategies need to be applied to the DRAs in a MIMO configuration in order to achieve the desired characteristics.

2.2. *Periodic Structures*

The mutual coupling between DRAs in a MIMO configuration can be significantly reduced by incorporating periodic structures such as metamaterials, EBGs, FSS, and metasurfaces. As opposed to the diversity techniques, these structures considerably enhance the isolation by limiting the coupling energy between the DRAs, though the design complexity is increased.

2.2.1. Circularly Polarized Designs

In [34], a two-element cylindrical MIMO dielectric resonator antenna for the sub-6 GHz band was presented. Mutual coupling was reduced using a carefully designed metasurface on top of the DRAs, which also generated circular polarization. The design operated at 3.81 GHz (3.86–4.23 GHz bandwidth), achieving over 20 dB isolation, 4 dBi gain, and over 90% radiation efficiency. The compact design allowed the close placement of DRAs, but with an increased profile and complexity in fabrication with a low gain.

In [35], a circularly polarized, wideband MIMO DRA was presented, featuring two rectangular DRAs on an FR4 substrate fed by microstrip lines. Wideband CP response was achieved by truncating the opposite edges of the DRAs. A cross-ring-shaped slot on the ground plane generated CP fields. Mutual coupling between the two DRAs was reduced using an electromagnetic band-gap surface etched on the ground plane as shown in Figure 5. Isolation exceeded 26 dB across the operating bandwidth. The design offered a wideband CP, simple feeding structure, and good isolation but a low gain.

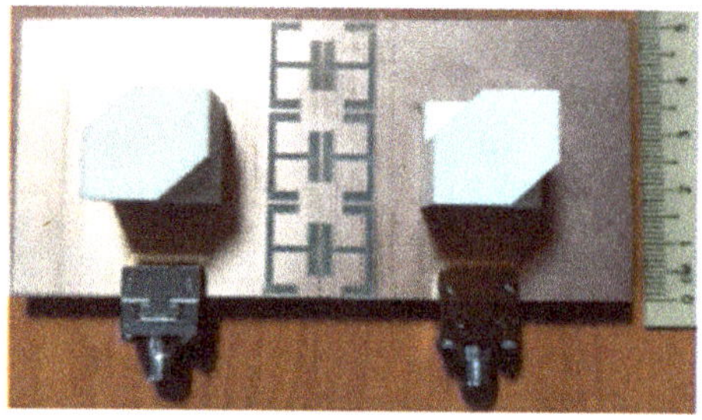

Figure 5. CP MIMO DRA with EBG on ground plane [35].

2.2.2. High Gain Designs

In [36], a triple-band MIMO DRA for sub-6 GHz 5G applications was presented. It featured six cylindrical DRAs on an FR-4 substrate, grouped with different heights but the same diameters to generate three resonant frequencies. CPW-fed conformal strips excited the DRAs, while a partially reflecting surface (PRS) reduced mutual coupling. The PRS increased the peak gain by 2 dB but decreased radiation efficiency (90% to 80%). Limitations included lower radiation efficiency, reduced impedance bandwidth, increased side lobe levels at higher frequencies, and an increased profile due to the minimum distance required for the PRS above the DRAs. Another design utilizing metamaterial for gain enhancement on top of the MIMO DRA for 5G mm-wave applications was proposed in [37]. The metamaterial served to reduce the mutual coupling between adjacent DRAs and assisted in achieving a gain greater than 7 dBi for the proposed design.

In [38], a MIMO DRA with an FSS wall for the 60 GHz mm-wave band was presented as shown in Figure 6. It included two cylindrical DRAs on a substrate, excited through ground plane slots fed by microstrip lines. The FSS wall reduced mutual coupling using Jerusalem-cross- and fan-shaped structures, achieving over 30 dB isolation, 1.5 dB increased gain, and over 90% radiation efficiency. A drawback was the tilted radiation pattern from the boresight direction.

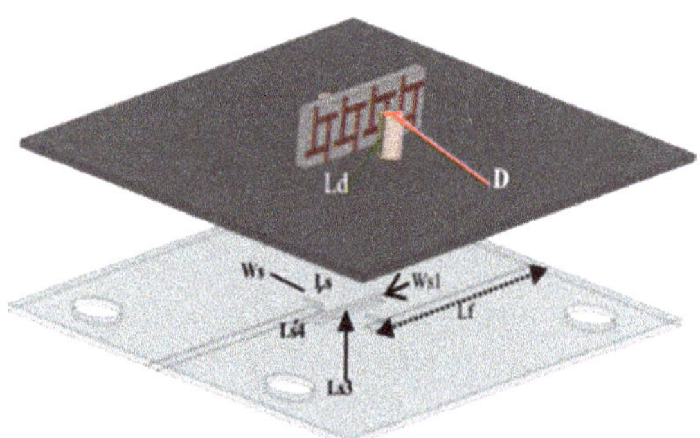

Figure 6. MIMO DRA with FSS wall [38].

In [39], a 60 GHz MIMO DRA with a metasurface was presented. It featured two cylindrical DRAs on an upper substrate, excited through rectangular slots and fed by microstrip feedlines on a lower substrate. A metasurface unit cell consisting of split ring resonators with gaps reduced mutual coupling. The design achieved high isolation (>18 dB), improved gain (7.9 dBi), and efficiency (91%) at 60 GHz, but with a tilted main beam in the H-plane.

2.2.3. High Isolation Designs

In [40], a hybrid isolator was presented for reducing mutual coupling in a two-element cylindrical MIMO DRA. The hybrid isolator, consisting of an EBG structure and a choke absorber, was placed between the DRAs as shown in Figure 7. Operating at 60 GHz (59.3–64.8 GHz bandwidth), the design achieved high isolation (>29 dB, up to 49 dB) in the entire operating band. The limitations of the proposed design are low radiation efficiency, an increased profile due to the choker absorber, and the shift of the direction of the main beam in the H-plane.

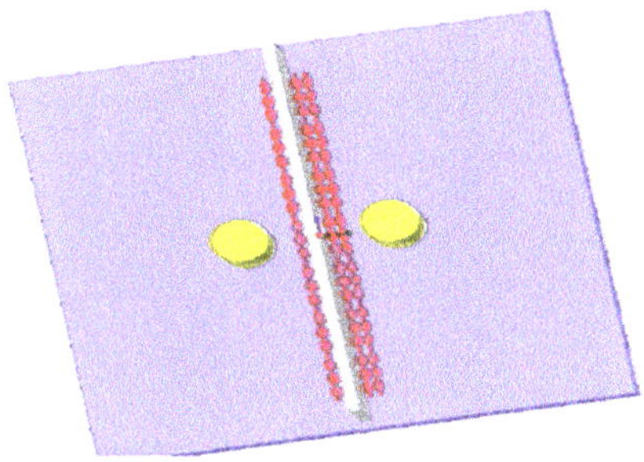

Figure 7. Hybrid-isolator-based MIMO [40].

In [41], a mutual reduction technique for a 60 GHz mm-wave MIMO DRA was proposed. It featured two cylindrical DRAs on a substrate, excited through a rectangular slot in the ground plane. A metamaterial polarization rotator (MPR) wall between the DRAs achieved high isolation by reducing mutual coupling on an average of 16 dB without affecting the radiation pattern. Limitations included an increased profile due to the multilayered structure, design complexity, and lower radiation efficiency (92% to 88%) when the MPR wall was employed.

In [42], a compact EBG structure was proposed to reduce mutual coupling between DRAs in a 60 GHz MIMO antenna. It featured two cylindrical DRAs on a substrate, excited through ground plane slots by microstrip feedlines. Six unit cells of the proposed EBG structure placed between the DRAs achieved an average isolation reduction of 13 dB, with the minimum and maximum isolation exceeding 15 dB and 30 dB, respectively. The design had the drawbacks of low gain and a 30° tilted radiation pattern in the E-plane. A comparison of several studies utilizing different periodic structures for mutual reduction has been given in Table 2.

Table 2. Comparison of reported designs utilizing periodic structures.

Ref.	Gain (dBi)	Bandwidth (%)	Isolation (dB)	Polarization	Decoupling Structure
[35]	4.83	25.9	>26	Circular	EBG
[39]	7.9	11.6	>18	Linear	Metasurface
[40]	8.1	8.8	29–49	Linear	EBG and choke absorber
[37]	>7.5	7.9	Up to 29.34	Linear	Metamaterial

The reported designs utilizing periodic structures for mutual coupling reduction indicate that high isolation can be achieved with such structures. Other desired performance characteristics including a high gain, the desired bandwidth, and circular polarization can also be achieved by carefully designing the periodic structures. However, a major drawback of such structures is the design complexity.

2.3. *Other Decoupling Structures*

The mutual coupling between the DRAs in a MIMO antenna can also be reduced by other decoupling structures which offer simpler designs as compared to periodic structures. Various such decoupling structures have been reported in the literature which will be discussed in this section.

2.3.1. Compact Designs

In [43], a MIMO DRA with port and radiation pattern decoupling was presented. The design utilized the principle of inducing a current in two adjacent DRAs through metallic posts. When a single DRA was excited, the induced current canceled out the fields in the adjacent DRA, achieving decoupling. Using the above concept, two designs were proposed: one with H-plane decoupling and the other with E-plane decoupling, as shown in Figure 8. The advantage of the proposed design is that it is very compact, as the two DRAs are placed without any spacing between them. A drawback, however, is the low bandwidth. In [44], a simulated substrate integrated MIMO DRA antenna for mm-wave applications was proposed in which the isolation between the DRAs was achieved using metallic vias and rectangular strips.

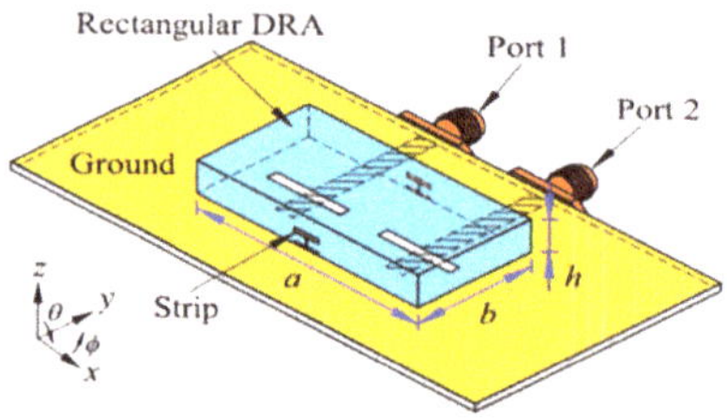

Figure 8. H-plane-decoupled MIMO DRA [43].

In [45], a decoupling method was proposed for a single-element MIMO DRA at 28 GHz. Techniques involving embedded metallic sheets (for E-plane coupling) and sidewall attachments (for H-plane coupling) reduced mutual coupling by suppressing EM wave concentration. The proposed design is very compact since a single-element DRA is utilized but with the drawback of limited bandwidth. In the simulated design of [46], decoupling was achieved in a closely spaced half-volume MIMO DRA by incorporating an inductor between the feeding microstrip lines.

In [47], conductive metallic strips were employed on the adjacent walls of the rectangular MIMO DRA to reduce mutual coupling in the H-plane as shown in Figure 9. The metallic strips suppressed the fields of the adjacent excited DRA from reaching the feed of the undriven DRA. The proposed design has a simple and compact design but with a smaller bandwidth. Similarly, Ref. [48] presented a decoupling method using metallic strips on adjacent and opposite walls for a mm-wave rectangular MIMO DRA to achieve an isolation greater than 50 dB.

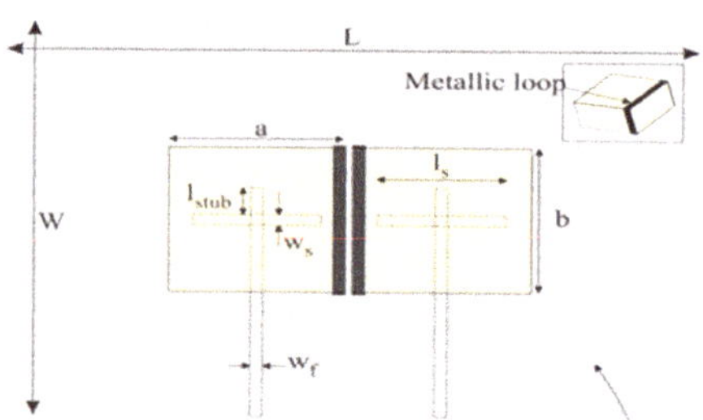

Figure 9. MIMO DRA with metallic strips on adjacent walls [47].

A starfish-shaped single-element compact MIMO was proposed in [49] with a degenerated ground structure for improved isolation and bandwidth. Another single-element cylindrical four-port MIMO DRA for mm-wave 5G applications employed a degenerated ground structure for isolation enhancement in [50]. Similarly, isolation was improved by utilizing slots in the ground plane in [51].

In [52], a technique to reduce mutual coupling in MIMO DRAs for 5G mm-wave applications was proposed. The design featured two rectangular DRAs on a Rogers substrate, excited by microstrip-fed slots as shown in Figure 10. Printed metallic strips on the DRAs' upper surface shifted the coupling field to the side, achieving over 24 dB isolation across the bandwidth without affecting the radiation pattern. The compact design required no additional space between the DRAs. However, the utilization of metallic strips decreased the gain and impedance bandwidth of the DRAs compared to the absence of the strips.

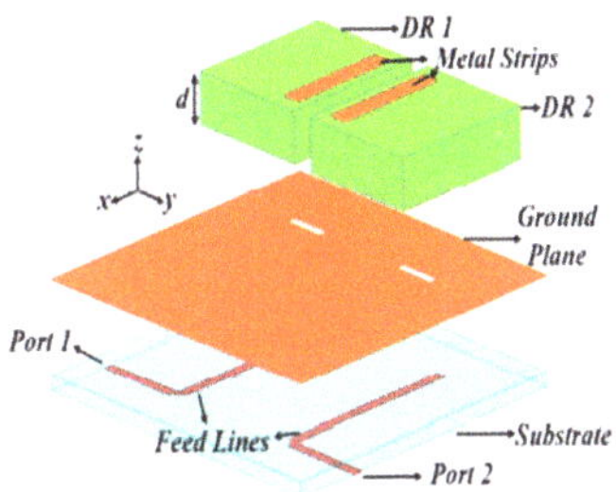

Figure 10. MIMO DRA with metal strips [52].

2.3.2. Circularly Polarized Designs

In [53], an A-shaped conformal metal strip achieved circular polarization while an S-shaped slot reduced mutual coupling in a sub-6 GHz rectangular MIMO DRA. The design operated at a resonant frequency of 3.72 GHz with an impedance bandwidth of 3.57–4.48 GHz. In [54], a two-element cylindrical MIMO DRA has been presented for 5G and X-band applications. The proposed design consists of a mirror S-shaped radiator that produces linearly polarized waves at 4.4–5.05 GHz and 7.85–8.43 GHz, while a cylindrical DRA placed in the center of the S-shaped radiator works to generate CP in 9.05–10.10 GHz band. For mutual coupling reduction between the two antennas, a partial ground plane with a slot has been used. The bottom side also consists of an elliptical radiator with a T-shaped slot to achieve an isolation greater than 20 dB in the operating bandwidth.

In [55], a dual-band circularly polarized MIMO DRA for sub-6 GHz 5G applications was presented. The design featured two ring DRAs on a substrate, excited by arc-shaped microstrip feed lines with conformal probes as shown in Figure 11. The probes generated orthogonal hybrid modes, enabling circular polarization. Improved isolation between ports was achieved by a rectangular slot in the ground plane. In [56], a MIMO DRA with CP agility for 5G applications was introduced. The design featured two DRAs on a hexagonal substrate, fed by L-shaped microstrip lines and conformal probes. The substrate reduced the antenna size, while the L-shaped feed lines and conformal probes enabled circular polarization.

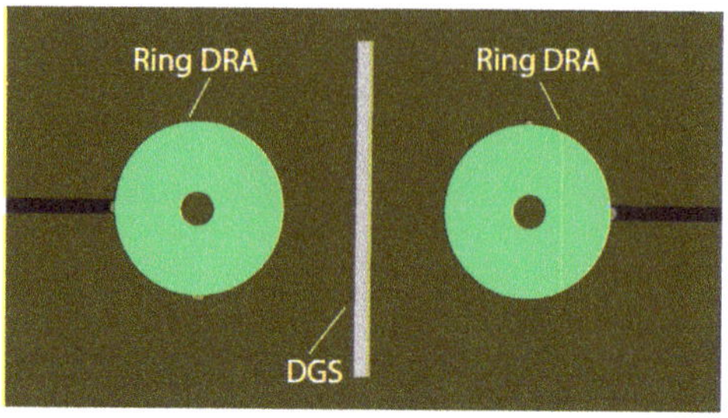

Figure 11. Geometry of dual-band CP MIMO DRA [55].

2.3.3. Broadband Designs

In [57], a four-element MIMO DRA was presented utilizing a linear array of 1 × 4 rectangular DRAs on small separated ground planes placed above a larger common ground plane. By adjusting the dimensions and distance of the small ground planes, mutual coupling between the DRAs was reduced. The design achieved a gain of 8.2 dBi and offered high isolation, an improved front-to-back ratio, and a wider bandwidth (3.37–4.10 GHz) compared to conventional DRAs. However, it has an increased profile and complexity.

In [58], a dual-element wideband triangular DRA for MIMO applications has been presented. The proposed design consists of two triangular DRAs that are fed by coaxial probes. Triangular DRAs have less size as compared to rectangular and cylindrical DRAs with the same permittivity and height operating at the same resonant frequency. The mutual coupling between the DRAs have been reduced by placing four metal vias between the DRAs. The position of the vias have been optimized to achieve maximum isolation and, at the same time, provide a wide bandwidth of 4.67–9.50 GHz (67.5%).

2.3.4. Multiband Designs

In [59], an on-chip MIMO DRA with multiband characteristics has been presented for mm-wave applications. The proposed design consists of two rectangular DRAs made from a silicon wafer and placed in a substrate of the same material. The energy to the DRAs was coupled by using coplanar waveguides. The mutual coupling between the DRAs was reduced by approximately 5–22 dB when a dielectric wall was placed between the DRAs. The resonant frequencies of the proposed design are 26, 36, 41, 47, and 52 GHz, with impedance bandwidths of 16%, 6.6%, 3.8%, 8.9%, and 5.5%, respectively.

In [60], a hybrid MIMO DRA operating in four bands was presented. The design comprised two cylindrical DRAs fed by a tapered rhombic ring-shaped feed, achieving quad-band operation. The resonant frequencies for the proposed design were 2.5 GHz, 5.09 GHz, 6.8 GHz, and 9 GHz, and the impedance bandwidth for these bands were 14.6%, 10.7%, 4.2%, and 5.3%, respectively. Mutual coupling between the radiators was reduced using a partial ground plane with a center slot and two L-shaped slots. The drawbacks were low bandwidth in all four bands, and decreased gain and efficiency due to degenerated ground. In [61], a quad-band MIMO antenna with a perforated DRA for achieving four bands and ground plane slots for enhancing isolation was simulated. The proposed design resonates at 2.95 GHz, 3.93 GHz, 4.8 GHz, and 5.8 GHz, with impedance bandwidths of 2.83–3.07 GHz, 3.79–4.04 GHz, 4.7–4.9 GHz, and 5.72–5.86 GHz, respectively.

The design and numerical analysis of a rectangular terahertz (THz) dielectric resonator (DR) antenna are presented in [62] recently. The study is presented with the DR aspect ratio chosen in a way that allows the antenna to operate in four modes with a triple-band response. The graphene strips have also been added to the DRs' top radiating surface. By carefully choosing the chemical potential of graphene, it is possible to manipulate the resonant modes of an antenna to produce either a large single-band or triple-band response. The antenna continues to provide the triple-band response operating at frequencies of 2.10–2.33, 2.56–2.65, and 2.77 THz.

2.3.5. High Isolation Designs

In [63], a decoupling mechanism was proposed to improve the isolation of DRAs in a MIMO arrangement as shown in Figure 12. A dielectric superstrate placed above the DRAs effectively weakened the interaction between adjacent DRAs, reducing mutual coupling. The proposed structure increased isolation by 25 dB, improved gain and efficiency, and maintained polarization purity without modifying the DRAs' resonant frequency. However, the design increased the MIMO profile and reduced the impedance bandwidth.

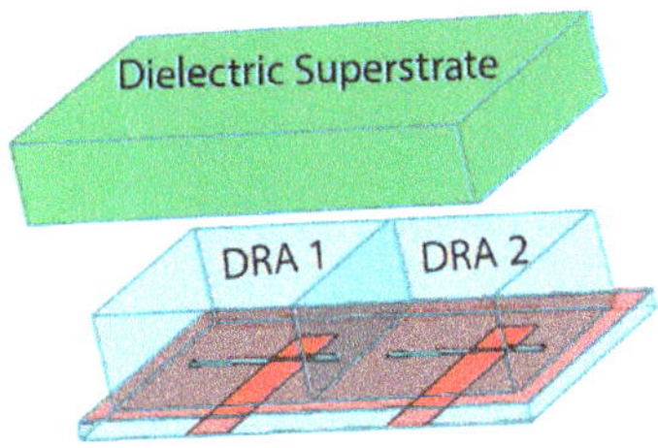

Figure 12. MIMO DRA with dielectric superstrate [63].

In [64], metallic vias were used to reduce mutual coupling in both H-plane- and E-plane-coupled MIMO DRAs for 5G mm-wave applications. The design achieved over 30 dB isolation, compactness, and improved gain, but required the optimization of the resonant frequency due to the vias' effect. A 60 GHz MIMO DRA with a dielectric superstrate for gain enhancement was presented in [65]. The mutual coupling between the DRAs was reduced by etching a slot from the ground plane and placing a metallic strip between the DRAs, achieving an isolation greater than 20 dB for the entire band of interest. A similar level of isolation was achieved by a four-port rectangular MIMO DRA utilizing vertical metal plates for minimizing mutual coupling in [66]. A comparison has been made for several decoupling structures with their performance parameters as shown in Table 3.

Table 3. Comparison of reported designs utilizing different decoupling structures.

Ref.	Gain (dBi)	Bandwidth (%)	Isolation (dB)	Polarization	Decoupling Structure
[47]	3.5	7.5	Up to 28	Linear	Metallic strips
[49]	2.97	79	>20	Linear	Defected ground structure
[53]	6	36.63	Up to 28.25	Circular	Defected ground structure
[57]	8.2	19.5	>25	Linear	Decoupling ground
[60]	2.18, 4.59, 4.11, 4.75	14, 10, 4, 5	Up to 22, 34, 30, 18	Linear	Defected ground structure
[63]	5.6	6.6	>25	Linear	Dielectric superstrate
[45]	6.4	3.4	Up to 23.45	Linear	Metallic sheets
[66]	4.09	10.4	>20	Linear	Vertical metal plates

From the designs reported in this section, it can be observed that a careful choice needs to be made when employing a particular decoupling structure for isolation enhancement, as some structures can have a pronounced effect on different performance parameters such as gain, bandwidth, and polarization.

2.4. Self-Decoupling Techniques

The techniques discussed so far enhance isolation through decoupling structures or the strategic placement of DRAs, but suffer from increased design complexity and size. This section explores mutual coupling reduction techniques that provide isolation without additional structures and with a built-in isolation mechanism. Most reported designs in this category are compact single-element DRAs, offering reduced complexity and size.

2.4.1. Single-Element Designs

In [67], a compact Y-shaped MIMO DRA was proposed for sub-6 GHz 5G applications. It consisted of three rectangular DRAs merged at a 120° angle, forming a Y-shape on a circular substrate with a ground plane as shown in Figure 13. Each arm was fed by a conformal strip, resulting in three ports. The mutual coupling was reduced by directing each port's main beam in a different direction. However, limitations included low radiation efficiency and isolation. In [68], the rectangular DRA was transformed into a plus-shaped design to improve isolation by removing portions of the DRA with strongly interfering fields.

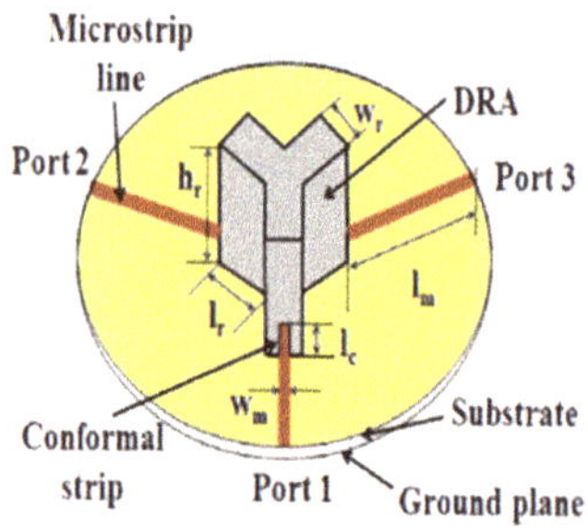

Figure 13. 3-D view of Y-shaped MIMO DRA [67].

In [69], a triple-band dual-polarized MIMO dielectric resonator antenna (DRA) was introduced for 5G and other wireless applications. The design featured a square DRA on an FR4 substrate, excited through rectangular slots by two orthogonal L-shaped monopole microstrip feeds as shown in Figure 14. Resonances were achieved at 1.7 GHz, 2.6 GHz, and 3.6 GHz, with respective impedance bandwidths of 1.63–1.84 GHz (12%), 2.43–2.71 GHz (10.89%), and 3.27–3.75 GHz (13.68%), respectively. The maximum isolation reached 20 dB for all the three bands. The design offered simplicity, dual-band operation, and ease of fabrication.

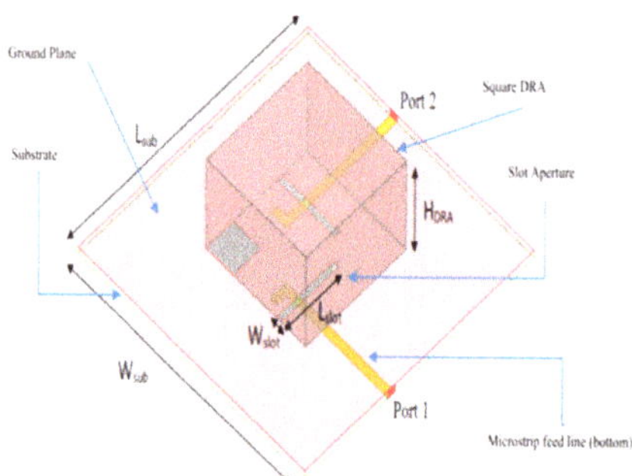

Figure 14. 3-D view of the proposed MIMO [69].

In [70], isolation between the ports of a single-element triangular MIMO DRA was improved by removing a semi-cylindrical portion of the DRA between the ports. A simulated two-port MIMO dielectric resonator antenna for 5G mm-wave applications presented in [71] consisted of two p-shaped dielectric resonators that were excited by a microstrip line through a slot, improving isolation and offering a compact size.

2.4.2. Multiple-Element Designs

In [72], isolation between the DRAs was achieved by exciting higher-order modes with the same radiation pattern as the active DRA. The proposed design had a resonant frequency of 5.25 GHz and an impedance bandwidth of 4.89–5.42 GHz, with a gain ranging from 7 to 7.8 dBi and radiation efficiency of 85–95%. The method offered a compact, low-loss, low-complexity, and cost-effective design, although the size of the antenna increased due to the utilization of higher-order modes.

The investigation of a unique self-decoupling technique for MIMO dielectric resonator antenna (DRA) arrays are presented in [73]. This approach is founded on the conformal strip's transmission properties that support the DRA. When the strip-fed DRA operates, it is discovered that a higher-order mode, TE_{113} mode, operates. This self-decoupling technique can be used with 2-D MIMO planar arrays, in addition to 1-D H- and E-plane coupled MIMO linear arrays. A 2 × 2 MIMO planar DRA array prototype is simulated, processed, and tested to confirm its viability. Figure 15 shows the design of the DRA structure and Figure 16 shows the corresponding S-parameter response. A comparison of several reported designs of self-decoupled MIMO DRAs for 5G applications has been provided in Table 4.

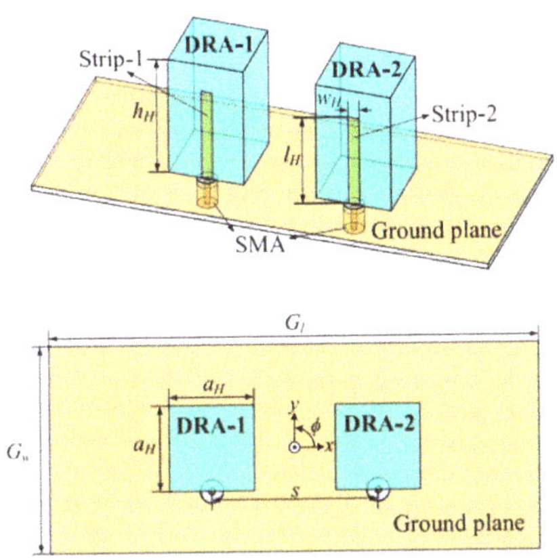

Figure 15. Configuration of the investigated coupled 1 × 2 MIMO DRA array [73].

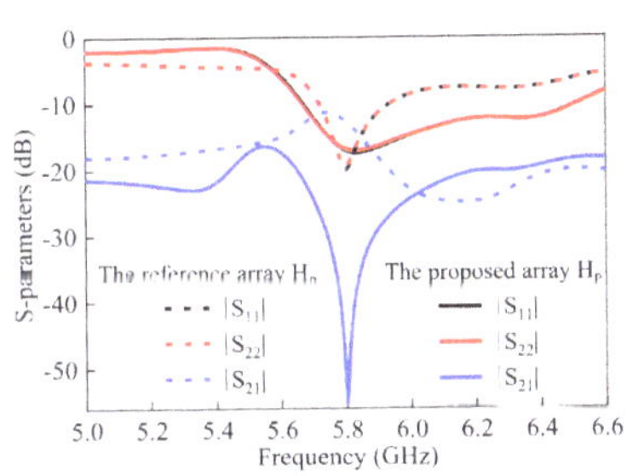

Figure 16. Simulated S-parameter response from the structure in [73].

Table 4. Comparison of self-decoupled designs.

Ref.	Gain (dBi)	Bandwidth (%)	Isolation (dB)	Polarization
[67]	8	64	>10	Linear
[68]	5.12	11.6	Up to 15	Linear
[69]	5.5, 5.9, 6.9	12, 10.89, 13.68	Up to 20	Linear
[72]	7–7.8	10.2	Up to 60	Linear

The self-decoupled designs have the advantage that no extra decoupling element is required for isolation, which reduces the design complexity and offers a compact design. A proper choice of the technique can provide high isolation, bandwidth, or gain, as per the designer's requirement. The authors, however, could not find any design that achieved circular polarization with a self-decoupling technique.

3. Discussion and Future Work Suggestions

The decoupling techniques reported for reducing mutual coupling in MIMO DRA for 5G have been compared in Table 5 below. Each technique has its own advantages and shortcomings, and, therefore, a careful decision needs to be made when choosing a decoupling technique for achieving the desired results.

Table 5. Comparison of various decoupling techniques.

Decoupling Technique	Advantages	Shortcomings
Diversity techniques	Easy to implement Less design complexity	Moderate isolation Increases the size and profile of MIMO
Electromagnetic band-gap structures	High isolation Can increase gain	Design complexity Shifts main beam Increases size of MIMO
Metamaterials	High isolation	Increases size of MIMO Design complexity
Metasurfaces	High isolation Can increase gain	Increases size of MIMO Shifts main beam from boresight direction Design complexity
Frequency selective surfaces	High isolation Can increase gain	Increases size of MIMO Design complexity Beam tilting from boresight direction
Conformal metal strips	Good isolation Compact design Ease of fabrication	Low bandwidth Low gain and efficiency Shifts the original resonance frequencies of DRAs
Dielectric superstrate	High isolation Can increase gain and efficiency Restores the tilt in main beam	Fabrication complexity Increases profile of MIMO
Metallic vias	Good isolation Does not occupy extra space	Lowers gain and efficiency Shifts resonance frequency Difficult to drill holes inside DRAs
Defected ground structures	Good isolation May improve bandwidth Does not increase profile	Lowers gain and efficiency due to backward radiation

The literature review has identified several research gaps that can be explored in future work:

- Limited circular polarization: Only a few MIMO dielectric resonator antenna (DRA) designs for 5G exhibit circular polarization, particularly in the mm-wave band.
- Increased edge–edge separation: When an isolating structure is inserted between the DRAs, the distance between their edges becomes large, impacting the overall compactness of the design.
- Lack of structural novelty: The majority of reported designs do not introduce novel structures for the DRAs, suggesting a need for more innovative approaches.
- Compactness challenges: It is challenging to achieve a compact structure when an isolating structure, which occupies more space, is placed between the DRAs.
- Limited bandwidth: Most designs have reported a small bandwidth, indicating the need for improvements to achieve wider frequency coverage.

- Profile reduction: Many designs have not made efforts to reduce the profile of MIMO antennas, highlighting a potential area for improvement.

The utilization of DRAs for 5G MIMO applications is still a developing area of research as evident from the number of publications in this area as compared to the proposed designs utilizing other antenna types for 5G MIMO arrangements. Addressing the above-mentioned research gaps can contribute to advancements in MIMO DRA designs for 5G applications.

4. Conclusions

An extensive literature review on multiple-input multiple-output (MIMO) dielectric resonator antennas (DRAs) designed for 5G applications was presented. The performance of the reported designs in terms of gain, isolation, size, bandwidth, profile, and radiation characteristics was discussed. The article also aimed to highlight the advancements in this research area, identify gaps in the existing literature, and suggest potential directions for future research.

Author Contributions: H.A., M.H.J. and F.C.S. were involved in the ideation of the study; H.A., M.H.J. and M.R. performed the literature search and data analysis; H.A., M.H.J., F.C.S. and M.R. drafted and revised the work. All authors have read and agreed to the published version of the manuscript.

Funding: This research received no external funding.

Data Availability Statement: Not applicable.

Conflicts of Interest: The authors declare no conflict of interest.

References

1. Sharawi, M.S. Printed multi-band MIMO antenna systems and their performance metrics [wireless corner]. *IEEE Antennas Propag. Mag.* **2013**, *55*, 218–232. [CrossRef]
2. Petosa, A.; Ittipiboon, A. Dielectric resonator antennas: A historical review and the current state of the art. *IEEE Antennas Propag. Mag.* **2010**, *52*, 91–116. [CrossRef]
3. Keyrouz, S.; Caratelli, D. Dielectric Resonator Antennas: Basic Concepts, Design Guidelines, and Recent Developments at Millimeter-Wave Frequencies. *Int. J. Antennas Propag.* **2016**, *2016*, 6075680. [CrossRef]
4. Mongia, R.K.; Bhartia, P. Dielectric resonator antennas—A review and general design relations for resonant frequency and bandwidth. *Int. J. Microw. Millim.-Wave Comput.-Aided Eng.* **1994**, *4*, 230–247. [CrossRef]
5. Singhwal, S.S.; Matekovits, L.; Kanaujia, B.K.; Kishor, J.; Fakhte, S.; Kumar, A. Dielectric Resonator Antennas: Applications and developments in multiple-input, multiple-output technology. *IEEE Antennas Propag. Mag.* **2022**, *64*, 26–39. [CrossRef]
6. Sahu, N.K.; Das, G.; Gangwar, R.K. Circularly polarized offset-fed DRA elements & their application in compact MIMO antenna. *Eng. Sci. Technol. Int. J.* **2022**, *28*, 101015.
7. Malhat, H.A.; Zainud-Deen, S.H.; El-Hemaily, H.; Hamed, H.A.; Ibrahim, A.A. Reconfigurable Circularly Polarized Hemispherical DRA Using Plasmonic Graphene Strips for MIMO Communications. *Plasmonics* **2022**, *17*, 765–774. [CrossRef]
8. Kumari, P.; Kumari, T.; Suman, K.K.; Gangwar, R.K.; Chaudhary, R.K. A Circularly Polarized Sub-6 GHz MIMO Antenna for 5G Applications. In Proceedings of the 2022 IEEE International Symposium on Antennas and Propagation and USNC-URSI Radio Science Meeting (AP-S/URSI), Denver, CO, USA, 10–15 July 2022.
9. Sahu, N.K.; Das, G.; Gangwar, R.K.; Rambabu, K. An arrangement for four-element MIMO DRA with complementary CP diversity. *IEEE Antennas Wirel. Propag. Lett.* **2021**, *20*, 1616–1620. [CrossRef]
10. Dwivedi, A.K.; Sharma, A.; Singh, A.K.; Singh, V. Circularly polarized two port MIMO cylindrical DRA for 5G applications. In Proceedings of the 2020 International Conference on UK-China Emerging Technologies (UCET), Glasgow, UK, 20–21 August 2020.
11. Patel, U.; Upadhyaya, T. Four-Port Dual-Band Multiple-Input Multiple-Output Dielectric Resonator Antenna for Sub-6 GHz 5G Communication Applications. *Micromachines* **2022**, *13*, 2022. [CrossRef]
12. Upadhyaya, T.; Park, I.; Pandey, R.; Patel, U.; Pandya, K.; Desai, A.; Pabari, J.; Byun, G.; Kosta, Y. Aperture-Fed Quad-Port Dual-Band Dielectric Resonator-MIMO Antenna for Sub-6 GHz 5G and WLAN Application. *Int. J. Antennas Propag.* **2022**, *2022*, 4136347. [CrossRef]
13. Girjashankar, P.R.; Upadhyaya, T.; Desai, A. Multiband hybrid MIMO DRA for Sub-6 GHz 5G and WiFi-6 applications. *Int. J. RF Microw. Comput.-Aided Eng.* **2022**, *32*, e23479. [CrossRef]
14. Alanazi, M.D.; Khamas, S.K. A Compact Dual Band MIMO Dielectric Resonator Antenna with Improved Performance for mm-Wave Applications. *Sensors* **2022**, *22*, 5056. [CrossRef]
15. Srivastava, K.; Dwivedi, A.K.; Sharma, A. Circularly polarized dielectric resonator-based multiple input multiple output antenna with pattern and polarization diversity for vehicular applications. *Int. J. Circuit Theory Appl.* **2021**, *49*, 3421–3433. [CrossRef]

16. Kulkarni, J.; Kulkarni, N.; Desai, A. A Two-Port Dual Band Microstrip Feed Based Cylindrical Dielectric Resonator Antenna Array for Sub-6 GHz 5G and Super Extended-C Band Applications. In Proceedings of the 2021 International Conference on Communication Information and Computing Technology (ICCICT), Mumbai, India, 25–27 June 2021.
17. Girjashankar, P.R.; Upadhyaya, T. Substrate integrated waveguide fed dual band quad-elements rectangular dielectric resonator MIMO antenna for millimeter wave 5G wireless communication systems. *AEU-Int. J. Electron. Commun.* **2021**, *137*, 153821. [CrossRef]
18. Mishra, S.; Das, S.; Pattnaik, S.S.; Kumar, S.; Kanaujia, B.K. Three-dimensional dual-band dielectric resonator antenna for wireless communication. *IEEE Access* **2020**, *8*, 71593–71604. [CrossRef]
19. Sun, Y.-X.; Leung, K.W. Substrate-Integrated Two-Port Dual-Frequency Antenna. *IEEE Trans. Antennas Propag.* **2016**, *64*, 3692–3697. [CrossRef]
20. Sharawi, M.S.; Podilchak, S.K.; Hussain, M.T.; Antar, Y.M. Dielectric resonator based MIMO antenna system enabling millimetre-wave mobile devices. *IET Microw. Antennas Propag.* **2017**, *11*, 287–293. [CrossRef]
21. Jadoon, A.; Sarfraz, H.; Owais, O.; Nasir, J. MIMO Dielectric Resonator Antenna Array for 5G Systems. In Proceedings of the 2022 19th International Bhurban Conference on Applied Sciences and Technology (IBCAST), Islamabad, Pakistan, 16–20 August 2022.
22. Varshney, G.; Gotra, S.; Chaturvedi, S.; Pandey, V.S.; Yaduvanshi, R.S. Compact four-port MIMO dielectric resonator antenna with pattern diversity. *IET Microw. Antennas Propag.* **2019**, *13*, 2193–2198. [CrossRef]
23. Sharma, A.; Sarkar, A.; Biswas, A.; Akhtar, M.J. A-shaped wideband dielectric resonator antenna for wireless communication systems and its MIMO implementation. *Int. J. RF Microw. Comput.-Aided Eng.* **2018**, *28*, e21402. [CrossRef]
24. Kavitha, M.; Shanthi, S.; Beno, A.; Arul Rajan, B.; Sathish, M. Design of 2× 2 MIMO-DRA antenna for 5g communication. *Int. J. Sci. Technol. Res.* **2020**, *9*, 7025–7029.
25. Hussain, M.T.; Sharawi, M.S.; Podilchack, S.; Antar, Y.M. Closely packed millimeter-wave MIMO antenna arrays with dielectric resonator elements. In Proceedings of the 2016 10th European Conference on Antennas and Propagation (EuCAP), Davos, Switzerland, 10–15 April 2016.
26. Hussain, M.T.; Hammi, O.; Sharawi, M.S.; Podilchak, S.K.; Antar, Y.M. A dielectric resonator based millimeter-wave MIMO antenna array for hand-held devices. In Proceedings of the 2015 IEEE International Symposium on Antennas and Propagation & USNC/URSI National Radio Science Meeting, Vancouver, BC, Canada, 19–24 July 2015.
27. Hussain, S.A.; Taher, F.; Alzaidi, M.S.; Hussain, I.; Ghoniem, R.M.; Sree, M.F.A.; Lalbakhsh, A. Wideband, High-Gain, and Compact Four-Port MIMO Antenna for Future 5G Devices Operating over Ka-Band Spectrum. *Appl. Sci.* **2023**, *13*, 4380. [CrossRef]
28. Das, G.; Sahu, N.K.; Gangwar, R.K. Dielectric resonator based multiport antenna system with multi-diversity and built-in decoupling mechanism. *AEU-Int. J. Electron. Commun.* **2020**, *119*, 153193. [CrossRef]
29. Kumari, T.; Das, G.; Gangwar, R.K. Spatially decoupled 8-port box shaped MIMO DRA with quad-directional pattern diversity. *J. Electromagn. Waves Appl.* **2021**, *35*, 1221–1234. [CrossRef]
30. Kiani, S.H.; Altaf, A.; Anjum, M.R.; Afridi, S.; Arain, Z.A.; Anwar, S.; Khan, S.; Alibakhshikenari, M.; Lalbakhsh, A.; Khan, M.A. MIMO antenna system for modern 5G handheld devices with healthcare and high rate delivery. *Sensors* **2021**, *21*, 7415. [CrossRef] [PubMed]
31. Sharma, A.; Sarkar, A.; Biswas, A.; Akhtar, M.J. Millimeter-Wave Quad-Port Multiple-Input Multiple-Output Dielectric Resonator Antenna Excited Differentially by TE 20 Mode Substrate Integrated Waveguide. In Proceedings of the 2019 URSI Asia-Pacific Radio Science Conference (AP-RASC), New Delhi, India, 9–15 March 2019.
32. Elfergani, I.; Rodriguez, J.; Iqbal, A.; Sajedin, M.; Zebiri, C.; AbdAlhameed, R.A. Compact millimeter-wave MIMO antenna for 5G applications. In Proceedings of the 2020 14th European Conference on Antennas and Propagation (EuCAP), Copenhagen, Denmark, 15–20 March 2020.
33. Sharma, A.; Sarkar, A.; Adhikary, M.; Biswas, A.; Akhtar, M. SIW fed MIMO DRA for future 5G applications. In Proceedings of the 2017 IEEE International Symposium on Antennas and Propagation & USNC/URSI National Radio Science Meeting, San Diego, CA, USA, 9–14 July 2017.
34. Dwivedi, A.K.; Sharma, A.; Singh, A.K.; Singh, V. Metamaterial inspired dielectric resonator MIMO antenna for isolation enhancement and linear to circular polarization of waves. *Measurement* **2021**, *182*, 109681. [CrossRef]
35. Chen, H.N.; Song, J.-M.; Park, J.-D. A compact circularly polarized MIMO dielectric resonator antenna over electromagnetic band-gap surface for 5G applications. *IEEE Access* **2019**, *7*, 140889–140898. [CrossRef]
36. Das, G.; Sharma, A.; Gangwar, R.K.; Sharawi, M.S. Performance improvement of multiband MIMO dielectric resonator antenna system with a partially reflecting surface. *IEEE Antennas Wirel. Propag. Lett.* **2019**, *18*, 2105–2109. [CrossRef]
37. Murthy, N. Improved isolation metamaterial inspired mm-Wave MIMO dielectric resonator antenna for 5G application. *Prog. Electromagn. Res. C* **2020**, *100*, 247–261. [CrossRef]
38. Karimian, R.; Kesavan, A.; Nedil, M.; Denidni, T.A. Low-mutual-coupling 60-GHz MIMO antenna system with frequency selective surface wall. *IEEE Antennas Wirel. Propag. Lett.* **2016**, *16*, 373–376. [CrossRef]
39. Dadgarpour, A.; Zarghooni, B.; Virdee, B.S.; Denidni, T.A.; Kishk, A.A. Mutual coupling reduction in dielectric resonator antennas using metasurface shield for 60-GHz MIMO systems. *IEEE Antennas Wirel. Propag. Lett.* **2016**, *16*, 477–480. [CrossRef]
40. Mabrouk, I.B.; E'qab, R.; Nedil, M.; Denidni, T.A. Hybrid isolator for mutual-coupling reduction in millimeter-wave MIMO antenna systems. *IEEE Access* **2019**, *7*, 58466–58474.

41. Farahani, M.; Pourahmadazar, J.; Akbari, M.; Nedil, M.; Sebak, A.R.; Denidni, T.A. Mutual coupling reduction in millimeter-wave MIMO antenna array using a metamaterial polarization-rotator wall. *IEEE Antennas Wirel. Propag. Lett.* **2017**, *16*, 2324–2327. [CrossRef]
42. Mu'Ath, J.; Denidni, T.A.; Sebak, A.R. Millimeter-wave compact EBG structure for mutual coupling reduction applications. *IEEE Trans. Antennas Propag.* **2014**, *63*, 823–828.
43. Tong, C.; Yang, N.; Leung, K.W.; Gu, P.; Chen, R. Port and Radiation Pattern Decoupling of Dielectric Resonator Antennas. *IEEE Trans. Antennas Propag.* **2022**, *70*, 7713–7726. [CrossRef]
44. Zhao, X.; Tong, C.; Yang, N.; Lu, K.; Wu, Z.; Long, Y. Radiation Pattern Decoupling of Low-Profile Substrate Integrated Dielectric Resonator Antennas. In Proceedings of the 2022 IEEE MTT-S International Wireless Symposium (IWS), Harbin, China, 12–15 August 2022; pp. 1–3.
45. Sahu, N.K.; Gangwar, R.K. Dual-Port Compact MIMO-DRAs: Exploiting Metallic Sheets to Increase Inter-Port Isolation at 28-GHz 5G-Band. *IEEE Trans. Circuits Syst. II Express Briefs* **2022**, *69*, 4814–4818. [CrossRef]
46. Liang, Y.-Z.; Chen, F.-C. Decoupling of Two Extremely Close Half-Volume Rectangular Dielectric Resonator Antennas. In Proceedings of the 2022 International Conference on Microwave and Millimeter Wave Technology (ICMMT), Harbin, China, 12–15 August 2022.
47. Elahi, M.; Altaf, A.; Almajali, E.; Yousaf, J. Mutual Coupling Reduction in Closely Spaced MIMO Dielectric Resonator Antenna in H-Plane Using Closed Metallic Loop. *IEEE Access* **2022**, *10*, 71576–71583. [CrossRef]
48. Wu, X.J.; Chen, Z.; Yuan, T.; Liu, S.Z. A Decoupling Method for E-Plane Coupled Millimeter-Wave MIMO Dielectric Resonator Antennas. In Proceedings of the 2021 Cross Strait Radio Science and Wireless Technology Conference (CSRSWTC), Shenzhen, China, 11–13 October 2021.
49. Yadav, M.; Sivadas, S.; Patel, P. Starfish-Shaped MIMO Dielectric Resonator Antenna. In Proceedings of the 2020 IEEE 17th India Council International Conference (INDICON), New Delhi, India, 10–13 December 2020.
50. Belazzoug, M.; Messaoudene, I.; Aidel, S.; Boualem, H.; Chaouche, Y.B.; Denidni, T.A. 4-port MIMO CDR Antenna based with a Defected Ground Structure for 5G mm-wave applications. In Proceedings of the 2020 IEEE International Symposium on Antennas and Propagation and North American Radio Science Meeting, Montreal, QC, Canada, 5–10 July 2020.
51. Abedian, M.; Khalily, M.; Xiao, P.; Kabiri, Y.; Tafazolli, R. Closely Spaced MIMO Dielectric Resonator Antenna for Sub 6 GHz Applications. In Proceedings of the 2022 International Symposium on Antennas and Propagation (ISAP), Sydney, Australia, 31 October–3 November 2022; pp. 237–238.
52. Zhang, Y.; Deng, J.-Y.; Li, M.-J.; Sun, D.; Guo, L.-X. A MIMO dielectric resonator antenna with improved isolation for 5G mm-wave applications. *IEEE Antennas Wirel. Propag. Lett.* **2019**, *18*, 747–751. [CrossRef]
53. Ali, A.; Tong, J.; Iqbal, J.; Illahi, U.; Rauf, A.; Rehman, S.U.; Ali, H.; Qadir, M.M.; Khan, M.A.; Ghoniem, R.M. Mutual Coupling Reduction through Defected Ground Structure in Circularly Polarized, Dielectric Resonator-Based MIMO Antennas for Sub-6 GHz 5G Applications. *Micromachines* **2022**, *13*, 1082. [CrossRef]
54. Kulkarni, J. Multi-Band Circularly Polarized Cylindrical Dielectric Resonator Multiple Input Multiple Output Antenna for 5G and X-Band Applications. In Proceedings of the 2021 IEEE Indian Conference on Antennas and Propagation (InCAP), Jaipur, Rajasthan, India, 13–16 December 2021.
55. Singhwal, S.S.; Kanaujia, B.K.; Singh, A.; Kishor, J.; Matekovits, L. Dual-band circularly polarized MIMO DRA for sub-6 GHz applications. *Int. J. RF Microw. Comput.-Aided Eng.* **2020**, *30*, e22350. [CrossRef]
56. Singhwal, S.; Kanaujia, B.; Singh, A.; Kishor, J.; Matekovits, L. Multiple input multiple output dielectric resonator antenna with circular polarized adaptability for 5G applications. *J. Electromagn. Waves Appl.* **2020**, *34*, 1180–1194. [CrossRef]
57. Song, S.; Chen, X.; Da, Y.; Kishk, A.A. Broadband Dielectric Resonator Antenna Array With Enhancement of Isolation and Front-to-Back Ratio for MIMO Application. *IEEE Antennas Wirel. Propag. Lett.* **2022**, *21*, 1487–1491. [CrossRef]
58. Li, R.-Y.; Jiao, Y.-C.; Zhang, Y.-X.; Wang, H.-Y.; Cui, C.-Y. A Wideband MIMO Triangular Dielectric Resonator Antenna with A Simple Decoupling Structure. In Proceedings of the 2020 9th Asia-Pacific Conference on Antennas and Propagation (APCAP), Xiamen, China, 4–7 August 2020.
59. Alanazi, M.D.; Khamas, S.K. On-Chip Multiband MIMO Dielectric Resonator Antenna for MillimeterWave Applications. In Proceedings of the 2021 30th Wireless and Optical Communications Conference (WOCC), Taipei, Taiwan, 7–8 October 2021; pp. 174–177.
60. Divya, G.; Babu, K.J.; Madhu, R. Quad-band hybrid DRA loaded MIMO antenna with DGS for isolation enhancement. *Int. J. Microw. Wirel. Technol.* **2022**, *14*, 247–256. [CrossRef]
61. Malhat, H.A.; Zainud-Deen, S.H. Low-profile quad-band perforated rectangular dielectric resonator antenna for wireless communications. *J. Eng.* **2017**, *2017*, 448–451. [CrossRef]
62. Vishwanath; Babu, R.; Sharma, V.; Sahana, B.C.; Varshney, G. Controlling the Resonant Modes/bandwidth using graphene strip and Isolation Enhancement in a Two-Port THz MIMO DRA. *Opt. Quantum Electron.* **2023**, *55*, 659. [CrossRef]
63. Li, M.; Cheung, S. Isolation enhancement for MIMO dielectric resonator antennas using dielectric superstrate. *IEEE Trans. Antennas Propag.* **2020**, *69*, 4154–4159. [CrossRef]
64. Pan, Y.M.; Qin, X.; Sun, Y.X.; Zheng, S.Y. A simple decoupling method for 5G millimeter-wave MIMO dielectric resonator antennas. *IEEE Trans. Antennas Propag.* **2019**, *67*, 2224–2234. [CrossRef]

65. Hagras, A.; Denidni, T.A.; Nedil, M.; Coulibaly, Y. Low-mutual coupling antenna array for millimeter-wave MIMO applications In Proceedings of the 2012 IEEE International Symposium on Antennas and Propagation, Chicago, IL, USA, 8–14 July 2012.
66. Mishra, M.; Chaudhuri, S.; Kshetrimayum, R.S. Four-port DRA Array for MIMO Applications. In Proceedings of the 2020 International Symposium on Antennas and Propagation (ISAP), Osaka, Japan, 25–28 January 2021.
67. Sanghmitra; Rajput, A.; Mukherjee, B.; Patel, P. Y-shaped dielectric resonator MIMO antenna for omnidirection radiation pattern *Electromagnetics* **2022**, *42*, 157–167. [CrossRef]
68. Iqbal, A.; Nasir, J.; Qureshi, M.B.; Khan, A.A.; Rehman, J.U.; Rahman, H.U.; Fayyaz, M.A.; Nawaz, R. A CPW fed quad-port MIMO DRA for sub-6 GHz 5G applications. *PLoS ONE* **2022**, *17*, e0268867. [CrossRef] [PubMed]
69. Anuar, S.U.; Jamaluddin, M.H.; Din, J.; Kamardin, K.; Dahri, M.H.; Idris, I.H. Triple band MIMO dielectric resonator antenna for LTE applications. *AEU-Int. J. Electron. Commun.* **2020**, *118*, 153172. [CrossRef]
70. Sharma, A.; Sarkar, A.; Biswas, A.; Jaleel Akhtar, M. Equilateral triangular dielectric resonator based co-radiator MIMO antennas with dual-polarisation. *IET Microw. Antennas Propag.* **2018**, *12*, 2161–2166. [CrossRef]
71. Abdullah, A.I.O.; Elshoshi, A.H.; Elmrabet, A.M.; Ahmed, S.M. Performance of Two-Port Dielectric Resonator Antenna Used for 5G mm-wave Applications. In Proceedings of the 2021 IEEE 1st International Maghreb Meeting of the Conference on Sciences and Techniques of Automatic Control and Computer Engineering MI-STA, Tripoli, Libya, 25–27 May 2021; pp. 782–786.
72. Pan, Y.M.; Hu, Y.; Zheng, S.Y. Design of low mutual coupling dielectric resonator antennas without using extra decoupling element. *IEEE Trans. Antennas Propag.* **2021**, *69*, 7377–7385. [CrossRef]
73. Lai, Q.X.; Pan, Y.M.; Zheng, S.Y. A Self-Decoupling Method for MIMO Linear and Planar Dielectric Resonator Antenna Arrays Based on Transmission Characteristics of Feeding Structure. *IEEE Trans. Antennas Propag.* **2023**, *71*, 5708–5716. [CrossRef]

Disclaimer/Publisher's Note: The statements, opinions and data contained in all publications are solely those of the individual author(s) and contributor(s) and not of MDPI and/or the editor(s). MDPI and/or the editor(s) disclaim responsibility for any injury to people or property resulting from any ideas, methods, instructions or products referred to in the content.

electronics

MDPI

rticle

Design of High-Gain and Low-Mutual-Coupling Multiple-Input–Multiple-Output Antennas Based on PRS for 28 GHz Applications

inkyu Jung [1], Wahaj Abbas Awan [1], Domin Choi [1], Jaemin Lee [1], Niamat Hussain [2] and Nam Kim [1,*]

1 Department of Information and Communication Engineering, Chungbuk National University, Cheongju 28644, Republic of Korea
2 Department of Intelligent Mechatronics Engineering, Sejong University, Seoul 05006, Republic of Korea
* Correspondence: namkim@chungbuk.ac.kr

Abstract: In this paper, a high-gain and low-mutual-coupling four-port Multiple Input Multiple Output (MIMO) antenna based on a Partially Reflective Surface (PRS) for 28 GHz applications is proposed. The antenna radiator is a circular-shaped patch with a circular slot and a pair of vias to secure a wide bandwidth ranging from 24.29 GHz to 28.45 GHz (15.77%). The targeted band has been allocated for several countries such as Korea, Europe, the United States, China, and Japan. The optimized antenna offers a peak gain of 8.77 dBi at 24.29 GHz with a gain of 6.78 dBi. A novel PRS is designed and loaded on the antenna for broadband and high-gain characteristics. With the PRS, the antenna offers a wide bandwidth from 23.67 GHz to 29 GHz (21%), and the gain is improved up to 11.4 dBi, showing an overall increase of about 3 dBi. A 2 × 2 MIMO system is designed using the single-element antenna, which offers a bandwidth of 23.5 to 29 GHz (20%), and a maximum gain of 11.4 dBi. The MIMO antenna also exhibits a low mutual coupling of −35 dB along with a low Envelope Correlation Coefficient and Channel Capacity Loss, making it a suitable candidate for future compact-sized mmWave MIMO systems.

Keywords: high gain; partially reflecting surface; MIMO antenna; 28 GHz; 5G; mmWave

Citation: Jung, J.; Awan, W.A.; Choi, D.; Lee, J.; Hussain, N.; Kim, N. Design of High-Gain and Low-Mutual-Coupling Multiple-Input–Multiple-Output Antennas Based on PRS for 28 GHz Applications. *Electronics* **2023**, *12*, 4286. https://doi.org/10.3390/electronics12204286

Academic Editor: Reza K. Amineh

Received: 13 September 2023
Revised: 12 October 2023
Accepted: 13 October 2023
Published: 16 October 2023

Copyright: © 2023 by the authors. Licensee MDPI, Basel, Switzerland. This article is an open access article distributed under the terms and conditions of the Creative Commons Attribution (CC BY) license (https://creativecommons.org/licenses/by/4.0/).

1. Introduction

Mobile communication data traffic has shown a rapid surge soon after the introduction of Long-Term Evolution (LTE) to the public, thanks to a significant increase in communication device numbers. To meet the present requirements, the popularity of 5G mobile communication technology has risen. Research and development for advanced mobile communication technologies are currently in progress under the 5G umbrella, primarily due to its ability to offer higher data transmission rates per user in comparison to 4G technology. The 5G frequency spectrum consists of two main frequency bands: sub-6 GHz (3.5 GHz) and millimeter wave (mmWave) [1]. Among these, the mmWave frequency bands, which include representative bands like 28 GHz, 39 GHz, and 60 GHz within the 24 GHz to 100 GHz range, offer better reliability, lower delay rates, and faster communication speeds compared to sub-6 GHz [2]. Notably, the 28 GHz frequency has garnered significant attention due to its exceptional performance in comparison to other spectrum bands. This has led to South Korea allocating the range from 26.5 GHz to 28.9 GHz for mobile connectivity, while the European Union has been assigned frequencies ranging from 24.5 GHz to 27.5 GHz by ITU. Additionally, the 3GPP standard designates 24.25 GHz to 27.5 GHz for use in the United States, while the range reserved for mobile communication is 27.5 GHz to 28.35 GHz [3], as shown in Figure 1.

Countries	24 – 28 GHz band	37 – 40 GHz band	64 – 71 GHz band
Korea	26.5 – 29.5 GHz		
Japan	27.5 29.5 GHz		
Singapore	24.25 -26.5 GHz 26.5 – 27.5 GHz		
Canada	27.5 – 28.35 GHz	37 – 40 GHz	64 – 71 GHz
USA	24.75 – 25.25 GHz 27.5 28.35 GHz	37 – 40 GHz 47.2 – 48.2 GHz	64 – 71 GHz
UK	26 GHz		
Germany	26 GHz		
France	26 GHz		
Italy	26.5 – 27.5 GHz		
Sweden	26.25 – 27.5 GHz		
Finland	25.1 – 27.5 GHz		
China	24.75 – 27.5 GHz	40 – 43.5 GHz	

Figure 1. Current status of mmWave frequency allocation in major countries [3].

Conversely, the trend towards smaller sizes for 5G devices has led to limited space for antenna systems. This is where millimeter-wave antennas have an edge over traditional sub-6 GHz antennas, as they do not require as much space. However, the mmWaves are not very resilient to the surrounding environment and have poor performance in non-line of-sight scenarios [4]. Researchers have found that this limitation of the mmWave band spectrum can be overcome by designing higher gain antennas along with Multiple-Input-Multiple-Output (MIMO) capabilities. The MIMO systems offer numerous advantages over Single-Input-Single-Output (SISO) that are not limited to lower power consumption improved signal range, reduced bit errors that result in reduced interference and enhanced NLOS connectivity [5]. Moreover, an effective MIMO antenna system should not only provide a compact size and high gain, but also excel in terms of factors like mutual coupling, efficiency, Envelop Correlation Coefficient (ECC), Channel Capacity Loss (CLL) and various diversity parameters including Diversity Gain [6–10]. As a result, numerous recent articles in the literature have explored different techniques to meet the growing demands for 5G millimeter-wave MIMO antenna systems.

As an example, a researcher developed an MIMO antenna enhanced with a ground stub to diminish mutual coupling [9]. They also employed a metasurface to mitigate mutual coupling between closely positioned elements. However, this approach led to increased dimensions of 31 mm × 48 mm. Another notable study is documented in [11], where researchers designed a series array antenna in a rectangular patch shape for 28 GHz applications. This antenna covers a broad frequency range of 27.6~29.1 GHz, boasting a high gain of 10.2 dBi and minimal mutual coupling of >−40 dB. Nevertheless, the utilization of a slot array technique makes the antenna larger in size.

A different approach involves the design of a low-mutual-coupling MIMO antenna employing orthogonally positioned elements in the infinity shell shape [12]. While this proposed antenna benefits from its compact size, it still faces various challenges, including low gain, limited bandwidth, and the absence of a common ground plane in its design, rendering it less suitable for practical applications. Similarly, [13] introduces an MIMO antenna based on a dielectric resonator. This design achieves a mutual coupling of >−24 dB within the frequency range of 25.2~28 GHz. However, the gain remains low for 5G applications.

In contrast to the works mentioned earlier, a flat-lens antenna is developed using an ultrathin Huygens' metasurface, aiming to achieve a high gain of 30 dBi [14]. However, the limitation of such antennas lies in their exceptionally large size, which confines their application to either measurement setups or military purposes where size constraints are not a concern.

On the other hand, a four-element MIMO antenna designed for a 28 GHz application employs a two-element array configuration as a single element [15]. This approach of using a two-element array aims to achieve a higher gain and broader bandwidth compared to a single-element design. Additionally, the antenna employs a defected ground structure technique along with orthogonal placement to achieve reduced mutual coupling between adjacent elements. The resulting antenna covers a wide frequency range of 25.5–29.6 GHz, exhibiting a moderate coupling of >-18 dB and a gain of 8.3 dBi.

Research efforts by [10,16] concentrate on enhancing the MIMO antenna by adding an extra layer of metasurface. For the four-element MIMO antenna with an orthogonal placement, introducing a metasurface at the back of the antenna enhances the gain from 6 dBi to 10 dBi [10]. However, the bandwidth and mutual coupling among elements remain consistent with the MIMO antenna lacking a metasurface. Conversely, [16] fully exploits the metasurface-loading technique to achieve both a wide bandwidth and high gain. This leads to a bandwidth improvement of 5.5 GHz, spanning 23.3–28.8 GHz and a gain increase of 4 dBi, resulting in a peak gain of 10.21 dBi. It is noteworthy that a maximum mutual coupling of -22.5 dB is observed at the operational frequency.

Another intriguing study is documented in [17], where a multilayered antenna based on a chiral metasurface is devised. This antenna offers the advantage of improved performance in terms of bandwidth, high gain, and reduced mutual coupling. However, it does face certain challenges, including structural complexity due to its multiple layers and elevated profile.

Based on the aforementioned considerations, it becomes evident that a pressing need exists for a mmWave MIMO antenna capable of delivering a superior performance across all parameters, while still maintaining a compact size, structural simplicity, and strong diversity performance. Therefore, this paper presents the design of an MIMO antenna characterized by its compact dimensions, low mutual coupling, high gain, and robust MIMO performance metrics.

The manuscript is organized as follows: Section 2 outlines the design process and outcomes for both the single antenna element and the MIMO antenna, along with their respective performance parameters. Section 3 details the design procedure for the metasurface and its integration with the antenna, along with an examination of various performance metrics. The concluding section, Section 4, provides a comparison between the proposed work and existing literature.

2. Antenna Design

The geometric representation of the proposed MIMO antenna system is depicted in Figure 2. This MIMO antenna comprises four circular patch antennas, each featuring a semi-circular slot and a pair of vias. In order to achieve minimal mutual coupling, these elements are positioned orthogonally to each other, as illustrated in Figure 2a. The MIMO antenna is backed by a full copper ground plane, as displayed in Figure 2b. For the simulation setup, the antenna is made using copper and the ROGERS 5880 substrate. The exact properties are assigned to all materials. The radiation box is placed at a quarter wavelength in all directions from the antenna system, where a central frequency of 28 GHz is chosen for the wavelength calculation. Air is assigned as a material for the radiation box.

For an additional enhancement of the proposed MIMO system's performance, a partially reflecting surface is incorporated. This surface consists of a 6×6 array of meta cells, where each meta cell is designed with a staircase-like structure on the upper side, as seen in Figure 2c, and a square loop on the reverse side, as depicted in Figure 2d. Placed atop the antenna with an airgap of G, this metasurface is shown in Figure 2e. Maintaining uniform dimensions of 24×24 mm^2 for both the antenna and metasurface facilitates the measurement setup.

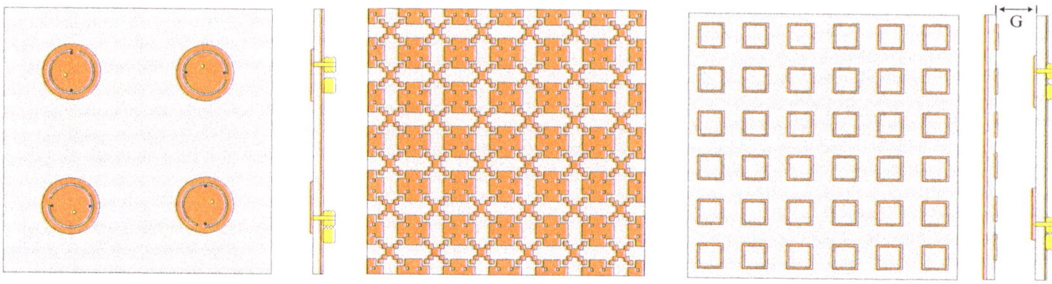

(a) (b) (c) (d) (e)

Figure 2. The proposed MIMO antenna with PRS. (**a**) Front view of MIMO antenna, (**b**) side view of MIMO antenna, (**c**) front view of PRS, (**d**) back view of PRS, (**e**) side view of MIMO antenna with PRS.

2.1. Single-Element Antenna Design

The proposed approach involves four sequential steps, commencing with the design of a circular patch antenna, followed by the creation of two distinct models utilizing a slot and a pair of vias. These models are subsequently merged and optimized to yield the ultimate broadband antenna.

The selection of the circular patch antenna stems from its numerous advantages, including its compact size, compatibility with array structures, and microwave monolithic integrated circuits. Furthermore, circular patch antennas can be seamlessly integrated into both planar and non-planar configurations.

The resonant frequency of the circular patch is determined by its radius (r), while impedance matching can be managed by optimizing the positioning of the feeding pin. The radius at the desired frequency can be extracted using the following formula [18]:

$$r = \frac{F}{1 + \frac{2h}{\pi \varepsilon_r F}\left[\ln \frac{\pi F}{2h} + 1.7726\right]^{\frac{1}{2}}} \tag{1}$$

Here, the ε_r refers to the dielectric constant of the substrate, h denotes the substrate thickness, while F is the constant, which can be estimated in terms of the resonating frequency by the mean of the following expression [18]:

$$F = \frac{8.791 \times 10^9}{f_r \sqrt{\varepsilon_r}} \tag{2}$$

The antenna is designed using RT5880 by ROGERS corp., having low dielectric value and low loss. The pin position is optimized for good impedance matching at a center frequency of 26 GHz, as shown in Figure 3. The circular patch offers bandwidth ranges 25.6–27.25 GHz, as depicted in Figure 4a.

In order to extend the bandwidth to encompass the entire globally allocated 5G frequency spectrum, the technique of etching slots and employing loading vias is employed. In the first scenario, a semi-circular slot is etched from the radius, as demonstrated in Figure 3. This leads to a redistribution of surface currents, resulting in a broader bandwidth of 25.37–27.05 GHz, as depicted in Figure 4.

In the second scenario, a pair of vias is loaded to reduce the length of the radiating structure in connection with the ground. This introduces added resistance to the radiator, which in turn facilitates impedance matching across a wider frequency range of 25.3–28.5 GHz, as displayed in Figure 4a. However, while both antenna-2 and antenna-3 exhibit a wider bandwidth compared to antenna-1, neither covers the desired frequency band. Consequently, both antennas were integrated, as presented in Figure 3.

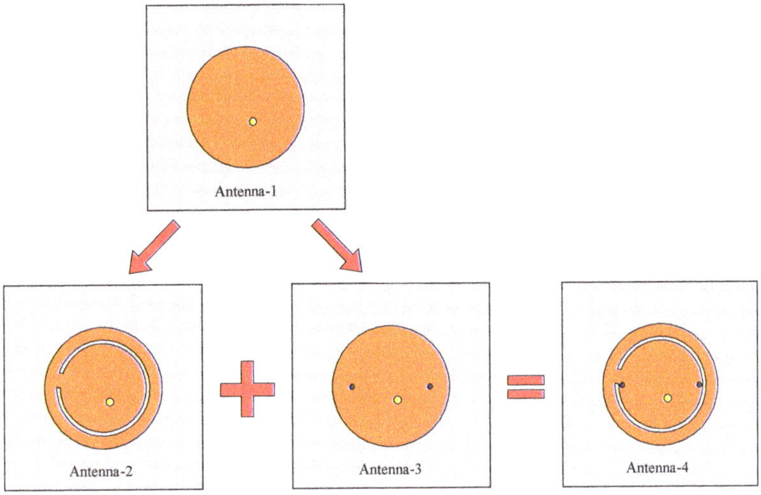

Figure 3. Design evolution of proposed circular patch antenna.

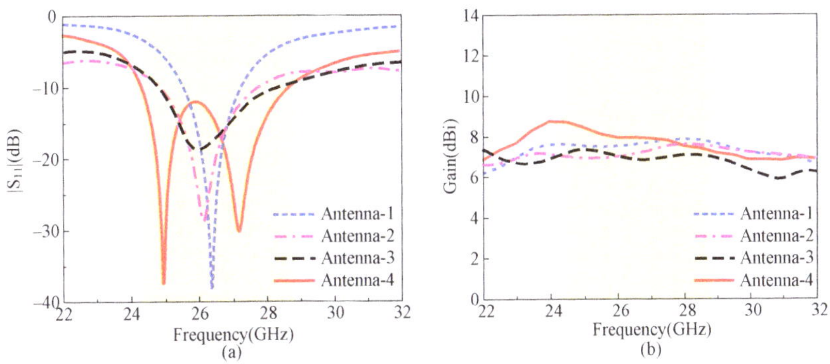

Figure 4. Performance comparison of the various antenna designs (**a**) $|S_{11}|$ and (**b**) gain.

This integrated antenna results in the generation of a broad frequency range spanning 24.29–28.45 GHz, featuring two resonances at 25 GHz and 27 GHz. The occurrence of these resonances is attributed to the interplay of the slot and the via, resulting in distinct electrical lengths with respect to the feeding point. Furthermore, Figure 4b illustrates a comparison between the basic antenna and the final integrated antenna, confirming a marginal enhancement in the peak gain of the antenna. To further understand the broad impedance bandwidth mechanism, the surface current distribution is illustrated in Figure 5. The current is distributed along the inner patch as well as the outer patch with respect to the slot for the 25 GHz band, which refers to a bigger electrical length. On the other hand, for 27 GHz, the current is mainly distributed along the inner patch while the small amount of current at the outer patch also shows that it takes part in achieving higher resonance. The antenna configuration is shown in Figure 6, while the values of various optimized parameters are listed in Table 1.

Table 1. Optimized parameters of proposed antenna.

Parameter	A	Pin x	G	R	h
Length (mm)	12	3.2	0.1	2.5	0.787

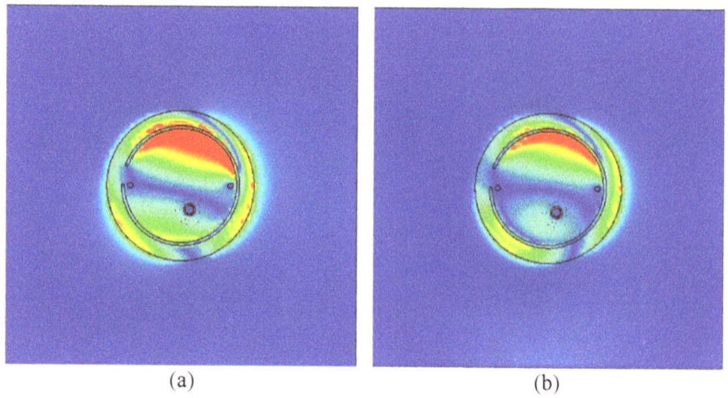

Figure 5. Current distribution of the optimized antenna at (**a**) 25 GHz and (**b**) 27 GHz.

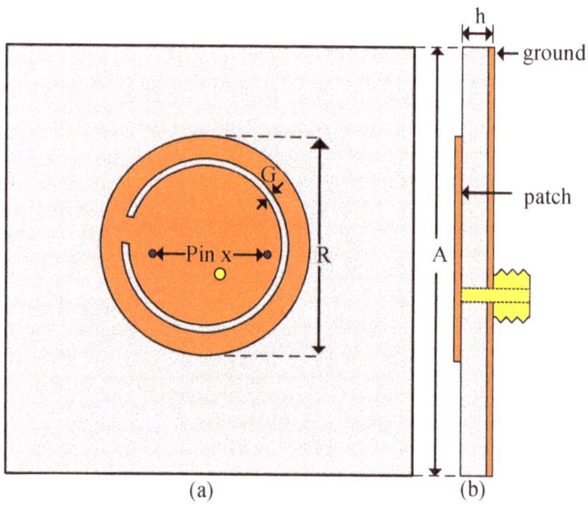

Figure 6. Geometrical illustration of proposed mmWave wideband single-element antenna. (**a**) Top view and (**b**) side view.

2.2. *Partial Reflective Surface Design*

Despite the antenna demonstrating commendable performance in terms of bandwidth and gain, there remains a need for performance enhancement due to the relatively low energy possessed by mmWave. This is particularly crucial to counteract the atmospheric absorption losses. To address this, the gain must be enhanced to mitigate such losses. In pursuit of this objective, metamaterials with unique attributes not commonly found in other materials are designed and integrated alongside the antenna.

Metamaterials are categorized into several types based on their working principles and behaviors, including frequency selective surfaces (FSSs), electronic bandgaps (EBGs), and partial reflective surfaces (PRSs). FSSs are typically paired with monopole antennas, positioned at the rear side to reduce rearward radiation. However, an FSS has limitations, such as its larger volume and constraints when applied in compact devices. Conversely, EBG structures find their use between MIMO antenna elements to diminish mutual coupling, thus enhancing the signal strength.

PRS refers to a reflecting surface with varying reflection characteristics based on the incident frequency angle, polarization, or radio wave band. It is often chosen for its significant impact on gain improvement. When employed in this manner, it is mainly positioned in front of the antenna, functioning as a resonator along with the antenna's ground surface, forming a Fabry–Perot (FP) resonator. For the antenna with a PRS antenna, a substantial gain can be achieved in the direction of propagation. An advantageous aspect of the FP resonance antenna is its relatively uncomplicated structure, offering the potential for high gains. However, the FP resonator also has its downside, featuring a narrow impedance and radiated bandwidth due to resonance conditions being met at a single frequency.

In light of these considerations, research has been directed toward maximizing the high-gain advantage of PRS while addressing the limitation of bandwidth narrowness.

Thus, instead of designing the PRS structure on the single side of the substrate, a dual sided structure is utilized to achieve a wideband operation. The back side of the PRS unit cell contains a square loop structure, while the front side is modified using various steps, as shown in Figure 7. Initially, a full copper-covered side is used, which is then converted to a marble-like structure by etching four rectangular slots from it. Afterwards, more structural changes are carried out by etching slots from the corner patches, and rotating the middle structure after etching slots from the middle patch. Finally, small square slots are extracted from the corner patches, while small rectangular patches are added at the corner of the middle structure.

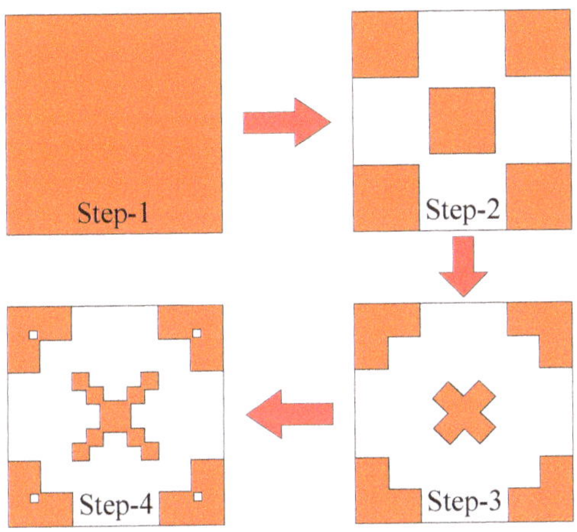

Figure 7. Design evolution of proposed PRS.

In the electromagnetic theory, the metasurface modifies electromagnetic waves through specific boundary conditions as opposed to the three-dimensional (3D) space's construction parameters, which are frequently used in meta materials and follow similar PRS principles. Using CST Studio's unit cell designing tool, the reflection coefficient and phase are analyzed considering general incidence and by using the open boundary condition on the xy plane of the cell.

Figure 8 shows the reflection coefficient and phase results of the optimized PRS. The reflection coefficient and phase are crucial key parameters in designing a wideband with a high-gain mmWave antenna. A higher reflective PRS leads to a higher gain, but narrower −3 dB bandwidth; to broaden this, a PRS with a positive phase gradient is used. The resulting reflection coefficient and phase of the optimized PRS have a positive gradient and almost exactly resemble the ideal phase of the PRS over a wide-bandwidth spectrum range.

Figure 9 presents the design of the proposed PRS unit cell, along with optimized parameters for various dimensions.

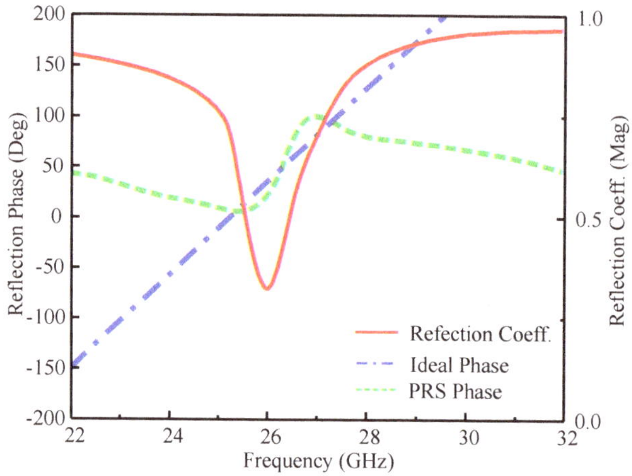

Figure 8. Reflection coefficient's magnitude and phase of the optimized PRS unit cell.

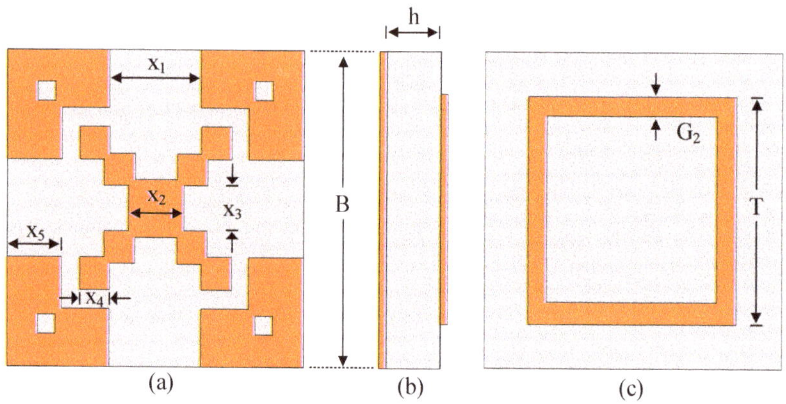

Figure 9. Proposed PRS unit cell: (**a**) top view, (**b**) side view, (**c**) bottom view. Parameter: B = 12; T = 3.2; G_2 = 0.1; h = 0.787; x_1 = 1.22; x_2 = 0.73; x_3 = 0.73; x_4 = 0.4; x_5 = 0.73 (units are in mm).

2.3. PRS-Loaded Single-Element Antenna

The PRS is loaded on the top side of the antenna, as shown in Figure 10. The Ray theory explains that the vector sum of the partially transmitted electromagnetic waves results in an electric field which is given by the following relation [19]:

$$\mathrm{TE} = \sum_{j=-\infty}^{\infty} f(\theta) E_j \tag{3}$$

where f (θ) denotes the radiated field pattern generated by the source having an angle (θ), while E_j denotes the reflected field as the collective sum of the vector field by the reflector and metasurface. To obtain the maximum power transmitted towards the broadside, the following resonance condition must be met.

$$\psi - \pi - \frac{4\pi G_3}{\lambda_0} = 2N\pi \tag{4}$$

where λ_0 denotes the free-space wavelength at the central frequency of the antenna; N will be an integer, while the G3 represents the gap among the antenna and PRS, which must satisfy the following relation:

$$G3 = \left(\frac{\psi}{360} - 0.5\right)\frac{\lambda_0}{2} - N\frac{\lambda_0}{2} \quad (5)$$

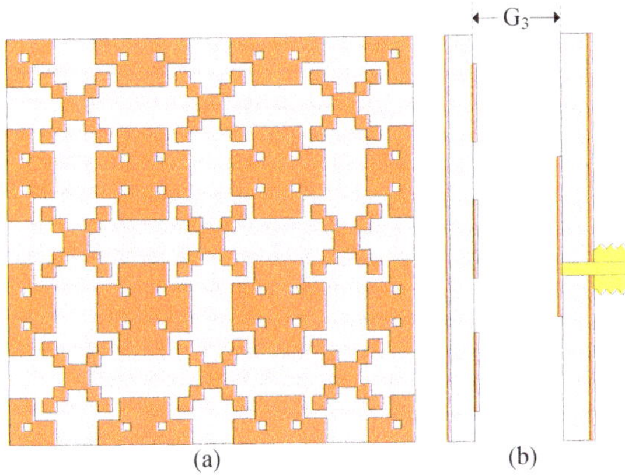

Figure 10. Single antenna with PRS: (**a**) top view, (**b**) side view.

However, due to a loss of materials, the gap between the PRS and antenna needs optimization. For said purpose, the parametric analysis is carried out and based upon the return loss, and the gain performance of the optimized gap is found to be 5.8 mm. Figure 11a shows the $|S_{11}|$ results of the antenna before and after loading PRS; the impedance bandwidth is improved from 4.16 GHz to 5.27 GHz with a range of 23.63–28.9 GHz. Similarly, the gain of the antenna significantly improved when loaded with the metasurface; an average gain improvement of 3 dBi is observed for the operational bandwidth, as shown in Figure 11b.

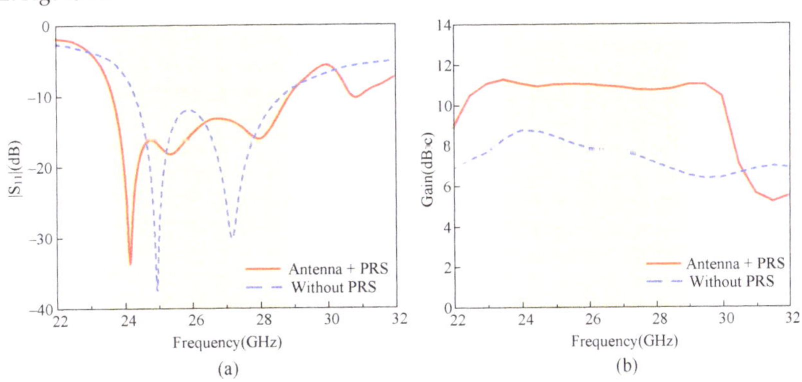

Figure 11. Single-element antenna with PRS performance; (**a**) S11 parameter, (**b**) gain.

2.4. PRS-Loaded MIMO Antenna

MIMO antennas leverage spatial diversity and spatial multiplexing techniques and offer multipath propagation through multiple antennas, resulting in elevated transmission rates; robust signals are less susceptible to attenuation or loss, ensuring stable communication [20]. However, the mutual coupling among MIMO antenna elements can disrupt system performance. As a result, MIMO antennas require minimal mutual coupling among elements to function effectively.

The fundamental antenna design from the preceding section serves as the foundation for constructing a four-element MIMO antenna. Orthogonal placement is employed to achieve low coupling among the elements, eliminating the need for additional decoupling structures. It is important to note that, due to measurement constraints and the size limitations of mmWave connectors, inter-spacing cannot be reduced.

Figure 12 shows the geometrical configuration of the MIMO antenna with interspace among adjacent antennas of w = 7 mm, referring to $\lambda/2$ at the lowest resonance. The total size of the antenna is 24×24 mm^2, which is twice the size of a single-element antenna. As stated earlier, the PRS is loaded to achieve a high gain; so, a 9×9 array of meta-cells is loaded on the topside of the MIMO antenna while keeping the same gap as that utilized for a single element, as shown in Figure 10.

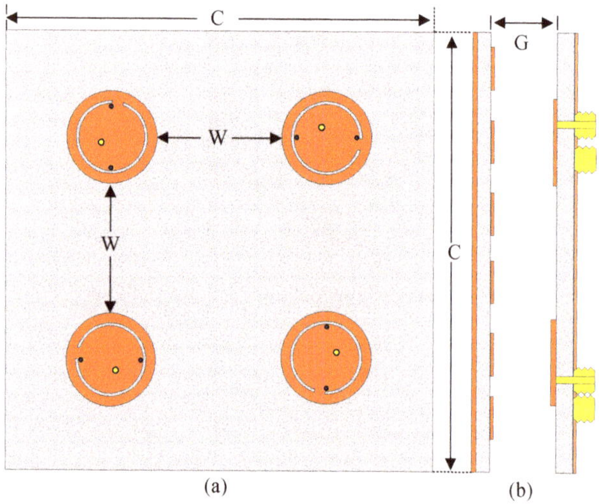

Figure 12. PRS-loaded MIMO antenna: (**a**) top view (**b**) side view.

Figure 13 shows the s-parameter performance of the proposed PRS-loaded MIMO antenna.

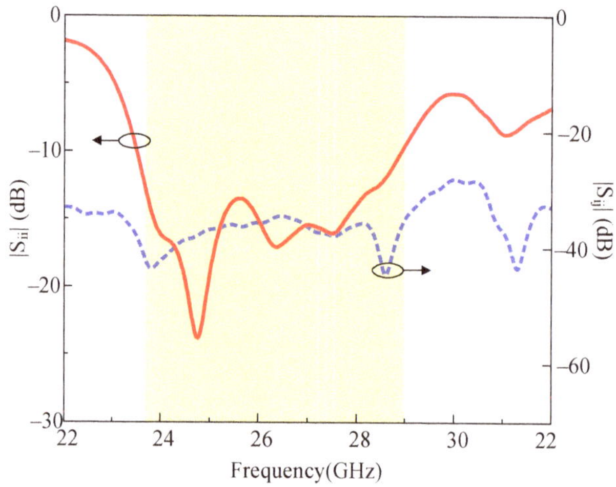

Figure 13. S-parameters of the proposed PRS-loaded MIMO antenna.

It is observed that the antenna offers an $|S_{11}| > -10$ dB impedance bandwidth of 5.5 GHz, starting from 23.5 GHz to 29 GHz. Moreover, the mutual coupling of > -33 dB is achieved throughout the operational region.

3. Results and Discussion

3.1. Fabrication and Measurment

To validate the findings presented in the preceding section, a prototype of the antenna and metasurface was fabricated and subjected to testing. In Figure 14a, the different components of the PRS-loaded antenna and its MIMO configuration are depicted, while the complete setup, including the use of polyurethane foam to create a gap between the PRS and the antenna, is illustrated in Figure 14b.

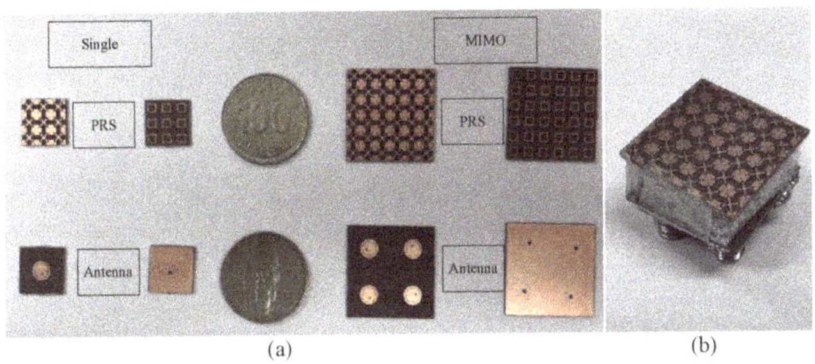

Figure 14. Manufactured 2 × 2 MIMO antenna photographs: (**a**) parts of each MIMO, (**b**) manufactured prototype.

The evaluation of radiation patterns and gain measurements was conducted within a newly established circular chamber, as shown in Figure 15. Further details regarding the functioning of the circular chamber and its superior efficiency compared to traditional mmWave far-field chambers can be found in [21].

Figure 15. Snaps of the proposed MIMO antenna with PRS in the anechoic chamber.

3.2. S-Parameter

The comparison among simulated and measured s-parameters is shown in Figure 16. The $|S_{11}|$ results of the single-element antenna loaded with PRS and that of MIMO remain changed; thus, they included are not included separately. The measured results offer wideband ranges 23.5 GHz to 30 GHz, while the simulation results have the operational band of 23.5 GHz to 29 GHz, as shown in Figure 16a. On the other hand, Figure 16b shows mutual coupling among the MIMO elements where both simulated and measured results offer a low mutual coupling of >−30 dB. A strong performance in terms of comparison is obtained; however, the discrepancy between the results may be due to fabrication and measurement setup tolerance [22].

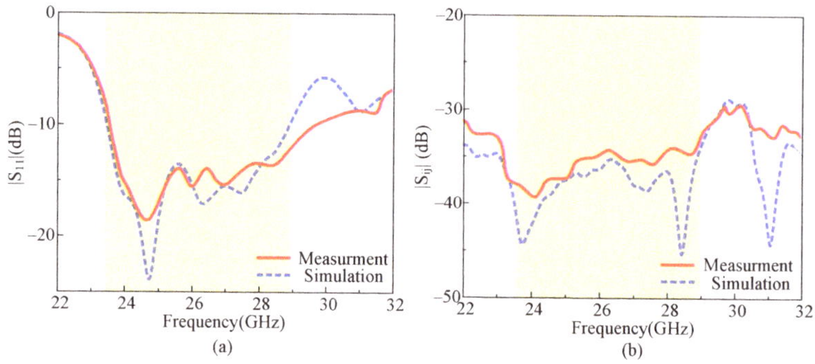

Figure 16. Comparison of simulation and measurement data of fabricated prototypes; (**a**) is $|S_{11}|$ and (**b**) is $|S_{12}|$.

3.3. *Radiation Pattern*

The radiation patterns of the proposed PRS-loaded MIMO antenna are measured; the comparison with simulated results is shown in Figure 17. Since measurement was conducted from 26.5 GHz due to the limitation of the measuring equipment, the result value of 26.5 GHz was attached, not the center frequency of 26 GHz. In both principal E- and H-planes, the antenna exhibits a broadside radiation pattern. At 26 GHz, the antenna exhibits a slightly titled radiation pattern in E-plane, which has the maximum value of radiation at θ = 15°, as shown in Figure 17a. Likewise, in H-plane, the antenna offers peak radiation at θ = 25° for 26 GHz, as shown in Figure 17b. On the other hand, the PRS-loaded antenna offers peak radiation toward θ = 0° for both E- and H-planes at the resonating frequency of 28 GHz, as shown in Figure 17a,b.

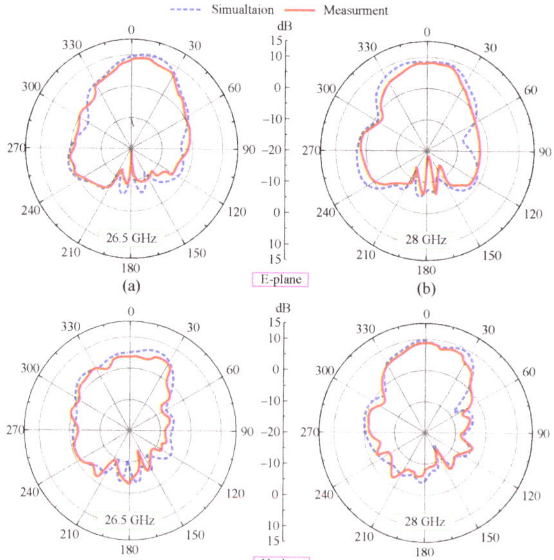

Figure 17. Comparison of measured and simulated results of the MIMO system: (**a**) 26.5 GHz and (**b**) 28 GHz at E-plane, and (**c**) 26.5 GHz and (**d**) 28 GHz at H-plane.

3.4. ECC

The Envelope Correlation Coefficient (ρ_{eij}) determines whether each individual antenna element phase is independent for the MIMO system. Ideally, the ECC must be zero for a perfect MIMO antenna system; however, due to the losses in materials and nature, an ECC value of less than 0.5 is acceptable. The ECC is determined by the means of S-parameters, so it is a unitless quantity. For any MIMO antenna system, the ECC can be calculated using the following expression, as explained in [23]:

$$\rho_{eij} = \frac{\left|S_{12}S_{11}^{*} + S_{22}S_{21}^{*}\right|^2}{\left[1-\left(|S_{11}|^2+|S_{21}|^2\right)\left[1-\left(|S_{22}|^2+|S_{21}|^2\right)\right]\right.} \tag{6}$$

The simulated and measured values of s-parameters are used to evaluate the ECC; the comparison of both results is shown in Figure 18. Both results offer a very low ECC value, which is close to 0 and less than 0.004.

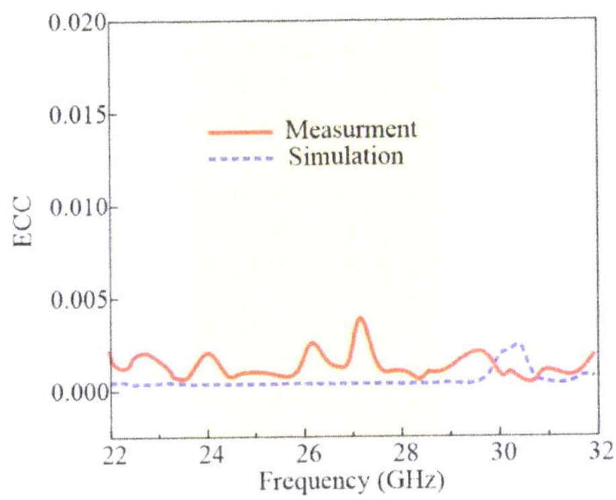

Figure 18. Simulated and measured ECC of the proposed PRS-loaded MIMO antenna.

3.5. Diversity Gain

Diversity gain is the loss of transmission power when diversity schemes are performed on modules for MIMO configurations; this is calculated from the above formula, as cited in the reference. The graph shown as ECC shows that DG is 9.99 dB over the entire communication band of the antenna, as shown in Figure 19. This shows that the antenna has excellent diversity performance.

$$DG = 10\sqrt{1-|\rho_{eij}|^2} \tag{7}$$

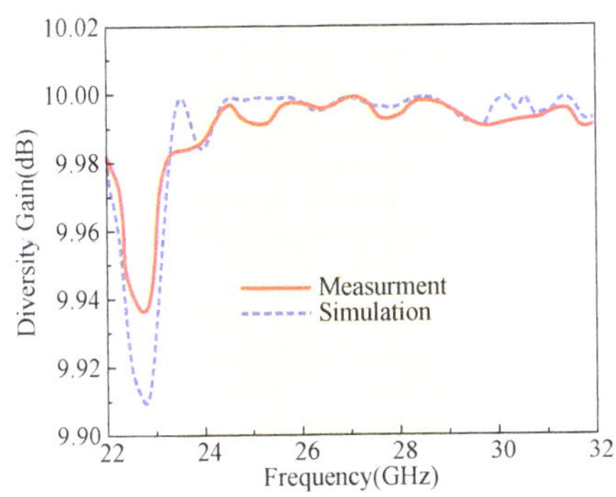

Figure 19. Simulated and measured DG of proposed PRS-loaded MIMO antenna.

3.6. CCL (Channel Capacity Loss)

CCL stands for Channel Capacity Loss, which occurs when a wireless communication system's receiver receives a signal. In an MIMO system, the receiver, equipped with multiple antennas, captures signals from various paths simultaneously. The presence of interference between antennas becomes evident through variations in the CCL. An antenna's performance directly influences CCL, with better antenna performance leading to lower CCL values. As the distance between the receiver and antenna decreases, CCL tends to increase. Additionally, the characteristics of the channel can also impact CCL. Consequently, CCL serves as a performance metric for evaluating MIMO antenna efficiency and should ideally be minimal to mitigate errors stemming from other factors. Mathematically, CCL is computed using the following expressions [24]:

$$C(loss) = -\log_2 \det(a) \quad (8)$$

$$a = \begin{bmatrix} \sigma_{11} & \sigma_{12} \\ \sigma_{21} & \sigma_{22} \end{bmatrix} \quad (9)$$

$$\sigma_{ii} = 1 - (|S_{ii}|^2 - |S_{ij}|^2) \quad (10)$$

$$\sigma_{ij} = -(S_{ii}^* S_{ij} + S_{ji} S_{jj}^*) \quad (11)$$

For the operational band, the proposed antenna offers a low CCL of <0.4 bits/s/Hz for both simulated and measured results, as shown in Figure 20.

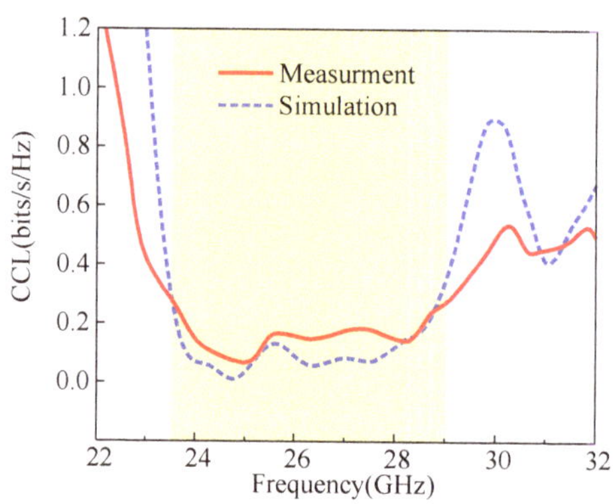

Figure 20. Simulated and measured CCL of proposed PRS-loaded MIMO antenna.

3.7. Comparison with Related Work

The performance comparison of the proposed work in terms of size, bandwidth, mutual coupling, −3 dB bandwidth and efficiency is performed with the recently reported work. Except from the work reported in [7,11], the proposed work overperformed in comparison with the other work in terms of size. Moreover, the work reported in [7,11] offers a narrow bandwidth, high mutual coupling and low gain as compared to the proposed work. Only the work reported in [8] offers the peak high-gain value of around 18 dBi, at the cost of twice the size of the proposed antenna. Thus, it can be concluded that the proposed PRS-loaded MIMO antenna system offers a good combination of various performance parameters by exhibiting a compact size, wideband and high gain, making it a potential candidate for future 5G devices operating using a 28 GHz band spectrum, as shown in Table 2.

Table 2. Comparison of proposed PRS-loaded MIMO antenna with related works.

Ref. No.	No. of Ports	Antenna Size (mm × mm)	$\lvert S_{11} \rvert$ BW	Isolation (dB)	Max Gain (dBi)	−3 dB Gain BW	Efficiency (%)
[9]	4 (2 × 2)	31 × 48	26~31 GHz	−21	10	26~31 GHz	-
[11]	4 (2 × 2)	32 × 32	27.6~29.1 GHz	−40	10.6	27.6~29.1 GHz	85
[12]	4 (2 × 2)	30 × 30	27~29 GHz	−29	6.1	27~29 GHz	90
[13]	4 (2 × 2)	25 × 25	25.2~28 GHz	−23.2	8.72	-	-
[15]	4 (2 × 2)	30 × 35	25.5~29.6 GHz	−10	8.1	26~29.6 GHz	82
[16]	4 (2 × 2)	22 × 22	23.3~28.8 GHz	−23	10.44	-	80
[25]	4 (2 × 2)	30 × 43	24.5~26.5 GHz	−45	10.27	24.5~26.5 GHz	-
This Antenna	4 (2 × 2)	24 × 24	23.5~29 GHz	−33	11.4	23.6~28.8 GHz	84

4. Conclusions

This paper introduces the design of a high-gain antenna based on PRS and its MIMO configuration, intended for 28 GHz applications. The basic radiator is derived from a conventional circular patch antenna through the incorporation of a pair of vias and a semi-circular slot. Subsequently, a PRS is developed for a similar operational band and positioned on the top side of the antenna with an air gap to enhance radiation characteristics.

The performance enhancement of the PRS-loaded antenna improves the bandwidth from 24.29–28.45 GHz to 23.6~28.8 GHz, accompanied by an elevation in peak gain from 8.3 to 11.4 dBi. To fulfill the requisites of contemporary and future devices utilizing MIMO systems, an MIMO antenna is constructed utilizing four-unit elements, strategically positioned orthogonally. The PRS is also extended to cover the MIMO antenna, maintaining the same distance as that of a unit element. The optimized PRS-loaded MIMO antenna offers a broad frequency range of 23.5–29 GHz and a peak gain of 11.4 dBi. A comprehensive exploration of various MIMO parameters reveals that the proposed work provides low-mutual-coupling and ECC values of less than −33 dB and 0.005, respectively.

Furthermore, a comparative analysis with related works highlights that the proposed PRS-loaded antenna outperforms other alternatives, positioning itself as a robust contender for future mmWave devices requiring high-gain MIMO antenna systems.

Author Contributions: Conceptualization, methodology, software, J.J. and W.A.A.; validation, formal analysis, investigation, resources, data curation, D.C., J.L., N.H. and N.K.; writing—original draft preparation, J.J., D.C. and J.L.; writing—review and editing, W.A.A., N.H. and N.K.; supervision, funding acquisition, N.K. All authors have read and agreed to the published version of the manuscript.

Funding: This work was supported by the Institute of Information & Communications Technology Planning & Evaluation (IITP) grant funded by the Korea government (MSIT) (No. 2022-0-01031, development of the measured EMF big data analysis and management platform).

Data Availability Statement: All data is included in the manuscript.

Conflicts of Interest: The authors declare no conflict of interest.

References

1. Ikram, M.; Sultan, K.; Lateef, M.F.; Alqadami, A.S.M. A Road towards 6G Communication—A Review of 5G Antennas, Arrays, and Wearable Devices. *Electronics* **2022**, *11*, 169. [CrossRef]
2. Dangi, R.; Lalwani, P.; Choudhary, G.; You, I.; Pau, G. Study and Investigation on 5G Technology: A Systematic Review. *Sensors* **2022**, *22*, 26. [CrossRef] [PubMed]
3. Hussain, N.; Jeong, M.J.; Abbas, A.; Kim, N. Metasurface-based single-layer wideband circularly polarized MIMO antenna for 5G millimeter-wave systems. *IEEE Access* **2020**, *8*, 130293–130304. [CrossRef]
4. Boccardi, F.; Heath, R.W.; Lozano, A.; Marzetta, T.L.; Popovski, P. Five disruptive technology directions for 5G. *IEEE Commun. Mag.* **2014**, *52*, 74–80. [CrossRef]
5. Lu, L.; Li, G.Y.; Swindlehurst, A.L.; Ashikhmin, A.; Zhang, R. An Overview of Massive MIMO: Benefits and Challenges. *IEEE J. Sel. Top. Signal Process.* **2014**, *8*, 742–758. [CrossRef]
6. Cuneray, K.; Akcam, N.; Okan, T.; Arican, G.O. 28/38 GHz dual-band MIMO antenna with wideband and high gain properties for 5G applications. *AEU-Int. J. Electron. Commun.* **2023**, *162*, 154553. [CrossRef]

7. Yang, B.; Yu, Z.; Dong, Y.; Zhou, J.; Hong, W. Compact tapered slot antenna array for 5G millimeter-wave massive MIMO systems *IEEE Trans. Antennas Propag.* **2017**, *65*, 6721–6727. [CrossRef]
8. Park, J.S.; Ko, J.-B.; Kwon, H.-K.; Kang, B.-S.; Park, B.; Kim, D. A tilted combined beam antenna for 5G communications using a 28-GHz band. *IEEE Antennas Wirel. Propag. Lett.* **2016**, *15*, 1685–1688. [CrossRef]
9. Wani, Z.; Abegaonkar, M.P.; Koul, S.K. A 28-GHz antenna for 5G MIMO applications. *Prog. Electromagn. Res. Lett.* **2018**, *78*, 73–79 [CrossRef]
10. Shoaib, N.; Shoaib, S.; Khattak, R.Y.; Shoaib, I.; Chen, X.; Perwaiz, A. MIMO antennas for smart 5G devices. *IEEE Access* **2018**, *6*, 77014–77021. [CrossRef]
11. Sufian, M.A.; Hussain, N.; Abbas, A.; Awan, W.A.; Choi, D.; Kim, N. A Series fed Planar Array-based 4-port MIMO Antenna for 5G mmWave IoT Applications. In Proceedings of the Asia-Pacific Microwave Conference (APMC), Yokohama, Japan, 29 November–2 December 2022; pp. 880–882.
12. Kamal, M.M.; Yang, S.; Ren, X.-C.; Altaf, A.; Kiani, S.H.; Anjum, M.R.; Iqbal, A.; Asif, M.; Saeed, S.I. Infinity shell shaped MIMO antenna array for mm-Wave 5G applications. *Electronics* **2021**, *10*, 165. [CrossRef]
13. Güler, C.; Bayer Keskin, S.E. A Novel High Isolation 4-Port Compact MIMO Antenna with DGS for 5G Applications. *Micromachines* **2023**, *14*, 1309. [CrossRef] [PubMed]
14. Pourgholamhossein, Z.; Jazi, M.N.; Denidni, T.A. Broadband flat-lens antenna design using ultrathin Huygens' metasurface for millimeter-wave applications. *IEEE Trans. Antennas Propag.* **2022**, *4*, 713–723. [CrossRef]
15. Khalid, M.; Iffat Naqvi, S.; Hussain, N.; Rahman, M.; Fawad; Mirjavadi, S.S.; Khan, M.J.; Amin, Y. 4-port MIMO antenna with defected ground structure for 5G millimeter wave applications. *Electronics* **2020**, *9*, 71. [CrossRef]
16. Sehrai, D.A.; Asif, M.; Shah, W.A.; Khan, J.; Ullah, I. Metasurface-based wideband MIMO Antenna for 5G millimeter-wave systems. *IEEE Access* **2021**, *9*, 125348–125357. [CrossRef]
17. Ferreira-Gomes, B.; Oliveira, O.N., Jr.; Mejía-Salazar, J.R. Chiral Dielectric Metasurfaces for Highly Integrated, Broadband Circularly Polarized Antenna. *Sensors* **2021**, *21*, 2071. [CrossRef]
18. Balanis, C.A. *Antenna Theory Analysis and Design*, 4th ed.; Wiley: New York, NY, USA, 2016.
19. Dong, G.; Huang, J.; Lin, S.; Chen, Z.; Liu, G. A compact dual-band MIMO antenna for sub-6 GHz 5G terminals. *J. Electromagn. Eng. Sci.* **2022**, *22*, 599–607. [CrossRef]
20. Mankong, U.; Chamsuk, P.; Nakprasert, S.; Potha, S.; Weng, Z.-K.; Dat, P.T.; Kanno, A.; Kawanishi, T. Millimeter Wave Attenuation Due to Wind and Heavy Rain in a Tropical Region. *Sensors* **2023**, *23*, 2532. [CrossRef]
21. Park, S.-H.; Jang, G.-H.; Seo, Y.-H.; Keum, H.-S.; Bang, S.-I. High-Speed Antenna Measurement System Using Multi-Probe Array Technique for 5G Applications. *Electronics* **2022**, *11*, 3435. [CrossRef]
22. Liu, L.; Weng, Y.F.; Cheung, S.W.; Yuk, T.I.; Foged, L.J. Modeling of cable for measurements of small monopole antennas. In Proceedings of the Loughborough Antennas & Propagation Conference, Loughborough, UK, 14–15 November 2011.
23. Lee, J.N.; Hyun, S.B.; Cho, Y.K. A Compact Ultra-Wideband Chip Antenna with Bandwidth Extension Patch and Simple Isolator for MIMO Systems for Mobile Handheld Terminals. *J. Electromagn. Eng. Sci.* **2022**, *22*, 272–282. [CrossRef]
24. Chae, S.H.; Oh, S.-K.; Park, S.-O. Analysis of Mutual Coupling, Correlations, and TARC in WiBro MIMO Array Antenna. *IEEE Antennas Wirel. Propag. Lett.* **2007**, *6*, 122–125. [CrossRef]
25. Tariq, S.; Naqvi, S.I.; Will, N.; Amin, Y. A Metasurface-Based MIMO Antenna for 5G Millimeter-Wave Applications. *IEEE Access* **2021**, *9*, 51805–51817. [CrossRef]

Disclaimer/Publisher's Note: The statements, opinions and data contained in all publications are solely those of the individual author(s) and contributor(s) and not of MDPI and/or the editor(s). MDPI and/or the editor(s) disclaim responsibility for any injury to people or property resulting from any ideas, methods, instructions or products referred to in the content.

electronics

MDPI

Article

Machine Learning-Inspired Hybrid Precoding for HAP Massive MIMO Systems with Limited RF Chains

Shabih ul Hassan [1], Talha Mir [2], Sultan Alamri [3,*], Naseer Ahmed Khan [4] and Usama Mir [5]

1 Information and Communication, University of Science and Technology of China, Hefei 230052, China
2 Department of Electronic Engineering, Baluchistan University of IT, Engineering and Management Sciences Pakistan (BUITEMS), Quetta 87300, Pakistan
3 College of Computing and Informatics, Saudi Electronic University, Riyadh 13316, Saudi Arabia
4 School of Computer Science, Northwestern Polytechnical University, Xi'an 710060, China
5 School of Computer Science, University of Windsor, Windsor, ON N9B 3P4, Canada
* Correspondence: salamri@seu.edu.sa

Abstract: Energy efficiency (EE) is the main target of wireless communication nowadays. In this paper, we investigate hybrid precoding (HP) and massive multiple-input multiple-output (MIMO) systems for a high-altitude platform (HAP). The HAP is an emerging solution operating in the stratosphere at an amplitude of up to 20–40 km to provide communication facilities that can achieve the best features of both terrestrial and satellite systems. The existing hybrid beamforming solution on a HAP requires a large number of high-resolution phase shifters (PSs) to realize analog beamforming and radio frequency (RF) chains associated with each antenna and achieve better performance. This leads to enormous power consumption, high costs, and high hardware complexity. To address such issues, one possible solution that has to be tweaked is to minimize the number of PSs and RFs or reduce their power consumption. This study proposes an HP sub-connected low-resolution bit PSs to address these challenges while lowering overall power consumption and achieving EE. To significantly reduce the RF chain in a massive MIMO system, HP is a suitable solution. This study further examined adaptive cross-entropy (ACE), a machine learning-based optimization that optimizes the achievable sum rate and energy efficiency in the Rician fading channel for HAP massive MIMO systems. ACE randomly generates several candidate solutions according to the probability distribution (PD) of the elements in HP. According to their sum rate, it adaptively weights these candidates' HP and improves the PD in HP systems by minimizing the cross-entropy. Furthermore, this work suggests energy consumption analysis performance evaluation to unveil the fact that the proposed technique based on a sub-connected low-bit PS architecture can achieve near-optimum EE and sum rates compared with the previously reported methods.

Keywords: energy efficiency; hybrid precoding; massive MIMO; machine learning; adaptive cross-entropy; Rician fading channel; HAP

Citation: Hassan, S.u.; Mir, T.; Alamri, S.; Khan, N.A.; Mir, U. Machine Learning-Inspired Hybrid Precoding for HAP Massive MIMO Systems with Limited RF Chains. *Electronics* **2023**, *12*, 893. https://doi.org/10.3390/electronics12040893

Academic Editor: Adao Silva

Received: 13 December 2022
Revised: 22 January 2023
Accepted: 31 January 2023
Published: 9 February 2023

Copyright: © 2023 by the authors. Licensee MDPI, Basel, Switzerland. This article is an open access article distributed under the terms and conditions of the Creative Commons Attribution (CC BY) license (https://creativecommons.org/licenses/by/4.0/).

1. Introduction

The capacity demand in fifth-generation (5G) wireless communication and beyond is facing major challenges, particularly in terms of "last mile" transmission. Line-of-sight (LOS) propagation paths are a bottleneck on the ground unless a significant number of base stations (BS) are deployed, while satellite systems have capacity limitations. To effectively provide high data rates, massive numbers of antennas, and consistent coverage, innovative technologies are required [1,2]. The high-altitude platform (HAP) is an emerging solution, operating at an amplitude of up to 20–40 km to provide communication facilities that can achieve the best features in both terrestrial and satellite systems. Relative to the terrestrial cellular base station (BS), the HAP has a wide coverage area and is more flexible and non-polluting. Furthermore, when compared with satellite communication technologies, it consumes far less energy and has low propagation, which means greater QoS for real-time users [3]. Massive MIMO is a technology used on HAPs that can improve energy efficiency and data rates by deploying a massive number of antennas, allowing multiple users to be served with the same time and frequency resources [4].

Massive MIMO technology allows high data rate transmissions by using the high number of antennas on the HAPs. Moreover, the enormous number of antennas in an array can provide sufficient gain by precoding, which can overcome the free space path loss [5]. However, in conventional fully digital (CFD) MIMO systems, each antenna requires a dedicated energy-intensive RF chain, which consumes a significant amount of energy (approximately 250 mW per RF chain). The energy consumption of CFD systems is too high due to the massive number of RF chains [6]. This allows researchers to consider hybrid (analog-to-digital) MIMO systems, which significantly reduce the number of RF chains by switching some operations to an analog domain [7].

A fully digital precoding technique demands a precise radio frequency (RF) chain consisting of a digital-to-analog converter (DAC), mixer, filter, and power amplifier for each antenna. When a large number of transmitting antennas is employed in the massive MIMO scenario, this results in prohibitively high hardware costs as well as significant energy consumption [8]. To solve this issue, recently, finite-resolution PS schemes have been developed. In this scheme, finite-resolution PSs are used directly instead of high resolution phase shifters. This can lower the energy consumption of a PS network without loss of performance, but it still demands a large number of PSs, each of which consumes a significant amount of energy [9].The other scheme is to use the switch network instead of the PS network [10]. This can considerably minimize hardware costs and energy utilization but it has a noticeable performance loss.

Few studies have specifically addressed the shortcomings of power consumption, SE and achievable sum rates for HAP networks, in contrast to the extensive research that has been undertaken in the spectrum sharing and theoretical analysis of HAP [11–14].

Due to the enormous amount of power consumed by base stations (BSs), energy efficiency (EE) has been viewed as an essential criterion for the development of future communication networks, both from an economic and environmental point of view. In this connection, the authors in [15] jointly optimized the transmission beamforming and covariance matrix in a MISO system in order to maximize the equivalent efficiency (EE). The authors of [16,17] addressed the EE maximization problem for terrestrial-aerial networks and proposed two array signal-processing approaches based on Dinkelbach's transformation in order to achieve suboptimal solutions. However, fully connected HP requires an extremely large number of active antennas and RF chains, which causes the existing techniques, such as those in [16,18,19], to have an extremely high level of hardware power consumption, which in turn results in a low level of system EE. In [20], the authors established beamforming techniques aimed at the mmWave HAP system. These algorithms were built for planar arrays that had thousands of antenna elements. The work in [21] presents HP for HAP massive MIMO systems in order to obtain the RF and the baseband precoder with limited RF chains. The authors applied duality to exploit the relation between the RF precoder and the statistical CSI, which is complicated to tackle. In [22], the authors proposed two stages of outer precoding design with the assumption of a zero-precoding (ZF) inner pre-beamformer and iterative algorithm to achieve the sum rate result for HAP. The authors of [23] used NOMA-based HAP communications and multiple antennas to satisfy the connectivity, dependability, and high data rate needs of 5G-and-beyond applications. Furthermore, a user selection and correlation-based user corresponding method for a NOMA-based multi-user HAPS system were proposed. In [24,25], the authors designed an outer beamformer to reduce the dimensional statistical eigen mode of the users and user grouping algorithm for HAP massive MIMO systems. The work in [26,27] proposed HAP to perform the beamforming technique. For this, the authors realized the interference alignment method for achieving the maximum sum rate of HAPs, respectively. In [28], the authors realized a decreased channel state information (CSI) overhead in a sub-connected RF precoding scheme, but this is incompatible with HAP massive MIMO systems. In [29], the initial RF precoder selection from a discrete codebook and the proposed algorithm iteration were utilized together to develop the RF precoder, avoiding an exhaustive search while maintaining a high level of complexity. The published literature did not focus on reducing the complexity of the architecture or energy consumption of the system, while in our proposed machine learning technique, we use limited RF chains and sub-connected HP with low-bit PSs for HAP massive MIMO systems, which reduces the power consumption and complexity and improves the overall sum rate, energy, and spectral efficiency (SE).

In the proposed work, we examined energy-efficient HP sub-connected one- and two-bit PSs for HAP massive MIMO systems, and our main contributions are the following:

- We investigate the energy consumption analysis to unveil the fact that the HPs based on a sub-connected low-bit PS architecture is substantially lower than that consumed by HR-PS-based HP. Moreover, utilizing one-bit PSs results in a slight and constant array gain loss.
- We propose 1- and 2-bit sub-connected PSs for an HAP massive MIMO system with significantly reduced hardware cost, complexity, and energy consumption in the Rician fading channel. The sub-connected HP design is expected to be easier to deploy and be more energy efficient. The sub-connected architecture is more practicable for antenna deployment due to its lower cost and lower hardware complexity.
- We propose ACE-based optimization with low-bit PSs to optimize the achievable sum rate, EE, and SE. First, our scheme randomly generates several candidate elements according to the probability distribution (PD) in HP, and then it weights them according to their sum rates, thereby improving the HP elements' PD by decreasing the cross-entropy. By repeating this process, we ultimately generate an HP with a sufficiently high probability close to the optimized level.
- We further examine the simulation results to show the fact that the HPs based on a sub-connected low-bit PS architecture can achieve near-optimum EE and sum rates compared with the traditional schemes.

The rest of this paper is arranged accordingly. Section 2 is about the energy consumption analysis of different hybrid precoding schemes. Section 3 is a brief description of the channel and system model, presenting ideas and steps for algorithms, ACE-based HPs with M-bit PSs, and complexity analysis. The experiment and discussion are in Section 4. The conclusions are in Section 5.

Notations: Lowercase and boldface $\mathbf{x}$ is used to denote the vector, and uppercase and boldface $\mathbf{X}$ is used to denote the matrix, while letters without boldface denote scalars. $(.)^T$ denotes a transpose, $(.)^{-1}$ denotes an inversion, $|.|$ denotes an absolute operator, $(.)^H$ denotes a conjugate transpose, $||.||_F$ denotes the Frobenius norm, $(.)^+$ denotes a pseudo inversion, $\otimes$ denotes the Kronecker product, and $\mathbf{I}_N$ denotes an $N \times N$ identity matrix

2. Energy Consumption Analysis of Different Hybrid Precoding Schemes

Here, we discuss a fully connected, high-resolution phase shifter (e.g., 4 bits) hybrid precoding-based architecture in which every RF chain is connected to each antenna through the network. The high-resolution (HR) PS-based hybrid precoding energy consumption can be represented as follows:

$$P_{HR-PSs} = P + N_{RF}P_{RF} + NN_{RF}P_{HR-PS} + P_B \tag{1}$$

In Equation (1), P represents the total transmission power, P_{RF} is the power consumed by the RF chain, P_{HR-PS} depicts the energy consumed by the PSs, P_B represents the baseband power consumption, N shows the number of PSs, and N_{RF} represents the number of RF chains. The energy consumed by the HR-PSs is (P_{HR-PS} = 40 mW) for the 4-bit phase shifters. The HR-PS-based HP architecture can attain the achievable sum-rate, but it requires a large number of PSs. As stated in [30], the power consumption by HR-PSs is quite high. Thus, to overcome this problem, a switch-based HP was introduced in [7], where instead of HR-PSs, switches were used, which could also decrease the hardware cost.

Nevertheless, this switch-based solution cannot actually obtain the array gain of mmWave massive MIMO systems. We describe the system in [7] as follows:

$$P_{Sw-PSs} = P + N_{RF}P_{RF} + N_{RF}P_{Sw} + P_B \tag{2}$$

where P_{Sw} is the power consumed by the switches, which is much lower than the HR-PSs (e.g., P_{Sw} = 5 mW) [31]. However, this design cannot achieve the array gain since only the N_{RF} antennas are active simultaneously, resulting in sum rate performance degradation.

In a sub-connected one-bit PS-based HP architecture, in which each RF chain is attached to the subantennas of an array with N/N_{RF}, the power consumption is much lower than in a fully connected architecture [7], and it provides the better trade-off between the two precoding schemes. The energy consumed by a sub-connected HP architecture can be express as

$$P_{SubC-PSs} = P + N_{RF}P_{RF} + NP_{SubC} + P_B \tag{3}$$

It is noted that energy consumed by an HP architecture sub-connected with one or two bits is low (P_{SubC} = 5 mW). Therefore, compared with HR-PSs, the sub-connected PS energy consumption is too low. Aside from that, the proposed sub-connected PS HP architecture is able to use all antennas to achieve the optimal array gain compared with SW-based HP [32]. Compared with the fullly connected HP architecture, the sub connected HP architecture with one- or two-bit PSs has a significantly lower number of phase shifters from $N_t \times N_t^{RF}$ to N_t, which will have more benefits and will excite the PSs. It can save energy and compensate the insertion loss of PSs. In addition, it also has less complexity due to its simplicity in its architecture and configurations.

3. Channel and System Model

We adopted a multi-user HAP massive MIMO system where an HAP was equipped with a uniform planner array (UPA) with C number of antennas in each column, while the number of antennas in each row was $N = C \times A$, as shown in Figure 1, and each RF chain was connected to a subset of the antenna serving K single-antenna users, as shown in Figure 2. The received signal at the kth user can be expressed as follows:

$$\mathbf{y} = \mathbf{H}\mathbf{F_A}\mathbf{F_D}\mathbf{s} + \mathbf{n} \tag{4}$$

where $\mathbf{s}$ denotes the transmission signal of a size $K \times 1$, while $\mathbf{F_A}$ is of a size $N \times N_{RF}$ and $\mathbf{F_D}$ is of a size $N_{RF} \times K$ for the analog and digital precoder, respectively. In addition, $\mathbf{n}$ represents the additive white Gaussian noise vector with a distribution $\mathcal{CN}(0, \mathbf{I}_K)$. $\mathbf{H} = [\mathbf{h_1}, \mathbf{h_2}, \ldots \mathbf{h_k}]$ denotes the overall downlink channel matrix of a size $K \times N$, where $\mathbf{h_k} \in \mathbb{C}^{N \times 1}$ is the channel vector for the kth user, which can be expressed as

$$\mathbf{h_K} = \sqrt{\alpha}\left(\sqrt{\frac{K_{r,k}}{K_{r,k}+1}}\mathbf{h_{LK}} + \sqrt{\frac{K_{r,k}}{K_{r,k}+1}}\mathbf{h_{NLK}}\right) \tag{5}$$

where K_r represents the Rician factor, $\alpha = (4\pi r_k/\lambda)^{-2}$ is the large scaling factor for the kth user, r_k represents the distance between the HAP and the kth user, as shown in Figure 1, and λ is the carrier wavelength, while $\mathbf{h_{LK}}$ and $\mathbf{h_{NLK}}$ are the line-of-sight (LOS) components and non-line-of-sight (NLOS) components, respectively, which can be expressed as [33]

$$\mathbf{h_{LK}} = \mathbf{a}(\theta_k, \phi_k) \otimes \mathbf{x}(\theta_k, \phi_k) \tag{6}$$

$$\mathbf{h_{NLK}} = \sqrt{\hat{\mathbf{W}}_k}[w_{1k}, w_{2k}, \ldots w_{Nk}]^T \tag{7}$$

where $\mathbf{a}(\theta_k, \phi_k) = [1, e^{(j2\pi d_h)}, \ldots \exp(j2\pi(N-1)d_h)]^T$, $\mathbf{x}(\theta_k, \phi) = [1, e^{(j2\pi d_v)}, \ldots \exp(j2\pi(N-1)d_v)]^T$, the angles $\theta_k \in [-\pi, \pi]$ and $\phi_k \in [0, \pi/2]$ are the angles of departure (AoDs), d is the distance between two adjacent antenna elements, and $\hat{\mathbf{W}}_{\mathbf{k}} \in \mathbf{C}^{N\times N}$ is the correlation matrix of the user K, which can be represented as follows:

$$[\hat{\mathbf{W}}_{\mathbf{k}}]_{a,b} = \int_{0}^{\pi/2}\int_{-\pi}^{\pi} f(\theta)f(\phi)\mathrm{e}^{(j\frac{2\pi}{\lambda}(d_1-d_2))}d\theta d\phi \tag{8}$$

where $d_2 = (a-b)d_d\cos(\phi)\cos(\theta)$, $d_1 = (a-b)d_v\cos(\phi)\sin(\theta)$, $f(\phi) = e^{(-\sqrt{2}|\phi-\phi_0|/\sigma)}$, in which σ and ϕ_0 are the mean AoDs, $f(\theta) = e^{\kappa\cos(\theta-\mu)/2\pi I_0(\kappa)}$, I_0 is the zero-order Bessel function of the first kind, $\mu = (-\pi,\pi)$ is the mean angle of departure (AoD) of user K, and κ is the angular spread.

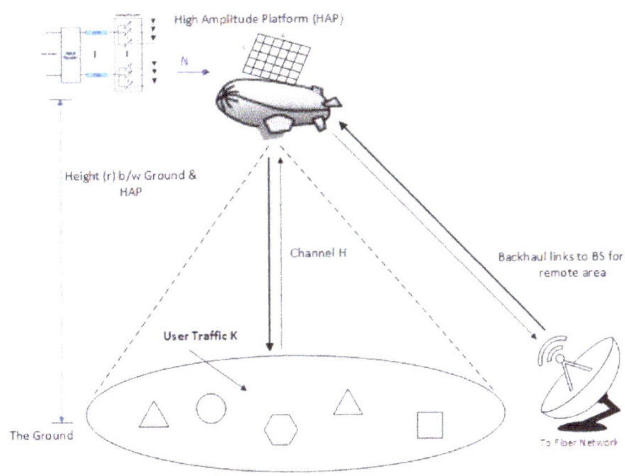

Figure 1. HAP massive MIMO system with user K.

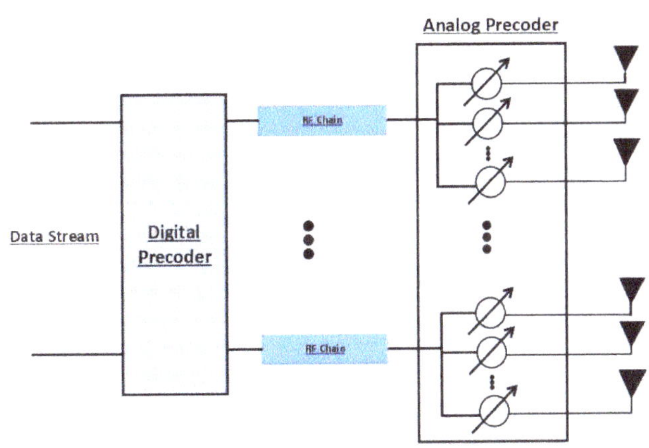

Figure 2. Sub-connected hybrid precoding architecture.

Furthermore, we need to explain the hardware constraints imposed by the proposed low-bit-based sub-connected HP, which differ from those imposed by traditional architectures. The first constraint is that the analog precoder $\mathbf{F_A}$ must be a block diagonal matrix rather than a full matrix:

$$\mathbf{F_A} = \begin{pmatrix} \mathrm{f}_1 & \cdots & 0 \\ \vdots & \ddots & \vdots \\ 0 & \cdots & \mathrm{f}_n \end{pmatrix} \tag{9}$$

where the analog weighting vector for the nth sub antenna array is $\mathbf{f_n}$ with a size $N \times 1$. These components have the same amplitude but different phases. All N nonzero elements $\mathbf{F_A}$ should belong to

$$\frac{1}{\sqrt{N}}[-1,+1] \tag{10}$$

Our main objective is to develop an efficient hybrid precoders $\mathbf{F_A^{opt}}$ and $\mathbf{F_D^{opt}}$ to maximize the sum rate R, which can be written as

$$\left(\mathbf{F_A^{opt}}, \{\mathbf{F_D^{opt}}\right) = \arg\max \mathbf{R}_{(sum)}$$
$$\mathbf{F_A^{opt}}, \mathbf{F_D^{opt}}$$
$$\text{s.t. } \mathbf{F_A^{opt}} \in \Re$$

$$\sum_{m=1}^{M} \left\| \mathbf{F_A^{opt}}, \mathbf{F_D^{opt}} \right\|_F^2 = \rho \tag{11}$$

where $\Re$ represents the set of all possible analog precoders that satisfy the constrains in Equations (9) and (10). The sum rate can be expressed as

$$R_k = \sum_{k=1}^{K} \log_2(1 + \eta_k) \tag{12}$$

where η_k represents the signal-to-interference-plus-noise ratio (SINR) for the kth user which can be expressed as follows:

$$\eta_k = \frac{\left|\mathbf{h_k^H F_A b_D^k}\right|^2}{\sum_{\bar{k} \neq k}^{K} \left|\mathbf{h_k^H F_A b_D^{\bar{k}}}\right|^2 + K\sigma^2} \tag{13}$$

Here, σ^2 represents the noise power, and $\mathbf{b_D^k}[m]$ represents the kth column of $\mathbf{F_D}$. It is worth mentioning that the constraints in Equations (9) and (10) on the analog precoder $\mathbf{F_A}$ are non-convex. This makes Equation (12) incredibly difficult to solve. The number of possible $\mathbf{F_A}$ values is finite, since all nonzero elements belong to the constraint in Equation (10). In order to tackle this challenging problem, we propose a machine learning (ML) adaptive cross-entropy optimization (ACE) with a low-bit, PS-based HP scheme.

3.1. Proposed ACE-Based Hybrid Precoding

To overcome the non-convex problem mentioned in Equations (9) and (10), we first decouple the joint design of the analog and digital precoders. As all of the analog precoder's N nonzero elements belong to Equation (10), the number of possibilities $\mathbf{F_A}$ is finite. As a result, we can consider the problem in Equation (11) to be a non-coherent combining problem. We can select an $\mathbf{F_A}$ candidate first and compute the optimal $\mathbf{F_D}$ with an efficient channel matrix $\mathbf{HF_A}$ without non-convex constraints. After searching for all possible $\mathbf{F_A}$ values, we can find the optimum analog precoder $\mathbf{F_A^{opt}}$ and digital precoder $\mathbf{F_D^{opt}}$. Unfortunately, such an exhaustive search involves filtering through all possible 2^N combinations, requiring excessively high complexity as the number of antennas N required is typically higher in massive MIMO systems, which in this case is N = 64, $2^{64} = 1.8 \times 10^{19}$. To tackle this challenging problem, we proposed an ACE algorithm with a one- or two-bit PS-based HP scheme, which is the advanced version of the CE algorithm [34]. In the CE algorithm, all elite contributions are treated as equal. Logically, when updating the PD, the elite contributions with better values should be more important. Therefore, if we can weight the elites as per their objective values, then we can expect better results. In ACE, multiple candidates' HP schemes are randomly generated based on the PD of the element in the HP. Then, they are weighted according to their sum rates for the calculated candidate HP. To enhance the probability distribution (PD) of an element $\mathbf{F_A}$, we reduce the CE between the two PDs.

We formulated the non-zero elements in $\mathbf{F_A}$ as an $N \times 1$ vector at the beginning, which can be expressed as $\mathbf{f} = \left[(\mathbf{f_1^A})^\mathbf{T}, (\mathbf{f_2^A})^\mathbf{T}, \ldots (\mathbf{f_{NA}^A})^\mathbf{T} \right]^T$, and the probability function $\mathbf{q} = [q_1, q_2, q_3, \ldots q_N]^T$ as the $N \times 1$ vector. The probability is denoted by $0 \le q_n \le 1$, where $f_n = \frac{1}{\sqrt{N}} f_n$ is the nth element of $\mathbf{f}$. First, we initialized $\mathbf{q}^{(0)}$ as $\mathbf{q}^{(0)} = \frac{1}{2} \times 1_{N \times 1}$ after setting the initial values of the parameters, namely the parameter for the ACE-based HP with sub-connected low-bit PSs. The updated method has the steps presented below.

In step 3, we produce the Z candidate analog precoder $\{\mathbf{F_A^z}\}_{z=1}^{Z}$, which depends upon the probability distribution function $\Xi(\Re; q^{(i)})$, and in step 4, we calculate the digital precoding $\mathbf{F_D}$ based on the efficient channel $(\mathbf{H}_{eq}^z)^H = \mathbf{H}\{\mathbf{F_A^{opt}}\}$. We employed a zero-forcing digital precoding scheme with low complexity and near-optimal performance in this paper, which can be expressed as

$$\mathbf{G}^z = (\mathbf{H}_{eq}^z)^H \left(\mathbf{H}_{eq}^z (\mathbf{H}_{eq}^z)^H \right)^{-1} \tag{14}$$

$$\mathbf{F_D^z} = \beta^z \mathbf{G}^z \tag{15}$$

where $\beta^z = \frac{\sqrt{\rho}}{\left\| \mathbf{F_A^z} \mathbf{G}^z \right\|_F}$ represents the power-normalized factor.

In step 5, we calculate the achievable sum rate by putting $\mathbf{F_A^z}$ and $\mathbf{F_D^z}$ in Equation (12). The achievable sum rates will be sorted in descending order in step 6. In step 7, the elite Z_{elite} values can be obtained with the highest sum rate. These candidates are utilized to update the probability distribution $\mathbf{q}^{i+1}$ and minimize the CE, which is equivalent to [34]

$$\mathbf{q}^{i+1} = \arg\max \frac{1}{Z} \sum_{z=1}^{Z_{elite}} \ln \Xi \left(\Re_A^{[z]} \mathbf{q}^{(i)} \right) \tag{16}$$

where $\Xi\left(\Re_A^{[z]} \mathbf{q}^{(i)} \right)$ is the probability to output $\mathbf{F_A^z}$ as stated in Equation (16). All elites have the same effects, and this leads to decreased performance. In order to fix this problem, we propose adaptively weighting the elites depending upon their sum rates. Thus, an auxiliary parameter U, which denotes the average of all sum rates, is introduced.

$$U = \frac{1}{Z_{elite}} \sum\nolimits_{z=1}^{Z_{elite}} R(\mathbf{F_A^{[z]}}) \tag{17}$$

The weight w_z of the elite $\mathbf{F_A^z}$ can be calculated in step 8 based on $\{w\}_{z=1}^{Z_{elite}}$, and Equation (15) can be modified to become

$$\mathbf{q}^{(i+1)} = \arg\max \frac{1}{Z} \sum_{z=1}^{Z_{elite}} w_z \ln \Xi \left(\Re_A^{[7]} \mathbf{q}^{(i)} \right) \tag{18}$$

where $\Xi\left(\Re_A^{[z]} \mathbf{q}^{(i)} \right)$ is the probability to calculate the $\mathbf{F_A^z}$. Notice that $\Xi\left(\Re_A^{[z]} \mathbf{q}^{(i)} \right) = \Xi\left(\mathbf{f}^{[z]}; \mathbf{q}^{(i)} \right)$, the nth element $f_n^{(z)}$ of $\mathbf{f}^{[z]}$, is the Bernoulli random variable, where $f_n^{(z)} = \frac{1}{\sqrt{N}}$ has a probability $q_n^{(i)}$, and similarly, $f_n^{(z)} = -\frac{1}{\sqrt{N}}$ has a probability $1 - q_n^{(i)}$. Thus, we have

$$\Xi\left(\Re_A^{[z]} \mathbf{q}^{(i)} \right) = \prod_{n}^{N} (q_n^{(i)})^{\frac{1}{2}(1+\sqrt{N} f_n^{(z)})} (1 - q_n^{(i)})^{\frac{1}{2}(1+\sqrt{N} f_n^{(z)})} \tag{19}$$

By putting Equation (19) into Equation (18), the first-order derivatives based on $q_n^{(i)}$ from Equation (18) is given as follows:

$$\frac{1}{Z} \sum_{z=1}^{Z} w_z \left(\frac{1 + \sqrt{N} f_n^{(z)}}{2 q_n^i} - \frac{1 + \sqrt{N} f_n^{(z)}}{2(1 - q_n^i)} \right) \tag{20}$$

For Equation (20), when set to zero, $1 - q_n^{(i)}$ is updated in step 9 as follows:

$$q_n^{(i+1)} = \frac{\sum_{z=0}^{Z_{elite}} w_z\left(\sqrt{N} f_n^{(z)} + 1\right)}{2\sum_{z=1}^{Z_{elite}} w_z} \tag{21}$$

We can further simplify Equation (21) to avoid the local optimum:

$$q_n^{(i+1)} = \Omega^{i+1} \times q_n^{(i+1)} + (1 - \Omega^{i+1}) \times q_n^{(i)} \tag{22}$$

where Ω^{i+1} in iteration $(i+1)$ is a smooth size for the step. We used steps 3–10 for the predefined iteration. Finally, the optimal analog $\mathbf{F}_{\mathbf{A}}^{\mathbf{opt}}$ and digital precoding $\mathbf{F}_{\mathbf{D}}^{\mathbf{opt}}$ are obtained.

3.2. ACE-Based Hybrid Precoding with M-Bit PS

We present a more general case (i.e., m-bit PSs) for the proposed ACE-based HP. We first generated the parametric sample distribution, which produced the applicant alternatives for the next iterations. The simple method of producing a random sample $\hat{\mathbf{b}} = \{\hat{\mathbf{b}}\}_{n=1}^{N}$ is to draw independently from $\{\hat{b}_1, \hat{b}_2, \dots \hat{b}_N\}$. Here, each sample belongs to a discrete distribution $\{\mathbf{q}_{m,n}^{(i)}\}$, and the mth quantized process of a set M is chosen to be $\hat{b}_n$. The elite candidates should then change the potential to minimize the CE:

$$\mathbf{q}_{m,n}^{(i+1)} = \arg\max_{q(i)} \frac{1}{Z} w_z \sum_{z=1}^{Z_{elite}} \ln \Xi\left(\Re_A^{[z]} \mathbf{q}_{m,n}^{(i)}\right) \tag{23}$$

where $\Xi\left(\Re_A^{[z]} \mathbf{q}_{m,n}^{(i)}\right)$ is given by

$$\Xi\left(\Re_A^{[z]} \mathbf{q}_{m,n}^{(i)}\right) = \prod_{n}^{N} \sum_{m=1}^{M} w_z (q_{m,n}^{(i)}) \mathbf{1}_{\{\hat{\mathbf{a}}^{(z)} \in \Re_{m,n}\}} \tag{24}$$

The indicator function is $\mathbf{1}_{\{.\}}$ if the statement is valid (i.e., = 1); otherwise, $\mathbf{1}_{\{.\}} = 0$. In addition, $\Re_{m,n} = \hat{\mathbf{b}} \in M^N : \hat{b}_n = \frac{2\pi m}{|M|}$, in which $M = \left\{\frac{2\pi m}{2^M}(z = 1, \dots, 2^M)\right\}$ refers to the set containing all possible analog precoders which satisfy the given constraint in Equation (9). Here, it is given that a single PS can be allocated to only one quantized process. The sum of every constrained probability is equal to one, and we implement the Lagrange multiplier to fulfill the constraints:

$$\mathbf{q}_{m,n}^{(i+1)} = \arg\max_{q(i)} \frac{1}{Z} w_z \sum_{z=1}^{Z_{elite}} \ln \Xi\left(\Re_A^{[z]} \mathbf{q}_{m,n}^{(i)}\right) + \sum_{n=1}^{N} L_n\left(\sum_{m=1}^{M} \mathbf{q}_{m,n}^{(i)} - 1\right) \tag{25}$$

When we take the first derivative of Equation (25) with respect to the probability $\mathbf{q}_{m,n}^{(i)}$ and then set the outcome to zero, we obtain the following result:

$$\frac{1}{Z} \sum_{z=1}^{Z_{elite}} \mathbf{1}\{\hat{\mathbf{a}}^{(z)} \in \Re_{m,n}\} + L_n \mathbf{q}_{m,n}^{(i)} = 0 \tag{26}$$

Finally, we add Equation (26) for $m = 1, 2, \dots, M$, and we modify the probability as given below:

$$\mathbf{q}_{m,n}^{(i+1)} = \frac{\sum_{z=1}^{Z_{elite}} w_z\{\hat{\mathbf{a}}^{(z)} \in \Re_{m,n}}{\sum_{z=1}^{Z_{elite}} w_z} \tag{27}$$

3.3. Complexity Analysis

According to Algorithm 1, the difficulty of the proposed ACE-based HP sub-connected low-resolution bit PS scheme initiates steps 4, 5, and 9. As in step 4, the effective channel matrices $\left\{\mathbf{H}_{\mathrm{eq}}^{Z}\right\}_{z=1}^{Z}$ for each candidate solution Z, as well as the related digital precoder $\{\mathbf{F}_{\mathrm{D}}^{z}\}_{z=1}^{Z}$, must be calculated as in Equations (15) and (16). As a result, this component has a complexity of $Z(NZK^2)$. In step 5, the sum rate for each candidate solution is computed. Here, we employ the traditional ZF precoder, which reduces the SINR of each user for each candidate to $\gamma^z = (\beta^z/\sigma)^2$. According to Equation (22), step 9 is the updating step of $\mathbf{q}_n^{(i+1)}$, which composes the complexity of $Z(NZ_{\mathrm{elite}})$. After I iterations, the overall computational cost of the proposed ACE-based HP sub-connected low-resolution bit system PS is $Z(ZINK^2)$. Meanwhile, I and Z are not essentially so big. This enables the complexity of the ACE-based HP sub-connected low-resolution bit PS to be reasonable and comparable to some current methods.

Algorithm 1 ACE-based hybrid precoding with sub-connected phase shifters (PSs)

Input:
Rician Channel-Matrix H, Iterations I, Candidates Z, Elites Z_{Elite}, Smoothing Step size Ω

Output:
Analog Precoder $\mathbf{F}_{\mathbf{A}}^{\mathbf{opt}}$, Digital Precoder $\mathbf{F}_{\mathbf{D}}^{\mathbf{opt}}$

1. initialization i = 0,$\mathbf{p} = 1/2 \times \mathbf{1}_{N\times 1}$
2. **for** *iteration* $= 1, 2, \ldots$
3. Generate Z candidates randomly as $\{\mathbf{F}_{\mathbf{A}}^{\mathbf{opt}}\}_{z=1}^{Z}$ based on $\Xi(\Re; q^{(i)})$
4. Calculate Z digital precoder $\{\mathbf{F}_{\mathbf{D}}^{\mathbf{opt}}\}_{z=1}^{S}$ based on effective channel by (15)
5. Calculate the sum-rate $\mathbf{R}\{\mathbf{F}_{\mathbf{A}}^{\mathbf{opt}}\}_{z=1}^{Z}$ by (12)
6. Sorting $\mathbf{R}\{\mathbf{F}_{\mathbf{A}}^{\mathbf{opt}}\}_{z=1}^{Z}$ in descending order as $\mathbf{R}(\mathbf{F}_{\mathbf{A}}^{[1]} \geq \mathbf{F}_{\mathbf{A}}^{[2]} \geq, \ldots, \mathbf{F}_{\mathbf{A}}^{[Z]})$
7. Select elite as $\mathbf{F}_{\mathbf{A}}^{[1]}, \mathbf{F}_{\mathbf{A}}^{[2]}, \ldots, \mathbf{F}_{\mathbf{A}}^{[Z_{Elite}]}$
8. Calculate weight for each elite $\mathbf{F}_{\mathbf{A}}^{[S]}$ according to their sum-rate
9. Update the probability $\mathbf{q}^{(i+1)}$ according to weight and $\{\mathbf{F}_{\mathbf{A}}^{\mathbf{opt}[z]}\}_{z=1}^{Z_{Elite}}$
10. **end for**

4. Experiment and Discussion

In this section, we will discuss the effectiveness of the ACE-based optimization that was proposed with low-bit PSs in terms of the achievable sum rate, EE, and SE. The following are the primary simulation parameters. We used the assumption that the users were dispersed in a 20-km radius under the HAP circle, and the height of the HAP was 20 km. It operated at a frequency of 2.4 gigahertz. The number of transmitting antennas $C = A = 10$, the transmitting power P was set to 20 dB, and we set the bandwidth to 10 MHz. The Rician factor K_r =10, and the transmitting antenna array was a UPA with antenna spacing $d = 2\lambda$. The number of RF chains was four. Furthermore, the number of candidate solutions, the number of elite candidates, and the number of iterations were set to Z = 200, Z_{elite} = 40, and I = 20, respectively.

Figure 3 presents that the proposed ACE sub-connected low-bit PS-based HP obtained a higher sum rate than the technique that is currently being used for HAP massive MIMO systems. For the ACE-Algorithm 1, the values were the same as above. The zero-forcing (ZF) precoding system is designed for use with the fully connected PS architecture, whereas the hybrid precoding system for conventional analog selection (AS) is designed for the switch-based architecture. When compared with CE, the ACE-based HP method has only one more step, but this step has a minimally increased level of complexity. The proposed ACE algorithm is, therefore, more effective. In Figure 3, we can see that the sum rate of the proposed ACE-based low-bit PSs was significantly higher than that of traditional AS-based HP and CE-based HP, and the difference between the ZF precoding and ACE-based HP remained constant. Nevertheless, the gap can be closed if the PSs continue to increase in terms of their resolution.

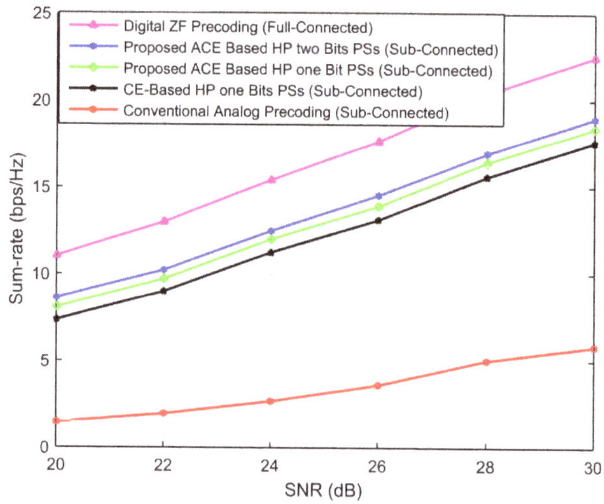

Figure 3. Sum-rate of ACE-based method with sub-connected PSs when C = A = 10 and users and RF chains are 4.

Figure 4 compares the sum rates that can be achieved by the proposed ACE-based HP with sub-connected low-bit PSs to the sum rates that can be achieved by the existing solutions for the HAP massive MIMO system. We used 8 for the number of RF chains and 8 for the number of users K, while the number of antennas C = A = 12. The remaining parameters are the same ones that were discussed above. It is evident from this that the proposed solution with sub-connected low-bit PSs was capable of achieving a significantly higher sum rate than the conventional HP, CEO-based HP with one-bit PSs, and fully connected ZF-based HP gap. In addition, a noticeable performance gap may be seen between the proposed solutions, which used 1-bit and 2-bit PSs, and the HR-based full-connected HP, which used 4-bit PSs. However, this gap could be narrowed by increasing the resolution of the PSs.

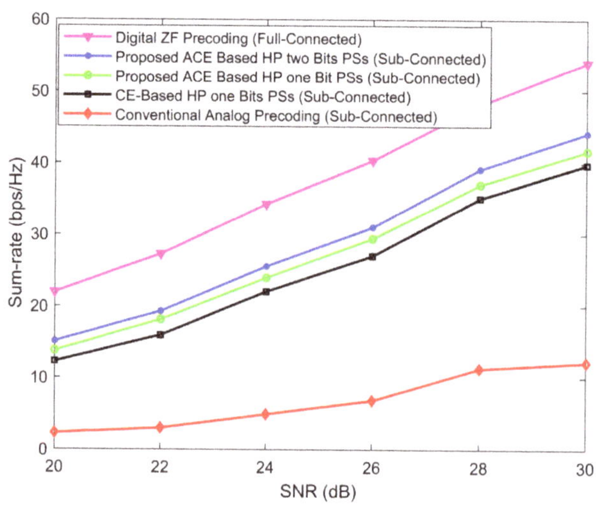

Figure 4. Sum rates of ACE-based method with sub-connected PSs when C = A = 12 and user count and RF chain are 8.

Figure 5 presents the energy efficiency attained by the proposed ACE-based HP with sub-connected PSs and some existing HP schemes mentioned above. The number of RF chains was equal to the user count K, and it varied from 1 to 32, while the other parameters were similar to those in Figure 2. In accordance with [35], the energy efficiency was calculated as the ratio of the sum rate to the total power consumption:

$$EE = \frac{R_{sum}}{p_{total}}(bps/Hz/W) \tag{28}$$

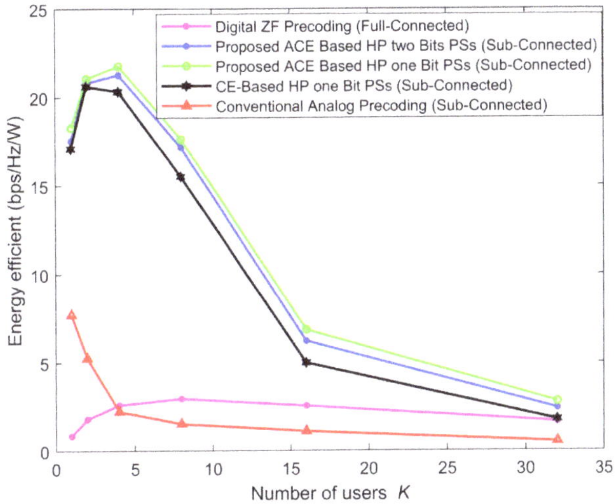

Figure 5. Energy efficiency of ACE-based method with sub-connected PSs (when C = A = 10), where SNR = 30.

The practical values for power consumption are as follows. The energy consumed by the sub-connected architecture where $p = 0.03$ W for the baseband power consumption is $P_B = 0.2$ W, the power consumed by the RF chain $P_{RF} = 0.3$ W, the one-bit PS power consumption $P_{ps} = 0.005$ W, and the typical valves of PS power consumption are 0.015 W, 0.045 W, 0.06 W, and 0.078 W for PSs of 3, 4, 5, and 6 bits, respectively [36,37]. Our proposed ACE-based HP scheme with sub connected PSs attained a high value of energy efficiency when compared with other existing HPs, as mention above.

Figure 6 presents that the proposed ACE-based algorithm with low-bit PSs obtained improved SE in comparison with other sub-connected schemes, especially when the number of users was not extremely large. The number of users K could range from 1 to 32, and the RF chains had the same configuration as the users. The bandwidth was fixed at 10 MHz, and the other parameters were comparable to those shown in Figure 4. In addition, it can be seen that when K equaled eight, the proposed ACE with two-bit PSs had a greater spectral efficiency than other sub-connected schemes, with the exception of the ZF precoding fully connected method. The reason for this was that when the number of users and RF chains increased, the number of PSs also increased, leading to a sharp rise in the total number of PSs in systems that were PS-based. In practice, the utilization of PSs with high resolutions that need a significant amount of energy is not appropriate. The gap between the proposed algorithm and the ZF-based HP is constant behavior. This can be reduced by increasing the resolution of the PSs.

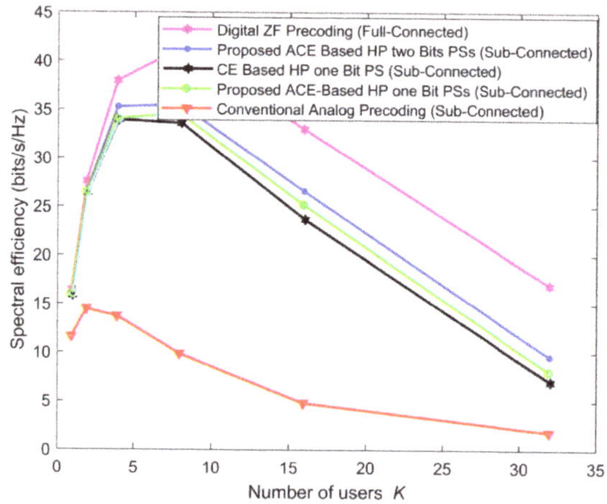

Figure 6. Spectral efficiency of ACE-based method with sub-connected PSs (when C = A = 10 and SNR = 30).

Finally, the impact of imperfect CSI was noticed in the proposed ACE-based HP, where $\hat{\mathbf{H}}$ represents the imperfect CSI as in [38]:

$$\hat{\mathbf{H}} = \varepsilon H + \sqrt{1-\varepsilon^2}\mathbf{E} \tag{29}$$

where $\varepsilon \in [0,1]$ represents the CSI accuracy, **H** is the original channel matrix, and **E** is the error matrix whose elements are distributed according to an independent and identically distributed distribution $\mathcal{CN}(0,1)$. Figure 7 depicts the ACE-based HP sum rates for various CSI scenarios when $A = C = 10$ and N_{RF} and $K = 4$. We considered a number of flawed CSI cases with varying values in addition to a perfect CSI scenario. CSI accuracy is not a factor in the proposed ACE-based HP with one-bit PSs. The proposed scheme's sum rate, which was reached with a value of $\varepsilon = 0.8$, was quite close to the perfect CSI situation. Furthermore, even when the CSI accuracy was extremely low, as when *varepsilon* = 0.4, the ACE-based proposal could still achieve 80 percent of the sum rate.

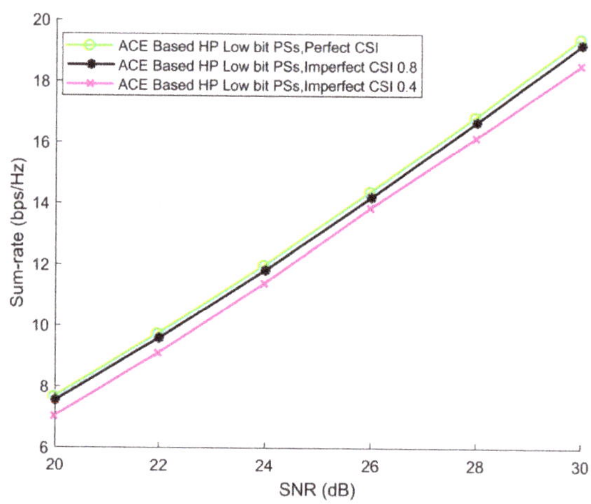

Figure 7. ACE-based HP with different CSI conditions (when C = A = 10 and users and RF chain = 4).

5. Conclusions

In this paper, we presented an EE design for HPs with a sub-connected low-bit PS HAP massive MIMO system. Initially, we focused on how to reduce power consumption without sacrificing performance. This energy utilization study showed that low-bit PS-based hybrid precoding consumes very little energy, and it also revealed that the array gain loss that is sustained by low-bit PSs is constrained and constant. Afterward, a low-complexity method that was based on ACE-based optimization with low-bit phase shifters was proposed in order to handle the sum rate maximization problem. The outcome of our simulation demonstrates that the proposed algorithm obtained a sufficiently high value for sum rate performance metrics as well as a higher value for energy efficiency compared with other algorithms.

Author Contributions: Study conception and design, preparation, data collection, analysis, and drafting, S.u.H.; conception, design, critical review for intellectual contents, and supervision, S.A.; data collection, analysis, and proofreading, T.M.; critical revision and drafting, N.A.K.; critical revision, U.M. All authors have read and agreed to the published version of the manuscript.

Funding: This research received no external funding.

Data Availability Statement: Codes and other relevant data are available and will be uploaded to https://github.com/Hassanbangash12/HAP-Massive-MIMO-with-Limited-RF-Chains (accessed on 10 December 2022) after the acceptance and publication of the manuscript. Interested authors can access and download the simulation code from the given link.

Acknowledgments: We thankfully acknowledge the kind support of the representative institutions for the utilization of common facilities during this work.

Conflicts of Interest: The authors declare that they have no known competing financial interests or personal relationships that could have appeared to influence the work reported in this paper.

References

1. Mumtaz, S.; Rodriguez, J.; Dai, L. *MmWave Massive MIMO: A Paradigm for 5G*; Academic Press: Cambridge, MA, USA, 2016.
2. Tozer, T.; Grace, D. High-altitude platforms for wireless communications. *Electron. Commun. Eng. J.* **2001**, *13*, 127–137. [CrossRef]
3. Jia, Z.; Wu, Q.; Dong, C.; Yuen, C.; Han, Z. Hierarchical Aerial Computing for Internet of Things via Cooperation of HAPs and UAVs. *IEEE Internet Things J.* **2022**. [CrossRef]
4. Zhang, G.; Jiang, L.; Ji, P.; Zou, S.; He, C.; He, D. A Modified K-means User Grouping Design for HAP Massive MIMO Systems. In Proceedings of the 2021 International Conference on Networking and Network Applications (NaNA), Lijiang, China, 29 October–1 November 2021; pp. 288–292.
5. Liu, X.; Li, X.; Cao, S.; Deng, Q.; Ran, R.; Nguyen, K.; Tingrui, P. Hybrid precoding for massive mmWave MIMO systems. *IEEE Access* **2019**, *7*, 33577–33586. [CrossRef]
6. Huang, Y.; Liu, C.; Song, Y.; Yu, X. Near-optimal hybrid precoding for millimeter wave massive MIMO systems via cost-efficient Sub-connected structure. *IET Commun.* **2020**, *14*, 2340–2349. [CrossRef]
7. Gao, X.; Dai, L.; Han, S.; Chih-Lin, I.; Heath, R.W. Energy-efficient hybrid analog and digital precoding for mmWave MIMO systems with large antenna arrays. *IEEE J. Sel. Areas Commun.* **2016**, *34*, 998–1009. [CrossRef]
8. Alkhateeb, A.; Leus, G.; Heath, R.W. Limited feedback hybrid precoding for multi-user millimeter wave systems. *IEEE Trans. Wirel. Commun.* **2015**, *14*, 6481–6494. [CrossRef]
9. Lu, A.Y.; Chen, Y.F. Adaptive Hybrid Beamforming Schemes in Millimeter Wave MIMO Systems. In Proceedings of the 2021 IEEE International Symposium on Radio-Frequency Integration Technology (RFIT), Hualien, Taiwan, 25–27 August 2021; pp. 1–3.
10. Yu, J.; Lee, H.; Eom, C.; Lee, C. A Switch Network-Based Adaptive Sub-Connected Hybrid Precoding Architecture in Sub-THz Communications. In Proceedings of the 2021 36th International Technical Conference on Circuits/Systems, Computers and Communications (ITC-CSCC), Jeju, Korea, 27–30 June 2021; pp. 1–4.
11. Jo, S.; Yang, W.; Choi, H.K.; Noh, E.; Jo, H.S.; Park, J. Deep Q-Learning-Based Transmission Power Control of a High Altitude Platform Station with Spectrum Sharing. *Sensors* **2022**, *22*, 1630. [CrossRef]
12. Tashiro, K.; Hoshino, K.; Nagate, A. Nullforming-Based Precoder for Spectrum Sharing between HAPS and Terrestrial Mobile Networks. *IEEE Access* **2022**, *10*, 55675–55693. [CrossRef]
13. Sun, Z.; Qi, F.; Liu, L.; Xing, Y.; Xie, W. Energy-efficient Spectrum Sharing for 6G Ubiquitous IoT Networks Through Blockchain. *IEEE Internet Things J.* **2022**. [CrossRef]
14. Lin, Z.; Lin, M.; Wang, J.B.; De Cola, T.; Wang, J. Joint beamforming and power allocation for satellite-terrestrial integrated networks with non-orthogonal multiple access. *IEEE J. Sel. Top. Signal Process.* **2019**, *13*, 657–670. [CrossRef]

15. Lu, Y.; Xiong, K.; Fan, P.; Ding, Z.; Zhong, Z.; Letaief, K.B. Global energy efficiency in secure MISO SWIPT systems with non-linea power-splitting EH model. *IEEE J. Sel. Areas Commun.* **2018**, *37*, 216–232. [CrossRef]
16. Huang, Q.; Lin, M.; Wang, J.B.; Tsiftsis, T.A.; Wang, J. Energy efficient beamforming schemes for satellite-aerial-terrestria networks. *IEEE Trans. Commun.* **2020**, *68*, 3863–3875. [CrossRef]
17. Niu, H.; Lin, Z.; Chu, Z.; Zhu, Z.; Xiao, P.; Nguyen, H.X.; Lee, I.; Al-Dhahir, N. Joint beamforming design for secure RIS-assistec IoT networks. *IEEE Internet Things J.* **2022**. [CrossRef]
18. Lin, Z.; Lin, M.; Champagne, B.; Zhu, W.P.; Al-Dhahir, N. Secrecy-energy efficient hybrid beamforming for satellite-terrestria integrated networks. *IEEE Trans. Commun.* **2021**, *69*, 6345–6360. [CrossRef]
19. Lin, Z.; An, K.; Niu, H.; Hu, Y.; Chatzinotas, S.; Zheng, G.; Wang, J. SLNR-based secure energy efficient beamforming ir multibeam satellite systems. *IEEE Trans. Aerosp. Electron. Syst.* **2022**. [CrossRef]
20. Srinivasan, M.; Gopi, S.; Kalyani, S.; Huang, X.; Hanzo, L. Airplane-aided integrated next-generation networking. *IEEE Trans Veh. Technol.* **2021**, *70*, 9345–9354. [CrossRef]
21. Ji, P.; Jiang, L.; He, C.; He, D. Hybrid Precoding for HAP Massive MIMO Systems. In *Proceedings of the International Conference or Space Information Network*; Springer: Berlin/Heidelberg, Germany, 2019; pp. 255–263.
22. Zong, D.; Jiang, L.; He, C.; Ji, P.; He, D. Two-stage precoding design in Rician channel for HAP massive MIMO systems. Ir Proceedings of the 2019 22nd International Symposium on Wireless Personal Multimedia Communications (WPMC), Lisbon Portugal, 24–27 November 2019; pp. 1–5.
23. Cumalı, İ.; Özbek, B.; Kurt, G.K.; Yanikomeroglu, H. User Selection and Codebook Design for NOMA-Based High Altitude Platform Station (HAPS) Communications. *IEEE Trans. Veh. Technol.* **2022**. [CrossRef]
24. Lian, Z.; Jiang, L.; He, C.; He, D. User grouping and beamforming for HAP massive MIMO systems based on statistical-eigenmode *IEEE Wirel. Commun. Lett.* **2019**, *8*, 961–964. [CrossRef]
25. Ji, P.; Jiang, L.; He, C.; Lian, Z.; He, D. Energy-efficient beamforming for beamspace HAP-NOMA systems. *IEEE Commun. Lett* **2020**, *25*, 1678–1681. [CrossRef]
26. Sudheesh, P.; Mozaffari, M.; Magarini, M.; Saad, W.; Muthuchidambaranathan, P. Sum-rate analysis for high altitude platform (HAP) drones with tethered balloon relay. *IEEE Commun. Lett.* **2017**, *22*, 1240–1243. [CrossRef]
27. He, Y.; Wang, D.; Huang, F.; Zhang, R.; Gu, X.; Pan, J. Downlink and Uplink Sum Rate Maximization for HAP-LAP Cooperated Networks. *IEEE Trans. Veh. Technol.* **2022**, *71*, 9516–9531. [CrossRef]
28. Teng, Y.; Wei, M.; Liu, A.; Lau, V.; Zhang, Y. Mixed-timescale per-group hybrid precoding for multiuser massive MIMO systems *IEEE Signal Process. Lett.* **2018**, *25*, 675–679. [CrossRef]
29. Yu, H.; Qu, W.; Fu, Y.; Jiang, C.; Zhao, Y. A novel two-stage beam selection algorithm in mmWave hybrid beamforming system *IEEE Commun. Lett.* **2019**, *23*, 1089–1092. [CrossRef]
30. El Ayach, O.; Rajagopal, S.; Abu-Surra, S.; Pi, Z.; Heath, R.W. Spatially sparse precoding in millimeter wave MIMO systems *IEEE Trans. Wirel. Commun.* **2014**, *13*, 1499–1513. [CrossRef]
31. Méndez-Rial, R.; Rusu, C.; Alkhateeb, A.; González-Prelcic, N.; Heath, R.W. Channel estimation and hybrid combining for mmWave: Phase shifters or switches? In Proceedings of the 2015 Information Theory and Applications Workshop (ITA), San Diego, CA, USA, 1–6 February 2015; pp. 90–97.
32. Mir, T.; Siddiqi, Z.; Mir, U.; MacKenzie, R.; Hao, M. One-bit hybrid precoding for wideband millimeter-wave massive MIMO systems. In Proceedings of the 2019 IEEE 89th Vehicular Technology Conference (VTC2019-Spring), Kuala Lumpur, Malaysia, 28 April–1 May 2019; pp. 1–6.
33. Michailidis, E.T.; Kanatas, A.G. Three-dimensional HAP-MIMO channels: Modeling and analysis of space-time correlation. *IEEE Trans. Veh. Technol.* **2010**, *59*, 2232–2242. [CrossRef]
34. Rubinstein, R.Y.; Kroese, D.P. *The Cross-Entropy Method: A Unified Approach to Combinatorial Optimization, Monte-Carlo Simulation and Machine Learning*; Springer Science & Business Media: Berlin/Heidelberg, Germany, 2013.
35. Liu, Y.; Feng, Q.; Wu, Q.; Zhang, Y.; Jin, M.; Qiu, T. Energy-efficient hybrid precoding with low complexity for mmWave massive MIMO systems. *IEEE Access* **2019**, *7*, 95021–95032. [CrossRef]
36. Méndez-Rial, R.; Rusu, C.; González-Prelcic, N.; Alkhateeb, A.; Heath, R.W. Hybrid MIMO architectures for millimeter wave communications: Phase shifters or switches? *IEEE Access* **2016**, *4*, 247–267. [CrossRef]
37. Huang, C.; Alexandropoulos, G.C.; Zappone, A.; Debbah, M.; Yuen, C. Energy efficient multi-user MISO communication using low resolution large intelligent surfaces. In Proceedings of the 2018 IEEE Globecom Workshops (GC Wkshps), Abu Dhabi, United Arab Emirates, 9–13 December 2018; pp. 1–6.
38. Li, X.; Fang, J.; Li, H.; Wang, P. Millimeter wave channel estimation via exploiting joint sparse and low-rank structures. *IEEE Trans. Wirel. Commun.* **2017**, *17*, 1123–1133. [CrossRef]

Disclaimer/Publisher's Note: The statements, opinions and data contained in all publications are solely those of the individual author(s) and contributor(s) and not of MDPI and/or the editor(s). MDPI and/or the editor(s) disclaim responsibility for any injury to people or property resulting from any ideas, methods, instructions or products referred to in the content.

electronics

MDPI

Article

DMCNet-Pro: A Model-Driven Multi-Pilot Convolution Neural Network for MIMO-OFDM Receivers

Pengyuan Li [1,2], Tianlin Zhu [2], Yutong Xin [3,4], Gang Yuan [2], Xiong Yu [2], Zejian Lu [4,*], Zili Liu [4] and Qing Yan [3]

1 Polytechnic Institute, Zhejiang University, Hangzhou 310058, China; pengyuan_li@163.com
2 Beijing Institute of Tracking and Telecommunications Technology, Beijing 100094, China; 13911966061@139.com (T.Z.); asqasq2022@163.com (G.Y.); 19800232763@163.com (X.Y.)
3 School of Artificial Intelligence, Beijing University of Posts and Telecommunications, Beijing 100876, China; xinyutong@bupt.edu.cn (Y.X.); qingyan@bupt.edu.cn (Q.Y.)
4 China Academic of Electronics and Information Technology, Beijing 100041, China; liuzili1@cetc.com.cn
* Correspondence: lzj_bupt@163.com

Abstract: Nowadays, wireless communication technology is evolving towards high data rates, a low latency, and a high throughput to meet increasingly complex business demands. Key technologies in this direction include multiple-input multiple-output (MIMO) and orthogonal frequency division multiplexing (OFDM). This research is based on our previous work DMCNet. In this article, we focus on studying the deep learning (DL) application of neural networks to solve the reception of single-antenna OFDM signals. Specifically, in multi-antenna scenarios, the channel model is more complex compared to single-antenna cases. By leveraging the characteristics of DL, such as automatic learning of parameters using deep neural networks, we treat the reception process of MIMO-OFDM signals as a black box and utilize neural networks to accomplish the signal reception task. Moreover, we propose a data-driven multi-pilot convolution neural network for MIMO-OFDM receivers (DMCNet). By incorporating complex convolution and complex fully connected structures, we design a receiver network to recover the transmitted signals from the received signals. We validate the accuracy and robustness of DMCNet under different channel conditions, comparing the bit error rates with different schemes. Additionally, we discuss the factors influencing various channel effects. At the same time, we also propose a model-driven scheme, DMCNet-pro, which has a higher accuracy and fewer parameters in some scenarios. The experimental results demonstrate that the DL-based reception scheme exhibits promising feasibility in terms of accuracy and interference resistance when compared to traditional approaches.

Keywords: MIMO; OFDM; multiple pilots; neural networks; complex convolution

Citation: Li, P.; Zhu, T.; Xin, Y.; Yuan, G.; Yu, X.; Lu, Z.; Liu, Z.; Yan, Q. DMCNet-Pro: A Model-Driven Multi-Pilot Convolution Neural Network for MIMO-OFDM Receivers. *Electronics* **2024**, *13*, 330. https://doi.org/10.3390/electronics13020330

Academic Editor: David A. Sánchez-Hernández

Received: 20 November 2023
Revised: 7 January 2024
Accepted: 8 January 2024
Published: 12 January 2024

Copyright: © 2024 by the authors. Licensee MDPI, Basel, Switzerland. This article is an open access article distributed under the terms and conditions of the Creative Commons Attribution (CC BY) license (https://creativecommons.org/licenses/by/4.0/).

1. Introduction

Orthogonal frequency division multiplexing (OFDM) and multiple-input multiple-output (MIMO) are key technologies used in current 5G and even future 6G wireless communications [1]. OFDM enables parallel transmission of signals by using overlapping orthogonal sub-channels to transmit data. The data streams are modulated onto various sub-channels. Its main advantages are the high spectrum efficiency and the ability to cope with adverse channel conditions without complex equalization filters. MIMO technology utilizes multiple antenna arrays at both the transmitting and receiving ends of the system. At the transmitting end, the serial data streams are parallelized through space-time transformations and are sent out through different transmitting antennas. At the receiving end, multiple antennas are used to receive the signals, and the received signal is processed through decoding modules. MIMO technology fully utilizes spatial and temporal diversity and multiplexing gains, significantly increasing the channel capacity and spectrum efficiency without increasing the occupied bandwidth [2].

However, the communication environment is becoming increasingly complex, and the transmission of signals is subject to various unfavorable factors, making it increasingly difficult to accurately receive signals. Traditional multi-antenna, multi-carrier signal receiver schemes often require channel estimation, symbol detection, and other steps. In multi-antenna scenarios, researchers have proposed optimal signal detection algorithms, also known as Maximum Likelihood (ML) detection algorithms [3]. These methods exhaustively search through all possible transmitted signals to find the optimal solution, providing the best performance in terms of solution accuracy, but they come with a high computational complexity. Commonly used linear signal detection algorithms include the Zero Forcing (ZF) detection algorithm [4] and the Minimum Mean Square Error (MMSE) detection algorithm [5]. The ZF algorithm is a linear precoding technique used in wireless communication. It aims to eliminate interference by designing a matrix that cancels out signals from unwanted users, improving system performance, especially in multi-user scenarios. The ZF algorithm has a simple solution but suffers from significant performance degradation when the noise power is high. The MMSE detection algorithm takes noise into account and performs better than the ZF algorithm at lower signal-to-noise ratios. However, in increasingly complex communication environments, these traditional methods exhibit poor robustness, susceptibility to interference, and limited accuracy. Therefore, in the field of wireless signal reception, there is a need for new technologies to complement and optimize traditional methods.

Fortunately, with the continuous development of computer hardware, the training capability of deep neural networks has become stronger. In recent years, the development of deep learning theory has been rapid, and related research has been applied to the field of wireless communications. Compared to traditional approaches, deep learning algorithms learn knowledge automatically from existing data and experimental results, surpassing the limitations of traditional mathematical approaches and exhibiting superior performance and adaptability in certain scenarios. Reference [6] first proposed the design of a single-antenna OFDM receiver using a five-layer fully connected structure. The fully connected network does not require explicit channel estimation and signal detection, treating the entire receiver as a black box. This breaks the modular structure of traditional approaches, and the experimental results demonstrate that the neural-network-based approach exhibits stronger robustness. Model-driven approaches combine deep learning with expert knowledge in the field of wireless communications. The receiver is divided into two sub-networks: channel estimation (CE) and signal detection (SD). A typical representative is ComNet [7] proposed by the team from Southeast University. Subsequently, SwitchNet [8], also based on a model-driven approach, was proposed to address performance issues in ComNet. Additionally, DCCN [8], based on a complex convolutional structure, includes a receiver, but its code is not open-source, making it difficult to reproduce the simulation links and training data. On the other hand, some individual achievements in researching deep learning MIMO receivers have been published in recent years. These studies focus on the design of single-carrier MIMO receivers. In 2018, Shi Jin et al. proposed OAMP-Net [9], which uses a multi-layer iterative neural network approach to design MIMO receivers.

In current practical applications, MIMO and OFDM are often combined, known as MIMO-OFDM. Compared to pure MIMO or OFDM techniques, the signal transmission process of MIMO-OFDM is more complex. Factors such as the number of antennas and the number of subcarriers and the specific characteristics of OFDM, such as cyclic prefix and peak clipping, increase the complexity of the communication process. Therefore, this paper aims to expand the aforementioned research by designing a small-scale MIMO-OFDM receiver that incorporates the principles of deep learning. The link of the DL-based MIMO-OFDM receiver is shown in Figure 1. Specifically, in this research, we start by studying the reception scheme for OFDM signals and extending it to MIMO-OFDM receiver design. Deep learning methods are employed to conduct research in this field, treating the entire receiver as a black box and utilizing deep neural networks to perform tasks such as channel estimation and signal detection. Furthermore, a deep-learning-based receiver called DMCNet is proposed [10], which combines complex convolution and complex, fully connected structures. This receiver network aims to recover the transmitted signals from the received signals and validate their accuracy and robustness under different channel conditions. Finally, based on the data-driven model DMCNet, a model-driven approach is proposed. Experimental results demonstrate that the model-driven approach, when combined with relevant algorithms from traditional communications, achieves a higher accuracy and requires fewer parameters in certain scenarios.

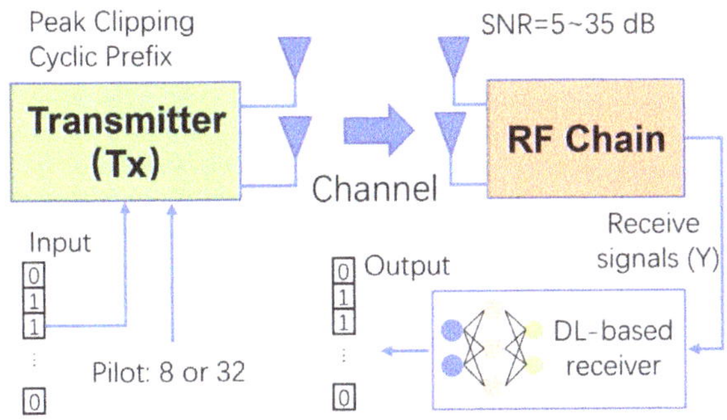

Figure 1. The link of a DL-based MIMO-OFDM receiver.

Motivated by the discussion above, we design a model-driven multi-pilot convolution neural network for MIMO-OFDM receivers (DMCNet-pro). Comparing to previous work, a signal denoising module is added to denoise the pilot frequency and data, respectively, in DMCNet-pro. The network we propose shows great performance regarding accuracy and robustness. Additionally, fewer parameters are required for this network in some scenarios.

The rest of this paper is organized as follows. Section 2 introduces the dataset and the channel model. Section 3 introduces the structure of DMCNet. Section 4 introduces the structure of DMCNet-pro. Section 5 introduces the experimental design and results. Section 6 introduces the conclusion of the experiment. We only consider a simplified single user scenario for a clear mathematical analysis, establishing fundamental principles before tackling complexities like multi-user interference in realistic settings.

2. Dataset and Channel Model

This paper focuses on the design of a 2×2 MIMO-OFDM wireless communication receiver. The 2×2 MIMO system strikes a balance between performance and simplicity with two transmit and two receive antennas. It effectively improves system capacity, reliability, and signal coverage, making it widely adopted across wireless communication standards. The required dataset for the experiments includes a 2×2 MIMO channel model and simulation programs for transmission and reception using the channel model. The wireless channel model follows the Wireless World Initiative New Radio (WINNER II) [11], and the channel scenario represents a typical urban street environment. In the simulation programs, the communication frequency was set to 1.85 GHz. The sampled channel data consist of 340,000 data samples, with each channel dataset containing 256 real numbers representing the channel matrix of a 2×2 wireless channel. Each element in the matrix consists of 32 complex numbers. The generated data are divided into training, testing, and validation datasets, consisting of 300,000, 20,000, and 20,000 samples, respectively.

Multi-carrier modulation is implemented using OFDM with 256 subcarriers. The transmitter includes QAM of the transmitted bits X, insertion of pilot symbols, IFFT, addition of a cyclic prefix (CP) with a length of 32, and optional peak clipping. The settings for cyclic prefix and peak clipping can be adjusted through flags in the simulation program.

Two transmit antennas generate random bit sequences of length 512. The combined bit stream of the two sequences is denoted as X, with a length of 1024. The random bit signals are input into the simulation link, and the receiver obtains the signal y after passing through the channel. Since pilot symbols are added at the transmitter, the received signal y consists of two time slots: one for pilot symbols and one for data. Taking eight pilot symbols as an example, the transmitter first reads the known pilot sequence from a pre-generated pilot file, which contains 32 random bits ($8 \times 2 \times 2 = 32$) in total. The pilot data remain unchanged throughout the training, testing, and validation processes. Then, a random channel sample is selected from the channel samples. The signal after passing through the channel is a result of the data sequence and pilot sequence combined, containing 2048 floating point numbers.

3. The Structure of DMCNet

3.1. The Data Flow of DMCNet

The designed deep-neural-network-based receiver in this paper, referred to as DMCNet, is shown in Figure 2. DMCNet consists of three main modules: signal denoising, pilot processing, and signal detection. In the network models for 32 pilot symbols and 8 pilot symbols, the dimensions of the output at each layer are different. Therefore, only the structure of the model is provided here, while the detailed dimensions of each layer and the specifics of each module will be discussed in the following sections.

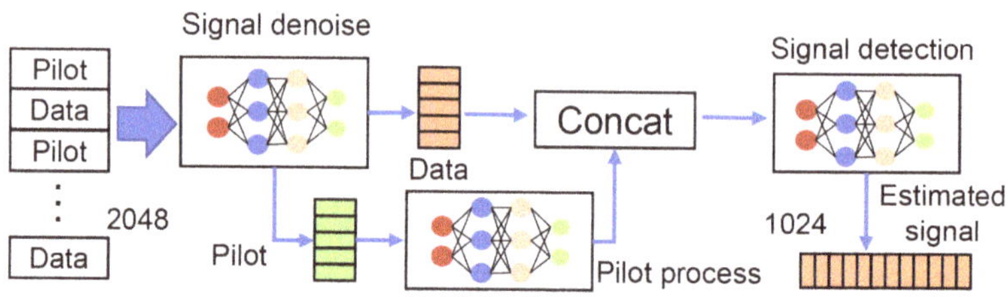

Figure 2. The structure of DMCNet.

Although DMCNet is divided into three modules, the main idea of DMCNet is to use a data-driven network to construct an end-to-end training model while enriching the model's computation by incorporating convolutional modules. As is well known, in the field of computer vision, images are often stored and processed as real numbers. However in the field of signal transmission, input signals often need to be mapped to the complex plane to perform operations that record both the magnitude and phase information. In this paper, the received signal is treated as a complex number, and complex neural network modules are used for information processing. We primarily employ complex convolution and complex fully connected structures. Subsequently, there have been some open-source implementations of this paper. In this paper, we primarily employ complex convolution and complex fully connected structures. Compared to traditional neural networks, complex neural networks have a more expressive power. The main difference between complex networks and real networks lies in the multiplication operation, as shown in the following equation:

$$(a + bi) \times (c + di) = (ac - bd) + (ad + bc)i. \tag{1}$$

For a complex fully connected network with one input neuron and one output neuron, assuming the input is $a + bi$ and the weight is $c + di$, according to the equation above, the complex fully connected network can be implemented using a real fully connected layer with an input size of 2 and an output size of 2. In this case, the four weights of the fully connected layer are $c, -c, d, -d$. Therefore, the complex fully connected layer can be implemented using a real fully connected layer with double the size. For the complex convolutional module, it can be implemented using higher-dimensional real convolutional layers. Let the input complex matrix be $\mathbf{W} = \mathbf{A} + i\mathbf{B}$ and the complex vector be $\mathbf{h} = \mathbf{x} + i\mathbf{y}$, where $\mathbf{A}$ and $\mathbf{B}$ are real matrices and $\mathbf{x}$ and $\mathbf{y}$ are real vectors. The convolution result of $\mathbf{W}$ and $\mathbf{h}$ can be represented as:

$$\mathbf{W}^*\mathbf{h} = (\mathbf{A}^*\mathbf{x} - \mathbf{B}^*\mathbf{y}) + i(\mathbf{B}^*\mathbf{x} + \mathbf{A}^*\mathbf{y}). \tag{2}$$

As we can see, the real and imaginary parts of the convolution result can be represented in matrix form as follows:

$$\begin{bmatrix} \Re(\mathbf{W}^*\mathbf{h}) \\ \Im(\mathbf{W}^*\mathbf{h}) \end{bmatrix} = \begin{bmatrix} \mathbf{A} & -\mathbf{B} \\ \mathbf{B} & \mathbf{A} \end{bmatrix} * \begin{bmatrix} \mathbf{x} \\ \mathbf{y} \end{bmatrix}. \tag{3}$$

Further information on the definition of activation functions and batch normalization (BN) layers in complex neural networks can be found in [12]. In this section, the experimental setups utilize the architecture of complex neural networks. Next, we will introduce the various data-processing modules in DMCNet separately.

3.2. Signal Denoising Module

The transmitted signal is subject to interference after passing through the channel, and the received signal contains varying levels of noise depending on the signal-to-noise ratio (SNR). A lower SNR corresponds to higher noise in the received signal, which significantly affects the overall signal detection performance. To mitigate the impact of channel noise, this paper introduces a denoising module in the first layer of DMCNet. The received signal is divided into data and pilot components, and each component undergoes denoising using their respective denoising networks. Two approaches were explored in the experiments. The first approach utilizes a two-layer fully connected network, where the input signal is initially expanded to twice its original size in the first layer and then compressed back to the original length in the second layer. The second approach employs a residual structure for the two-layer fully connected network, which incorporates a shortcut connection that subtracts the output of the fully connected layer from the original signal. This approach aligns better with the definition of the noise model. The structure of the denoising module is shown in Figure 3:

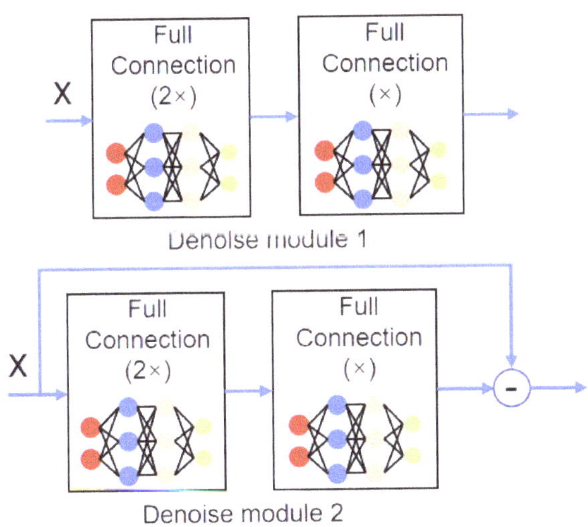

Figure 3. Denoising module structure.

3.3. Pilot Processing Network

In [6], it is mentioned that pilot signals and data signals are processed using the same fully connected network. However, as explained in Section 2, the pilot signals and data signals are not symmetrical regarding their positions during communication, even though they are transmitted through the same channel. The pilot signals are typically shorter in length compared to the data signals, and the original sequence of the pilot signals is known, allowing for channel estimation. To handle the pilot signals more precisely, this paper introduces a separate pilot processing and analysis step. The pilot signals are first processed and analyzed independently before being merged and processed together with the data signals. To accomplish this, a pilot processing module is designed; the diagram of the pilot processing module is shown in Figure 4 and the pilot processing module's details are shown in Table 1.

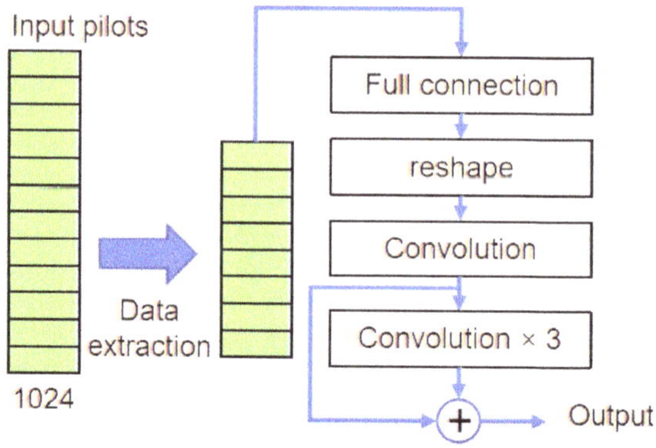

Figure 4. Pilot processing structure.

Table 1. Pilot processing module details.

	Module Type	Output Shape (8 Pilots)	Output Shape (32 Pilots)
Input pilot	–	1024	1024
Data extraction	–	64	256
Full connection layer	full connection	1024	2048
Reshape	reshape	[1, 256, 4]	[1, 256, 8]
Convolution layer 0	1 × 9	[1, 256, 8]	[1, 256, 16]
Convolution layer 1	1 × 9	[1, 256, 16]	1, 256, 32]
Convolution layer 2	1 × 9	[1, 256, 32]	[1, 256, 32]
Convolution layer 3	1 × 9	[1, 256, 8]	[1, 256, 16]
Convolution layer 4	1 × 9	[1, 256, 8]	[1, 256, 16]

After denoising, the received signal with a length of 2048 is divided into data and pilot components, each with a length of 1024. For an 8-pilot link data scenario, this means that pilot data are inserted every 32 subcarriers, and the remaining subcarriers have null pilot data, which does not contain useful information and are outputted through the communication link. In the approach described in [6], regardless of the number of pilot signals, all the data are inputted to the input layer of the neural network and the output result is obtained through a fully connected network. However, in the context of this paper's research scenario, there are more antennas and the number of OFDM subcarriers has doubled, resulting in the output data from the link increasing from 256 to 2048. Therefore, in this paper, some of the unused pilot positions are discarded. Since the data are shifted by half a pilot position, the model first extracts a set of pilot output data every 16 subcarriers. In the case of eight pilot signals, a total of $(256/16) \times 4 = 64$ floating point numbers are extracted, which represent the known pilot data through the channel link. The extracted pilot output data are initially processed through a residual module for information extraction. They are first dimensionally expanded using a fully connected network to the size of 1024 and then reshaped into a 1-row, 256-column dataset, with the third dimension of the tensor representing the number of channels. In the case of eight pilot signals, the reshaped tensor has a shape of (1, 256, 4). The dataset is then passed through a complex convolution with a size of 1×9 to expand the channels, resulting in an output size of (1, 256, 16).

A residual network structure is used here, where the pilot data, after being dimensionally expanded and reshaped, are further processed through a residual module for purification. The residual module is similar to the one described in Section 3 and consists of three layers of complex convolution. The output of the third layer is added to the original input of the first layer. The output of the residual module serves as the pilot result.

3.4. Signal Detection Network

The signal detection module's details are shown in Table 2. For the data part, a tensor of length 1024 is also extracted. These 1024 data points represent the information of the data part after passing through the channel, so no further extraction is needed. Similarly, these data are reshaped into a (1, 256, 4) tensor and concatenated with the implicit channel estimation results of the pilot part in the third dimension. This combined tensor serves as the input to the signal detection network. For eight pilots, the dimensions of the merged data are (1, 256, 12). In this data with 12 channels, the first 8 channels contain the results of the pilot part through the pilot module and the last 4 channels contain the original data part after reshaping. The merged tensor of size (1, 256, 12) is then expanded in the channels through two consecutive convolutions, resulting in a tensor of size (1, 256, 100). It is then processed through three Res-Block residual refinement modules. Each Res-Block consists of three cascaded convolutional layers, and the output after the first convolution is directly connected to the final output in a shortcut manner. Finally, the intermediate results are dimensionally reduced to a (1, 256, 4) tensor through convolution, flattened into 1024 floating point data, and the final estimation result is obtained through a bit decision function. It is worth noting that due to the relatively large number of channels in the data of the Res-Block part, a channel attention (CA) module is added to each Res-Block to weight the multi channel data and improve system performance.

Table 2. Signal detection module details.

	Module Type	Output Shape (8 Pilots)	Output Shape (32 Pilots)
Merge data	–	[1, 256, 12]	[1, 256, 20]
Convolution layer 0	1×9	[1, 256, 50]	[1, 256, 128]
Convolution layer 1	1×9	[1, 256, 100]	[1, 256, 256]
Res-Block 0	Res-Block	[1, 256, 100]	[1, 256, 256]
Res-Block 1	Res-Block	[1, 256, 100]	[1, 256, 256]
Res-Block 2	Res-Block	[1, 256, 100]	[1, 256, 256]

Table 2. *Cont.*

	Module Type	Output Shape (8 Pilots)	Output Shape (32 Pilots)
Convolution layer 2	1×9	$[1, 256, 4]$	$[1, 256, 4]$
Sigmoid	–	$[1, 256, 4]$	$[1, 256, 4]$
Bit decision	Threshold function	$[1, 1024]$	$[1, 1024]$

4. The Structure of DMCNet-Pro

Model-Driven Receiver Network Structure

In Section 3, a data-driven scheme is adopted to treat the whole receiver as a black box for end-to-end model training. In some previous work, such as [13], a model-driven scheme was used to solve the problem of a single-antenna OFDM signal receiver. Based on the research results in Section 3, a model-driven receiver design is also used in this paper. DMCNet, designed in Section 3, includes several network components to process different input data and perform end-to-end model training. This section attempts to split and improve it, and design a model-driven deep learning receiver scheme.

The model-driven network model is shown in Figure 5, which is named DMCNet-pro in this paper. The model-driven network DMCNet-pro includes signal denoising, a channel estimation network, and a signal detection network. In Section 3, it was shown that the model is also divided into several modules; however, in the training process, as a whole an end-to-end training method is adopted. The final output result of the receiver and the transmitted signal are used to calculate the loss function, and then the optimizer uses this loss function to adjust the parameters of the entire network to reduce the loss of the next iteration, and the process is repeated to guide the specified training cycle. In the model-driven scheme, the loss function is calculated and trained in stages by using some known information.

In DMCNet, a signal denoising network is added to initially remove the noise of the received signal passing through the channel. In DMCNet-pro, a signal denoising module is also added to denoise the pilot frequency and data, respectively. The difference is that the denoising network needs to be trained separately at this time. The input signal of the signal denoising module is the received signal, the label is the result after the same input passes through the noiseless channel, and the loss function is the mean square error loss between the actual received signal y and the result $\dot{y}$ after the same signal passes through the noiseless channel, which can be expressed as:

$$loss = \frac{1}{n}\sum_{i=1}^{n}(y_i - \dot{y}_i)^2 \tag{4}$$

where n represents the length of signal y. This loss function is passed to the Adam optimizer, which adjusts the model parameters.

The second step is required to train the channel estimation network. As mentioned above, the pilot frequency can actually be understood as a known sequence, which experiences the same channel propagation path as the actual data sequence to be sent; that is, it undergoes the same channel operation, and then obtains a string of output sequences. Through the output sequence and the known pilot sequence, the distribution of the channel can be estimated and the data can be received. Therefore, the received data and the known transmit pilot sequence can be used for channel estimation, and the least-squares (LS) channel estimation algorithm is used for initialization, that is:

$$\hat{H}_{LS} = \frac{y_P}{x_P} \tag{5}$$

where y_p represents the pilot part of the received signal and x_p represents the pilot sequence known at both ends of the transceiver. The initialized channel estimate $\widehat{H}_{LS}$ serves as the input to the channel estimate network. After initialization by the LS algorithm, the size of $\widehat{H}_{LS}$ is a vector with a length of 256, and the input channel estimation network is further purified. Since precise estimation of the output channel matrix is required, a convolution is added after Res-Block, and the output dimension of the convolution is $[1, 256, 1]$. The output $\widehat{H}$ is computed with the actual sample matrix H for the mean square error value, which is used as input to the model optimizer to adjust the network parameters.

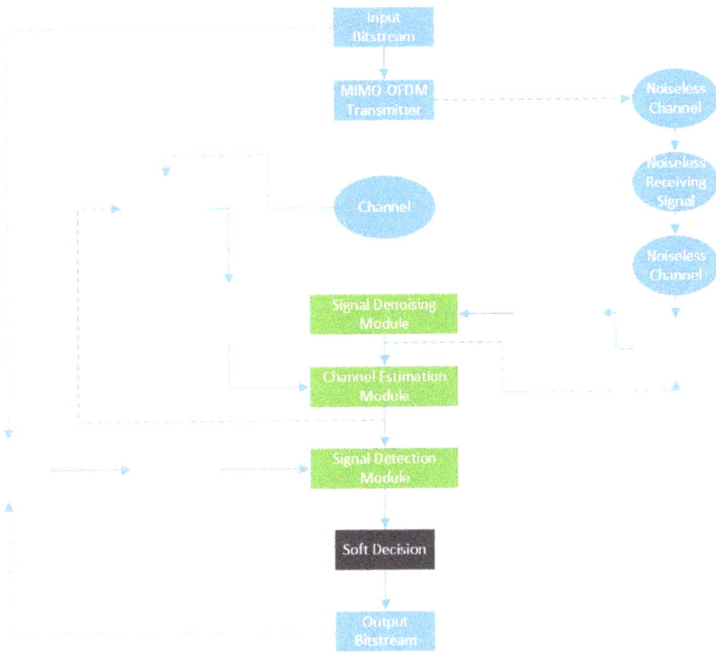

Figure 5. Model-driven MIMO-OFDM receiver.

The third step is required to train the network of the signal detection module. The structure of channel estimation and signal detection model is shown in Figure 6. First, the signal is initialized using the Zero Forcing (ZF) method. The ZF method can improve signal quality and system performance, particularly in multi-user scenarios.

$$\widehat{x}_{ZF} = \frac{y_D}{\widehat{H}} \tag{6}$$

where y_D is the data part of the received signal, $\widehat{H}$ is the output result of the channel estimation module, and the output $\widehat{x}_{ZF}$ is the input of the signal detection network. Comparing to DMCNet, the input of the signal detection network is the result of superposition of data and the pilot processing results. Here, $\widehat{H}$ and $\widehat{x}_{ZF}$ are also input to the signal detection network at the same time and superimposed. Similar to the signal monitoring network in DMCNet, the signal detection network structure in DMCNET-Pro is:

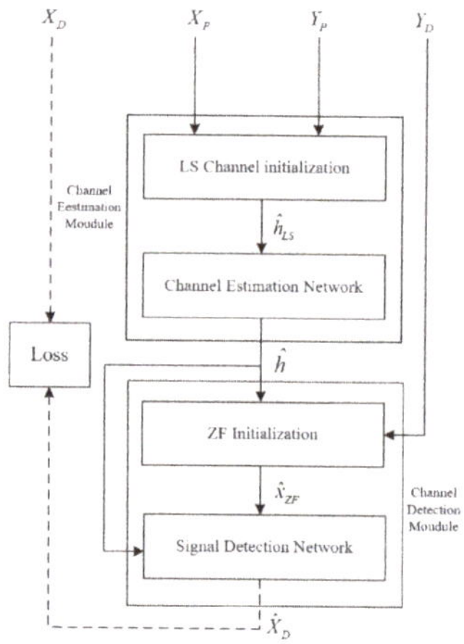

Figure 6. Channel estimation and signal detection model.

As we can see, the structure of the model-driven network is largely the same as that of DMCNet, but in this network, there are fewer output channels for each layer of convolution. The structure of signal detection model is shown in Figure 7. At eight pilots, $\widehat{H}$ and $\widehat{x}_{ZF}$ synthesize a feature map of size [1, 256, 5], which is first enhanced by convolution and then further purified by several refine blocks, while with the 32-pilot Res-Block, the number of convolutional output channels is 100, 100, and 50, respectively. Compared with DMCNet, it has fewer channels and fewer parameters, but the experimental results show that its performance is similar.

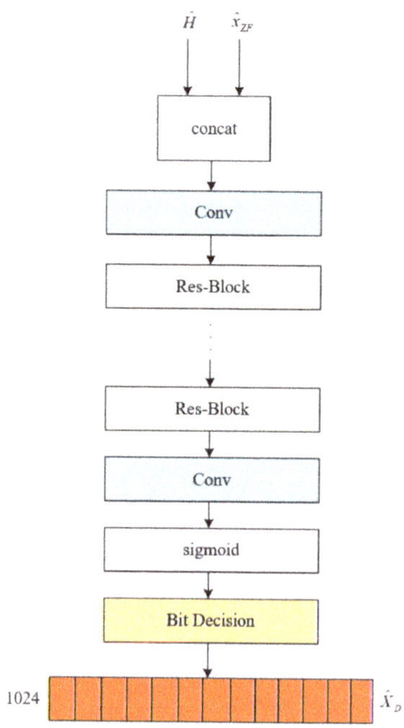

Figure 7. Signal detection model.

5. Experimental Design and Results

5.1. Experimental Design

The experiments in this section were conducted on a laboratory server platform. The server used had two NVIDIA GeForce RTX 2080 Ti GPUs, 16 GB of RAM, and 64 GB of GPU memory. The neural network was implemented using the Keras deep learning framework. The channel dataset used during the training process was collected from a wireless channel model following the Wireless World Research Forum's New Radio [11] initiative (WINNER II). The experimental process included two scenarios: 8 pilots and 32 pilots. In the eight-pilot scenario, there were 8 known complex training sequences for each of the two signals, totaling 16 known complex values. Additionally, 32 random bits (0 s and 1 s) were generated using random numbers. Similarly, for the 32-pilot scenario, 128 random bits were generated. The generated random bits were written to files named "Pilot8" and "Pilot32", respectively, and the pilot sequences remained unchanged throughout the subsequent experiments.

The channel model was the same as described in Section 2. Two antennas at the transmitter generated random data of 512 bits each, which were then transmitted through an Additive White Gaussian Noise (AWGN) simulated channel. The output of the AWGN channel serves as the input to the network. During the training process, the batch size was set to 128. Batches of 128 samples were generated using a Python generator for training, and the number of steps per epoch was set to 1000, meaning that 1000 batches of samples were input per epoch for training. The total number of epochs was set to 200. The Adam optimizer was used to optimize the model parameters. Since the final output is a binary sequence, binary cross-entropy was used to compute the model loss. It can be mathematically represented by the following equation:

$$\text{Loss} = -\frac{1}{N}\sum_{i=1}^{N} y_i \cdot \log(p(y_i)) + (1 - y_i) \cdot \log(1 - p(y_i)). \tag{7}$$

In this equation, y represents the binary label (0 or 1) and $P(y)$ denotes the probability of the output label. The learning rate follows a step-wise decay approach, where it is reduced by 0.75 times every 30 epochs.

For the model-driven receiver network DMCNet-pro, model training is carried out in a phased way, which is divided into four stages: training the denoising module (100 epochs); fixing the denoising module parameters and training the channel estimation module (100 epochs); training the parameters of the fixed-signal denoising module and the cardio estimation module to the signal detection module (100 epochs); and end-to-end training with a small learning rate (0.0001), fine-tuning the model, and training (200 epochs). The other settings for the experiment are the same as for DMCNet.

5.2. Performance and Parameter Count of Fully Connected Networks

In [6], the authors employed a fully connected network (DNN) as the receiver for single-antenna OFDM signals. For the MIMO-OFDM system studied in this paper, the experiments were also conducted using a fully connected network. The number of fully connected layers used in the experiments was 3, with a gradually increasing number of neurons in each layer. The output data were fixed as the 16 bits from the 128th to the 143rd positions. The last layer of the fully connected network utilized a sigmoid activation function to ensure the output fell within the [0, 1] range. Finally, a bit decision function was employed to convert the neural network output into binary (0 or 1) bits:

$$output = \begin{cases} 0 & x <= 0.5 \\ 1 & \text{else} \end{cases}. \tag{8}$$

The bit error ratio (BER) performance of the fully connected network under 8 pilot tones and 32 pilot tones is shown in Figures 8 and 9.

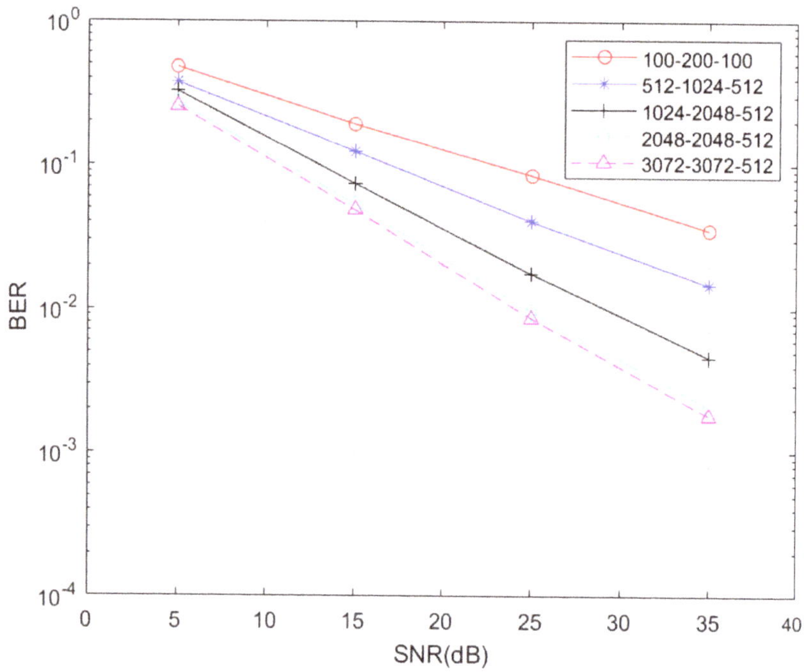

Figure 8. BER of MIMO-OFDM fully connected receiver under eight pilot tones.

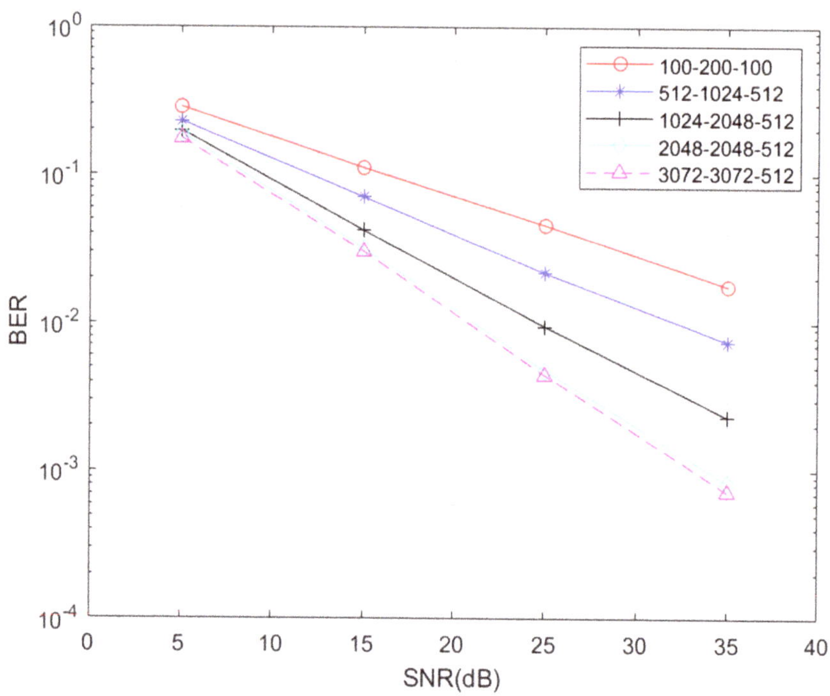

Figure 9. BER of MIMO-OFDM fully connected receiver under 32 pilot tones.

It can be observed that in the MIMO-OFDM scenario, the receiver performance of the fully connected network is significantly worse compared to the single-antenna OFDM scenario. By increasing the number of neurons in the fully connected layers, the BER performance can be gradually improved. However, once the number of neurons in the fully connected layers reaches a certain value, the network's bit error rate reaches a bottleneck, and further increasing the network size does not yield significant benefits.

In [6], the authors employed a scheme where each fully connected network estimates only 16 input bits, resulting in a total of 512 input bits. Therefore, 32 parallel fully connected networks are required for estimation, and their output results are subsequently concatenated. However, for the MIMO-OFDM receiver, the number of input bits is higher, necessitating a larger number of sub-networks. To investigate the impact of the output bit number on the receiver's BER performance, we conducted experiments with output lengths of 8 bits, 16 bits, 32 bits, 64 bits, and 128 bits under eight pilot symbols. The predicted data corresponded to the input starting from the 128th data point, which, respectively, covered the ranges of 128–135 bits, 128–142 bits, 128–159 bits, 128–171 bits, and 128–255 bits. The results are shown in Figure 10:

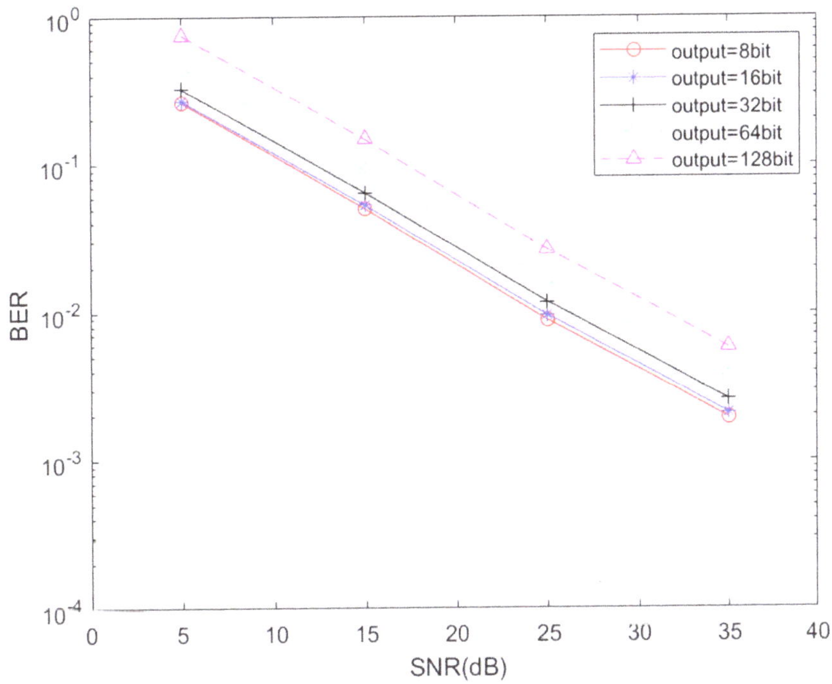

Figure 10. Relationship between prediction length and BER of an MIMO-OFDM fully connected receiver.

Based on the experimental results, it can be concluded that the performance of a single fully connected network deteriorates as the number of predicted bits increases. As the prediction length decreases, the BER gradually decreases, but at a slower rate. When the output length is 16 bits, further reducing the output length has little impact on performance as it reaches saturation. Additionally, a smaller number of predicted bits per network requires a larger number of sub-networks. In this study, a prediction length of 16 bits was used, which means a total of 64 sub-networks are needed to concatenate the final output.

Furthermore, the parameter count is a significant concern in the fully connected solution. In the case of MIMO-OFDM receivers with a large number of input bits, using a fully connected approach requires multiple large sub-networks to combine the outputs The parameter count for each parameter configuration in the graph is already substantial, and employing a fully connected approach with a high parameter count yields a mediocre performance. Therefore, the fully connected signal reception solution suitable for single-antenna scenarios is not applicable to MIMO-OFDM scenarios. The following fully connected experiments employ a 2048-2048-512 network configuration.

5.3. Results and Analysis of the DMCNet

This study compares the results between the 8-pilot and 32-pilot scenarios. As shown in Figure 11, it can be observed that in the eight-pilot scenario, with fewer pilots, the traditional approach struggles to utilize pilot data for channel information recovery and signal detection. Therefore, the BER performance saturates and is the worst when the SNR exceeds 15. On the other hand, in the deep-learning-based approach, both fully connected and convolutional networks show continuous BER reductions as the SNR increases. The conclusion drawn from this is that the neural-network-based approach exhibits stronger robustness regardless of the number of pilots, while the traditional approach heavily relies on pilot data for accurate channel estimation. The superiority of neural networks in this aspect mainly stems from their ability to approximate any function, and the characteristics of wireless channels can be seen as complex functions that can be learned from training data and network models.

In the case of 32 pilots, the traditional LMMSE-MMSE receiver can leverage more pilots to recover channel information, and the performance of the neural network algorithm is comparable to the traditional method.

Comparing the fully connected and convolutional receiver schemes, it can be observed that the convolutional receiver outperforms the fully connected receiver. In the single-antenna OFDM scenario, where the input–output signal length is short, the fully connected network can approximate the characteristics of the channel. However, in the multi-antenna scenario, with longer input signal lengths and where the channel state is represented as a multidimensional matrix, the operation of the input signal through the channel becomes more complex and the limitations of the fully connected network operation are insufficient to fit the channel model. The introduction of convolutional modules and residual modules in DMCNet enriches the network's operation and allows for a better fit to the channel model, resulting in a lower BER compared to the fully connected approach.

Furthermore, DMCNet does not require input data segmentation, and its output length is 1024 bits, allowing for the reception of the entire input signal through a single network. In terms of the parameter count, in the eight-pilot scenario, DMCNet has a parameter count of 4,202,496, while the fully connected network has a parameter count of 7,348,224 $\times$ 64. The parameter count of DMCNet is only 9% of the fully connected network, making it more suitable for practical deployment and usage.

Based on the experimental results, it can be concluded that the proposed DMCNet is suitable for receiving 2 $\times$ 2 MIMO-OFDM signals and outperforms fully connected networks in terms of performance with fewer parameters. It exhibits a stronger robustness in scenarios with fewer pilots, being able to fit results with limited data. Its performance in scenarios with more pilots is comparable to traditional approaches.

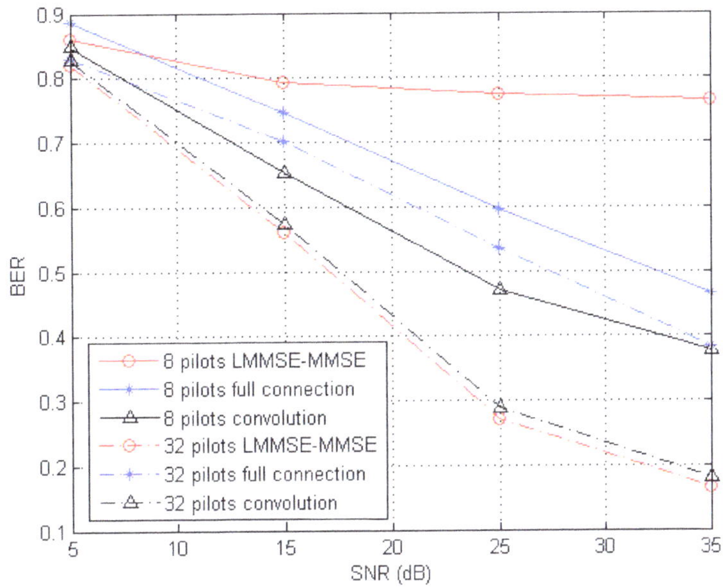

Figure 11. Accruacy comparison of different signal detection algorithms.

5.4. Results and Analysis of DMCNet-Pro

It can be seen that at 32 pilots, the accuracy of the model-driven scheme is slightly improved compared with the data-driven method and the traditional method, but at 8 pilots, the accuracy of the model-driven scheme is very close to that of DMCNet. The main reason for this is that when there is a large number of pilot sequences, the receiver can use traditional methods to perform channel estimation and signal detection comparatively. Therefore, LS initialization and ZF initialization provide guidance to the neural network, and the neural network can gradually adjust parameters in the right direction to reduce errors. However, in eight-pilot mode, due to the small pilot frequency, it is difficult for traditional methods to perform signal estimation and signal detection, and the accuracy of LS initialization and ZF initialization is not high, so the gain brought by changing to a model-driven model is not large.

During the experiment, it is found that in the model-driven receiver, the number of parameters of the model can be greatly reduced and the accuracy is close to that of the data-driven model. This is reflected in the number of output channels of Res-Block three-layer convolution. For this reason, an experiment was designed to explore the influence of the number of output channels of Res-Block on the performance. The BER of each group is shown in Figure 12, where 512-512-256* indicates the BER of the data-driven model DMCNet at 32 pilot frequency. It can be seen from the results that in the model-driven network, when the number of Res-Block output channels reaches 200-200-100, the accuracy is already better than that in the data-driven network with more parameters. When the number of parameters continues to increase, the gain in accuracy is not obvious. Therefore, the set of parameters of 200-200-100 is selected at 32 pilot frequency to set the number of channels in each layer of Res-Block. By comparing the parameter number and FLOPs of the 32-pilot model-driven network DMCNet, it can be seen that the model-driven scheme can achieve a better or similar receiving accuracy with less computation, which indicates that in the data-driven scheme, a large number of parameters and operations is needed to fit the signal transmission process. By making full use of the known information, part of the network parameters can effectively be replaced, and the method used for sub-module training also makes the training of each module's parameters more efficient. The same experiment was also performed at eight pilots. Due to the low accuracy of traditional algorithms for channel estimation and signal detection at eight pilots, the model-driven scheme does not bring significant accuracy or parameter number gains at eight pilots.

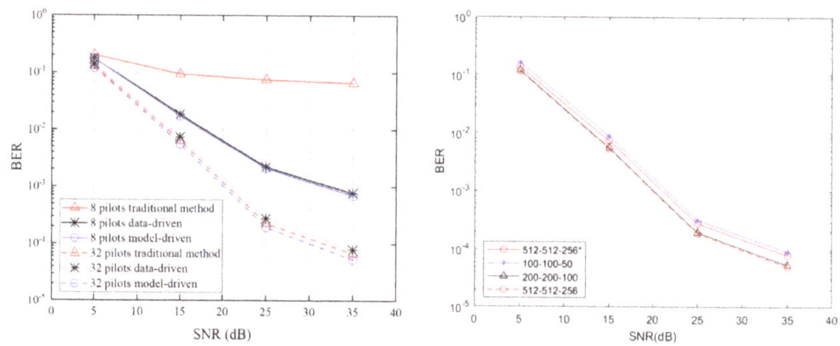

Figure 12. Comparison of data-driven and model-driven results (**left**) and model-driven Res-Block channel numbers (**right**).

6. Conclusions

This paper discusses the problem of signal receivers based on deep learning in an MIMO-OFDM scenario. Compared with a single antenna, the output signal length in the multi-antenna scenario is longer and the receiver design is more complex. The proposed receiver based on deep learning, DMCNet, combines complex convolution and a complex full connection structure, which has a better performance and fewer parameters than previous fully connected networks.

Compared with traditional receiver algorithms, the receiver scheme based on deep learning is more robust, and the accuracy is less affected when nonlinear noise (such as cyclic prefix and peak clipping) is introduced, which also indicates that the deep learning receiver has a certain ability to fit the channel changes. In addition, the influence of some key modules on the accuracy of the neural network is compared. Based on the data-driven model DMCNet, this work expands the model-driven scheme to DMCNET-Pro. Combined with the relevant algorithms in traditional communication, the model-driven scheme has a higher accuracy and fewer parameters in some scenarios.

In summary, the MIMO-OFDM receiver based on deep learning has more advantages than traditional methods in some aspects, and has certain research and application prospects.

Author Contributions: Conceptualization, T.Z.; Methodology, P.L. and Y.X.; Formal analysis, X.Y. and Z.L. (Zili Liu); Investigation, Q.Y.; Writing—original draft, G.Y.; Writing—review & editing, Z.L. (Zejian Lu). All authors have read and agreed to the published version of the manuscript.

Funding: This work is supported by National Key Technologies R&D Program (2022YFB4300400), China Scholarship Council (202106475002), Belt and Road Initiative Innovative Talent Exchange Foreign Expert Project (DL2023107003L), and in partly by National Natural Science Foundation of China (61801035).

Data Availability Statement: Data are contained within the article.

Conflicts of Interest: The authors declare no conflict of interest. The funders had no role in the design of the study; in the collection, analyses, or interpretation of data; in the writing of the manuscript; or in the decision to publish the results.

References

1. Qin, Z.; Ye, H.; Li, G.Y.; Juang, B.H.F. Deep learning in physical layer communications. *IEEE Wirel. Commun.* **2019**, *26*, 93–99. [CrossRef]
2. Cao, B.; Yang, Y.; Ran, P.; He, D.; He, G. ACCsiNet: Asymmetric convolution-based autoencoder framework for massive MIMO CSI feedback. *IEEE Commun. Lett.* **2021**, *25*, 3873–3877. [CrossRef]
3. Liu, H.; Li, G.; Cheng, X.; Li, D. Performance analysis framework of ML MIMO receiver over correlated Rayleigh fading channel. In Proceedings of the 2006 IEEE International Conference on Communications, Guilin, China, 25–28 June 2006; Volume 9, pp. 4137–4142.
4. Wang, C.; Au, E.K.; Murch, R.D.; Mow, W.H.; Cheng, R.S.; Lau, V. On the performance of the MIMO zero-forcing receiver in the presence of channel estimation error. *IEEE Trans. Wirel. Commun.* **2007**, *6*, 805–810. [CrossRef]
5. Eraslan, E.; Daneshrad, B.; Lou, C.Y. Performance indicator for MIMO MMSE receivers in the presence of channel estimation error. *IEEE Wirel. Commun. Lett.* **2013**, *2*, 211–214. [CrossRef]
6. Ye, H.; Li, G.Y.; Juang, B.H. Power of deep learning for channel estimation and signal detection in OFDM systems. *IEEE Wirel. Commun. Lett.* **2017**, *7*, 114–117. [CrossRef]
7. Jiang, P.; Wang, T.; Han, B.; Gao, X.; Zhang, J.; Wen, C.K.; Jin, S.; Li, G.Y. Artificial intelligence-aided OFDM receiver: Design and experimental results. *arXiv* **2018**, arXiv:1812.06638.
8. Zhao, Z.; Vuran, M.C.; Guo, F.; Scott, S.D. Deep-waveform: A learned OFDM receiver based on deep complex-valued convolutional networks. *IEEE J. Sel. Areas Commun.* **2021**, *39*, 2407–2420. [CrossRef]
9. He, H.; Wen, C.K.; Jin, S.; Li, G.Y. A model-driven deep learning network for MIMO detection. In Proceedings of the 2018 IEEE Global Conference on Signal and Information Processing (GlobalSIP), Anaheim, CA, USA, 26–29 November 2018; pp. 584–588.
10. Xin, Y.; Peng, J.; Lu, Z.; Lee, Y.; Yang, Y. Dmcnet: Data-driven multi-Pilot convolution neural network for MIMO-OFDM receiver. In Proceedings of the 2023 IEEE International Conference on Network Intelligence and Digital Content (IC-NIDC), Beijing, China, 3–5 November 2023; pp. 1–5.
11. Winner, I. *WINNER II Channel Models*; IST-4-027756 WINNER II, D. 1. 1. 2 V1. 2; MathWorks: Natick, MA, USA, 2007.
12. Howard, A.G.; Zhu, M.; Chen, B.; Kalenichenko, D.; Wang, W.; Weyand, T.; Andreetto, M.; Adam, H. MobileNets: Efficient Convolutional Neural Networks for Mobile Vision Applications. *arXiv* **2017**, arXiv:1704.04861.
13. Gao, X.; Jin, S.; Wen, C.K.; Li, G.Y. ComNet: Combination of Deep Learning and Expert Knowledge in OFDM Receivers. *IEEE Commun. Lett.* **2018**, *22*, 2627–2630. [CrossRef]

Disclaimer/Publisher's Note: The statements, opinions and data contained in all publications are solely those of the individual author(s) and contributor(s) and not of MDPI and/or the editor(s). MDPI and/or the editor(s) disclaim responsibility for any injury to people or property resulting from any ideas, methods, instructions or products referred to in the content.

electronics

Article

Compact Design Method for Planar Antennas with Defected Ground Structures

Won Jun Lee [1], Won-Sang Yoon [2], Dal Ahn [1] and Sang-Min Han [1,*]

[1] Department of ICT Convergence, Soonchunhyang University, Asan 31538, Chungnam, Republic of Korea; xp3589@naver.com (W.J.L.); dahnkr@sch.ac.kr (D.A.)
[2] Department of Electronic Engineering, Hoseo University, Asan 31499, Chungnam, Republic of Korea; wsyoon@hoseo.edu
* Correspondence: smhan@sch.ac.kr

Abstract: In this paper, a compact antenna design method is proposed for microstrip patch antennas using a double-layered defected ground structure (DGS) configuration. While a conventional single-layered defected ground structure yields a lower resonant frequency and Q-factor, a smaller circuit size can be achieved using an additional substrate with a higher dielectric constant. The size reduction obtained from the additional resonant LC elements is analytically explained using the equivalent circuit model. The characteristics of the additional substrates are investigated for various dielectric constants and thicknesses. From the experimental results, the proposed design method leads to a total size reduction of up to 51.7% and a miniaturized design for planar antennas with ground apertures. The proposed design method can be applied to various antenna designs with any DGS pattern. Furthermore, the size reduction method can maintain the structure of the resonant patch element and its radiation characteristics. Therefore, the proposed method is applicable to the design of microwave devices on microstrip-based configurations.

Keywords: defected ground structures (DGSs); dielectric substrates; microstrip antennas; minimization; multi-layer circuits

Citation: Lee, W.J.; Yoon, W.-S.; Ahn, D.; Han, S.-M. Compact Design Method for Planar Antennas with Defected Ground Structures. *Electronics* **2023**, *12*, 2226. https://doi.org/10.3390/electronics12102226

Academic Editors: Djuradj Budimir, Syed Muzahir Abbas and Yang Yang

Received: 19 March 2023
Revised: 10 May 2023
Accepted: 11 May 2023
Published: 14 May 2023

Copyright: © 2023 by the authors. Licensee MDPI, Basel, Switzerland. This article is an open access article distributed under the terms and conditions of the Creative Commons Attribution (CC BY) license (https://creativecommons.org/licenses/by/4.0/).

1. Introduction

Mobile devices have been rapidly developed toward minimization and integration in order to converge various wireless systems. Whereas active front-end circuits can be integrated on a single die with a one-chip solution, microwave devices are still limited in their compactness due to the wavelength-dependent design of the resonant circuits, such as filters, duplexers, and antennas. Since a microstrip patch antenna is one of the representative planar resonant antennas designed to be integrated into mobile systems, its miniaturization has been researched using various approaches [1,2].

The defected ground structure (DGS) was invented for microstrip resonators [3] and is implemented on a ground plane to replace microstrip line resonators. Its versatile usages include effective wavelength reduction [4–7], current flow control, signal isolation, and so on. In addition, the defected patterns are directly mounted on the backsides of microstrip antenna resonators. Despite the advantages of the DGS transmission line, it is hard to implement due to the floating DGS ground plane. The electromagnetic field is fringed with the DGS aperture on a ground plane, which causes radiation loss. In general, microwave devices with DGSs need to be implemented with some air space on the ground side to avoid interferences from other materials. When a conducting material makes contact with or approaches the DGS aperture, the original characteristics of the DGS transmission line cannot be retained due to field reflections or impedance changes. Therefore, microwave devices using DGS transmission lines have been implemented only as stand-alone modules with limited multiple and composite designs [8]. Since the DGS is a resonator itself, it is difficult to analytically design a resonant antenna, because the double resonators of an antenna and a DGS lead to unexpected resonance effects. Therefore, the applications of DGSs to antenna designs are limited in features such as compactness [1,2], polarization control [9–11], and inter-element isolation improvement [12,13]. Furthermore, the backside aperture of the DGS can contribute to bandwidth enhancement by reducing the Q-factor of the resonant antenna [14].

A microstrip patch antenna is popularly designed for a simple planar structure because it has omni-directional radiation patterns oriented toward a radiating direction. As the patch size is dependent on the resonant frequency, it requires a relatively large area on a microstrip substrate within low frequency ranges. Therefore, various modified design technologies have been researched to minimize the patch area on a planar substrate, such as patches with a shorting via pins [15,16], slots [17–21], folded patches [22,23], and fractal patch patterns [24,25]. These patch designs require specific design processes for each application because one has to modify the radiating element of the patch itself.

In this paper, a compact microstrip patch antenna with a DGS implemented on a double-layered (DL) substrate is proposed. The size reduction design procedure is introduced using an original patch antenna with a conventional single-layered (SL) DGS and a proposed DL-DGS. The double-layered substrate is designed with an SL-DGS attached to a dielectric substrate with a high dielectric constant. The suspended substrate increases the higher effective dielectric constant on the DGS surface, which can reduce the resonant antenna size. As the value of the LC increases using a DGS and an additional substrate, the resonant frequency of the antenna can be reduced because the resonant frequency is inversely proportional to the values of L and C. The proposed DL-DGS patch antenna is also modeled, verified for any additional LC resonator effects, and analyzed for its additional substrate characteristics. The proposed method can reduce the size of a planar patch antenna without any modification of the resonant patch. In addition, the patch antenna's characteristics can be maintained for the additionally attached dielectric substrate. Furthermore, the proposed design technique can be applied to any planar antenna design with ground slots to reduce the size of the radiating elements.

This paper is organized as follows. Section 2 presents the configuration of the proposed DL-DGS patch antenna and the potential to reduce the patch size. An equivalent circuit model and characterization are described for the proposed double-layered structure in Section 3. Section 4 discusses the size reduction and antenna performance based on the experimental results. In addition, a comparison with the other size reduction techniques for microstrip patch antennas with defected ground structures is provided. Finally, the feasibility and applications of the new DL-DGS patch antenna are outlined in Conclusions.

2. Double-Layered Design for Planar Patch Antennas with DGSs

A DL-DGS microstrip patch antenna with an H-shaped DGS was designed as shown in Figure 1a. The antenna pattern in the top layer and the DGS pattern in the middle layer are shown in Figure 1b. The original patch was designed without a DGS on an upper substrate, while the SL-DGS patch was designed with a DGS on an upper substrate, which is the same as the structure of the conventional DGS antenna. The proposed DL-DGS patch was configured with an additional lower substrate beneath the SL-DGS patch. For a convenient comparison of the resonator size, a microstrip square patch was designed with a resonant frequency of 2.4 GHz with l = 41 mm. An impedance transformer was designed with a quarter wavelength of j = 22.8 mm and a 135 Ω width of m = 0.3 mm. The middle layer has a ground plane with an H-shaped DGS pattern with a = 20 mm, b = 14 mm, c = 1 mm, and d = 7 mm. The bottom layer has no metal plane. The two-layered substrate has an upper substrate with ε_{r_1} = 2.2 and t_1 = 31 Mils, and the lower substrate has ε_{r_2} = 10.2 and t_2 = 50 Mils. The DGS pattern determines the additional L and C values of the resonator and the effect of the lower substrate. Since the H-shaped DGS pattern has a dominant inductance, a capacitance variance corresponding to the lower substrate material is expected.

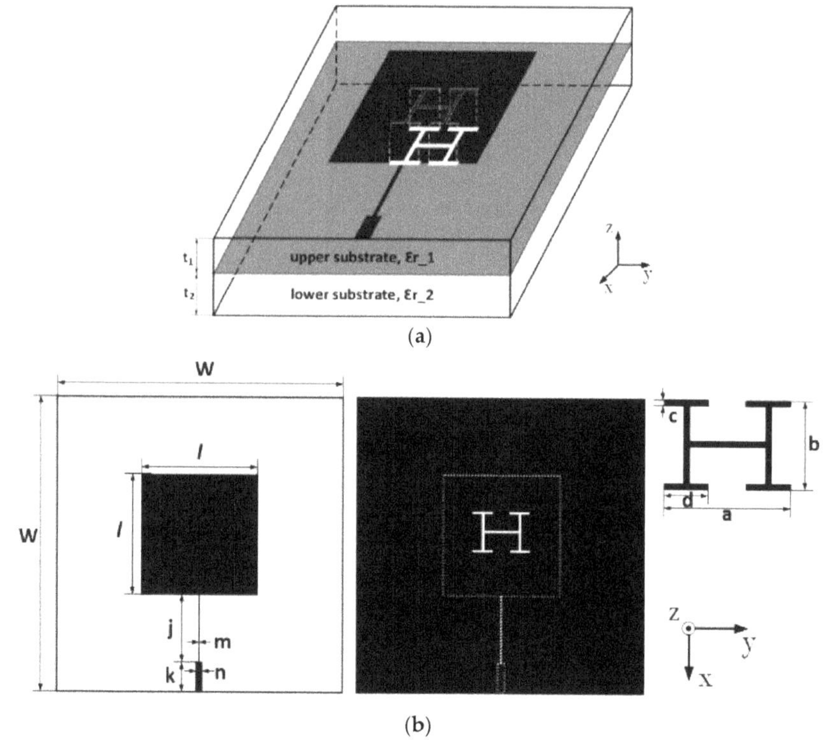

Figure 1. Configuration of the proposed DL-DGS microstrip patch antenna. (**a**) Double-layered configuration. (**b**) A top-layer view and a middle-layer view.

The effects of the DGS and additional substrate on the antenna size reduction were evaluated using an EM simulator from ANSYS HFSS (High-Frequency Structure Simulator). Figure 2 presents the S11 of an original, an SL-DGS, and a DL-DGS microstrip patch antenna. The original patch antenna was simulated without a DGS and a lower substrate, while the SL-DGS and DL-DGS patch antennas were simulated with an H-shaped DGS and an additional lower substrate, respectively. The original patch antenna has a resonant frequency (f_r) of 2.403 GHz, while the SL-DGS patch antenna has an f_r of 1.833 GHz. Through the DGS effect, the size of the patch antenna can be effectively reduced by 42.4%. Due to the small aperture size of the DGS, the two antennas show almost the same bandwidth. The DL-DGS microstrip patch antenna has an f_r of 1.661 GHz. The lower substrate yields an additional size reduction of the patch antenna by 10.9%. Compared to the original patch antenna, the proposed DL-DGS patch antenna is expected to achieve a size reduction of 53.3%.

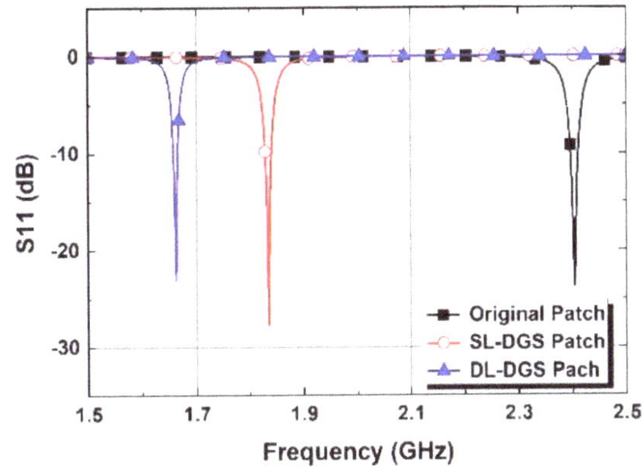

Figure 2. Simulated S11 for an original, an SL-DGS, and a DL-DGS patch antenna.

3. Design and Analysis of the Double-Layered DGS Patch Antenna

The proposed double-layer design of the microstrip patch antenna was analyzed for resonant operations using an equivalent circuit model. Figure 3 shows the equivalent circuit of the DL-DGS patch antenna. Each patch antenna design model is separately shown for the original, SL-DGS, and DL-DGS patch antennas, respectively. The original patch is modeled using an open-circuit half-wave transmission line model [26,27], while the SL-DGS antenna is modeled based on the original patch model, adding a DGS resonator of a parallel-connected L_s and C_s [8,27]. The DL-DGS antenna is modeled with an additional series-connected L_d and C_d [8].

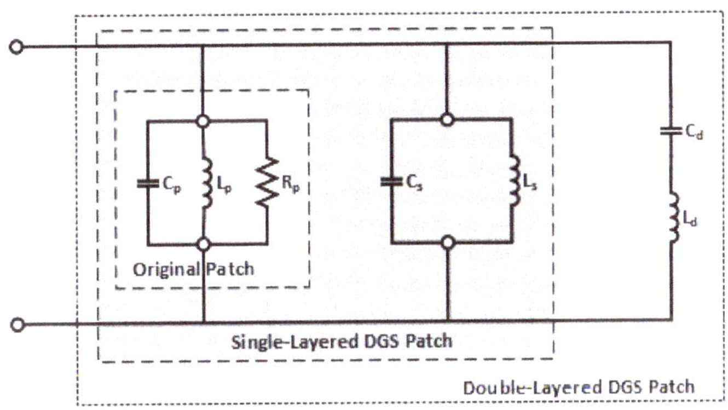

Figure 3. Equivalent circuit model of the proposed DL-DGS microstrip patch antenna.

In light of the equivalent circuit of the DL-DGS transmission line with a dumbbell-shaped DGS [8], the equivalent circuit of the DL-DGS antenna has a reasonable configuration. The equivalent circuit of the DL-DGS transmission line consists of a DGS resonator of C_s and L_s and lower substrate elements of C_d and L_d. The radiation loss parameters of R_s and R_d in the DL-DGS transmission line are ignored for an antenna equivalent circuit because the antenna itself has a radiation function. Whereas the parallel RLC resonator comprises a conventional SL-DGS model of L_s and C_s, the proposed structure has another series-resonant circuit induced by an additional ground current path. The parasitic parameters from the lower substrate of L_d and C_d account for the effect on the resonant frequency shift and Q-factor improvement because the total inductance decreases and the total capacitance increases after attaching a lower substrate. In the case of the equivalent circuit of the DL-DGS antenna, a resonant circuit of C_p, L_p, and R_p is added as an original patch function.

In order to verify the equivalent model and characterize the effect of the additional substrate applied to the DL-DGS patch antenna, various dielectric constants of the lower substrate were applied. Figure 4A presents the resonant frequency variation in the dielectric constant of the lower substrate with a thickness of 50 Mils. To verify the effect of the double-layered structure, an SL-DGS patch antenna was compared in the case of $\varepsilon_{r\_2} = 1$ (air). As the dielectric constant increases from 1 to 10.2, the resonant frequency decreases from 1.833 GHz to 1.661 GHz. The solid lines show the EM simulated results, while the dashed lines present the circuit simulated results based on the equivalent circuit models. The circuit simulations were performed using an ADS (Advanced Design System) from Keysight Technologies, Ltd. The equivalent circuit models are well-matched with every substrate environment together with the EM simulated results. For the effect of the thickness of the lower substrate, resonant frequency variations were applied. Figure 4B shows the resonant frequency variations for various thicknesses with a dielectric constant of the lower substrate of 10.2. As the thickness increases from 0 Mil to 50 Mils, the resonant frequency decreases from 1.883 GHz to 1.661 GHz, which indicates an effective size reduction. A thickness of 0 Mil indicates that there is no lower substrate (SL-DGS) as a reference antenna.

The equivalent circuit elements were extracted by matching the EM simulated results with the specified antenna structure and substrate material environment. The original patch antenna model was designed with L_p = 0.03 nH, C_p = 146.1 pF, and R_p = 57.0 Ω, as shown in Figure 3. Then, the resonant elements of the SL-DGS were designed with L_s = 1.46 nH and C_s = 110.0 pF.

For the various characteristics of lower substrates, the additional elements of the DL-DGS of L_d and C_d are presented in Table 1. The inductances yielded by the lower substrates are found to have almost the same value as L_d = 0.03 nH, whereas the capacitances increase as the dielectric constant increases. The H-shaped DGS has a relatively large inductance and small aperture size, and the material characteristics affect the additional capacitance increments. In addition, the capacitances increase as the lower substrate thickness increases.

Table 1. Characteristics of the DL-DGS patch antennas for various dielectric constants and thicknesses of the lower substrate.

ε_{r_2}	L_d (nH)	C_d (pF)	f_r (GHz)	BW (MHz)	Q-Factor
1.0 (air)	0.00	0.00	1.833	10.9	168.2
2.0	0.03	8.24	1.803	9.7	185.8
4.0	0.03	17.7	1.770	9.3	190.3
6.0	0.03	27.5	1.734	8.7	199.3
8.0	0.03	37.2	1.699	8.2	207.1
10.2	0.03	47.6	1.661	8.0	207.7
Thickness (Mils)	**L_d (nH)**	**C_d (pF)**	**f_r (GHz)**	**BW (MHz)**	**Q-factor**
0 (air)	0.00	0.0	1.833	10.9	168.2
10	0.03	19.4	1.765	9.8	180.1
20	0.03	32.0	1.718	9.0	190.9
30	0.03	38.5	1.695	8.6	197.1
40	0.03	43.9	1.674	8.2	204.2
50	0.03	47.6	1.661	8.0	207.7

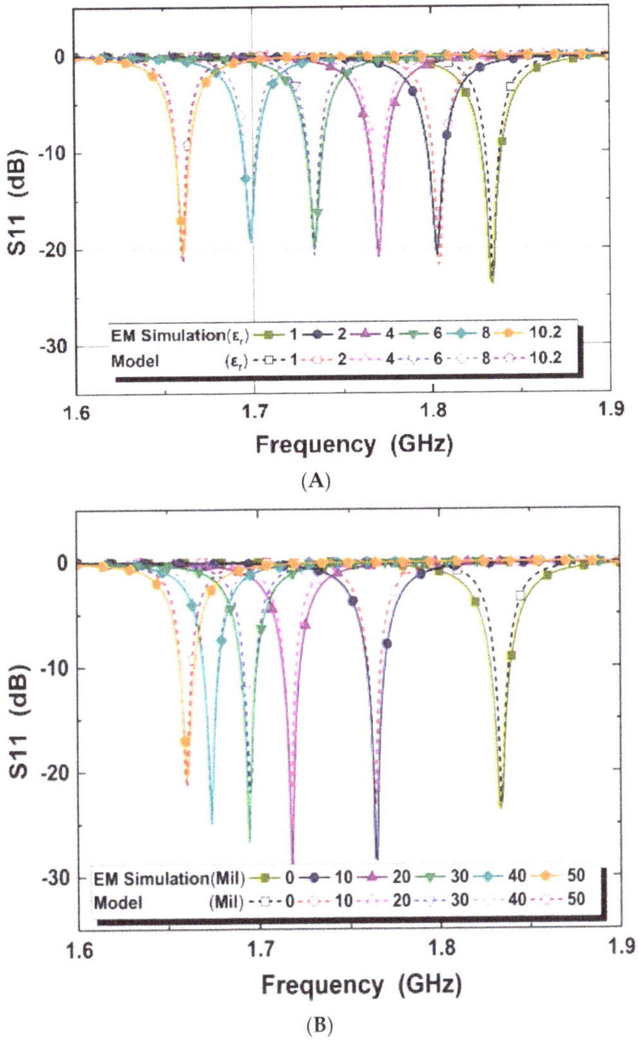

Figure 4. Resonant frequencies of the DL-DGS antenna with a lower substrate. (**A**) Dielectric constant variations for t_2 = 50 Mils. (**B**) Thickness variations for ε_{r_2} = 10.2.

Table 1 summarizes the characterization of the DL-DGS patch antenna. As the dielectric constant of the attached lower substrate increases, the total amount of LC products increase; therefore, the resonant frequency is decreased. From the equivalent circuit analysis, as shown in Figure 3, the proposed structure presents as a parallel resonator model. As the general resonator model of the patch antenna of C_p, L_p, and R_p expands with another resonant circuit of C_d and L_d due to the H-shaped DGS resonator, the total capacitance increases with $C_p + C_d$. In addition, the total inductance decreases with $L_p // L_d$. When the additional lower substrate is attached, the additional series-resonant elements of C_d and L_d contribute to the additional capacitance increment and inductance decrement. Because the equivalent circuit model of the proposed DL-DGS patch antenna has a parallel resonator model, the Q-factor can be expressed as follows:

$$Q = \frac{R}{\omega_r L} = \omega_r RC \quad (1)$$

where $\omega_r = 2\pi f_r$ and f_r indicate a resonant frequency. Therefore, the total capacitance increases and the total inductance decreases, as designed from an original patch to create the SL-DGS and DL-DGS patches, which means that the Q-factor can be increased. According to Table 1, the DL-DGS patch has a higher Q-factor than the SL-DGS patch antenna for $\varepsilon_{r\_2} = 1$. Additionally, as the dielectric constant increases in the lower substrate, the Q-factor increases, because a substrate with a higher dielectric constant generates a higher additional capacitance.

For the thickness variations, as for the dielectric constant variations, the same mechanism was investigated. The SL-DGS case shows a zero thickness of the lower substrate. The thicker dielectric substrate results in a higher parallel capacitance, which presents a lower resonant frequency and higher Q-factor, as shown in Table 1.

From the total capacitance increase, the resonant frequency of the DL-DGS patch is decreased while maintaining the physical patch size, which means that this technique can be used to design a smaller patch with the same resonant frequency. Therefore, the proposed method of adhering a dielectric slab with a higher dielectric constant and a greater thickness is verified.

4. Experiments and Evaluations of the Proposed Compact DL-DGS Patch Antenna

The proposed DL-DGS microstrip patch antenna was implemented on the upper substrate of an RT Duroid 5880 with $\varepsilon_{r\_1} = 2.2$ and $t_1 = 31$ Mils and on the lower substrate of an RT Duroid 6010 with $\varepsilon_{r\_2} = 10.2$ and $t_2 = 50$ Mils. Figure 5 shows photographs of the implemented DL-DGS antenna. While Figure 5a shows the top layer of the original patch on the upper substrate, the backside ground plane with an H-shaped DGS pattern on the upper substrate is presented in Figure 5b. Figure 5c shows the photo of the lower substrate without any metal on either side that is adhered to the bottom surface of the upper substrate, as shown in Figure 5b. The implemented substrate size is W × W = 95 mm × 95 mm.

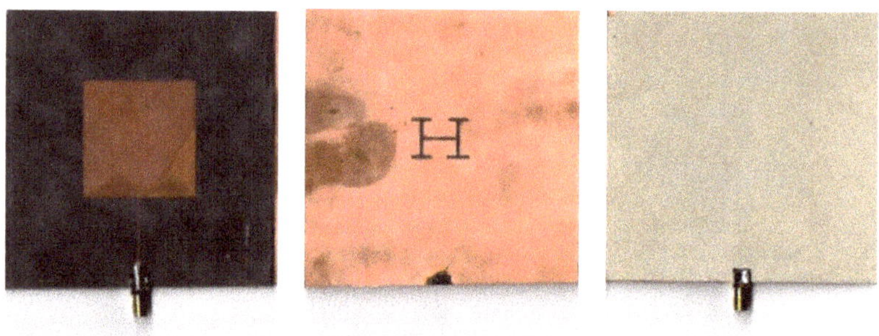

(a) (b) (c)

Figure 5. Photographs of the DL-DGS microstrip patch antenna. (**a**) Top view. (**b**) Middle-layer view. (**c**) Bottom view.

The resonant frequencies and Q-factors were measured as shown in Figure 6. The original, SL-DGS, and DL-DGS patch antennas present measured resonant frequencies of f_r = 2.399 GHz, 1.801 GHz, and 1.69 GHz, respectively, while the simulated results show values of 2.404 GHz, 1.834 GHz, and 1.662 GHz, respectively. The Q-factor is decreased from 243.5 for the original patch to 133.4 for the SL-DGS patch by the DGS aperture and then slightly increased to 149.5 for the DL-DGS patch after attaching a high-dielectric slab to the aperture.

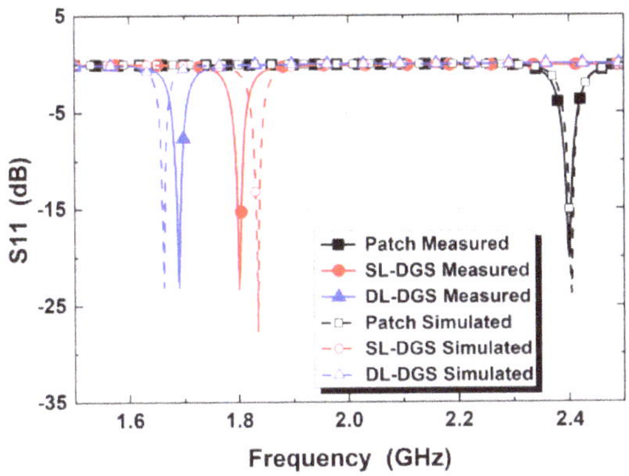

Figure 6. Measured and simulated S11 of an original patch, a conventional SL-DGS patch, and the proposed DL-DGS patch.

To compare the radiation characteristics, the antenna radiation patterns were measured in an anechoic chamber. In general, DGS antennas show a relatively poor linear polarization performance due to current direction changes induced by ground defection. In addition, the amount of backside radiation through the DGS aperture needs to be checked before application to normal patch antennas, because the electromagnetic energy may radiate through the DGS aperture. Figure 7a,b presents the radiation patterns of the SL-DGS and DL-DGS patch antennas, respectively. Each pattern was measured for an E-plane (x-z plane) and an H-plane (y-z plane) and compared to a cross-polarization pattern. While the SL-DGS patch antenna presents a gain of 3.9 dBi, an HPBW of 68.0°, and a cross-pol. Level of −27.6 dB, the DL-DGS patch antenna shows a gain of 2.4 dBi, an HPBW of 92.0°, and a cross-pol. Level of −24.1 dB. Regarding the effect of the DGS aperture radiation, the proposed DL-DGS patch antenna maintains good linear polarization characteristics for the defected ground. Because the high-dielectric substrate absorbs the electromagnetic field toward the backside substrate of the antenna, the front-to-back ratio (F/B) of 7.3 dB for the SL-DGS antenna decreases to 4.5 dB for the DL-DGS antenna. As the dielectric material ($\varepsilon_r > 1$) attracts the electromagnetic field, the additional high-dielectric substrate absorbs the backward field, which reduces the F/B compared to the SL-DGS antenna with an air backside boundary ($\varepsilon_r = 1$).

Based on the proposed compact design method, the circuit size reductions were compared. At first, the original patch antenna was designed with 2.4 GHz using a basic design theory stipulating that the square patch is designed with $l \times l = 0.49\lambda/\sqrt{\varepsilon_r} \times 0.49\lambda/\sqrt{\varepsilon_r}$ at the resonant frequency, with the expected patch size of 1681 mm^2. The design results show a resonant frequency of 2.399 GHz with a physical size of 1681 mm^2. With the same physical patch size, the SL-DGS patch was designed. The SL-DGS patch operates at 1.801 GHz, which should require the expected original patch size of 3025 mm^2. Therefore, the patch antenna operating at 1.801 GHz is designed with a size of 1681 mm^2 instead of 3025 mm^2, with a size ratio of 55.6%. The DL-DGS patch antenna operating at 1.690 GHz can be designed with a physical size of 1681 mm^2, instead of the expected original patch size of 3481 GHz, with a size ratio of 48.3%. Therefore, the antenna sizes can be compared as shown in Table 2. The conventional SL-DGS patch antenna can reduce the antenna size by 44.4%, while the DL-DGS patch can reduce the antenna size by 51.7%.

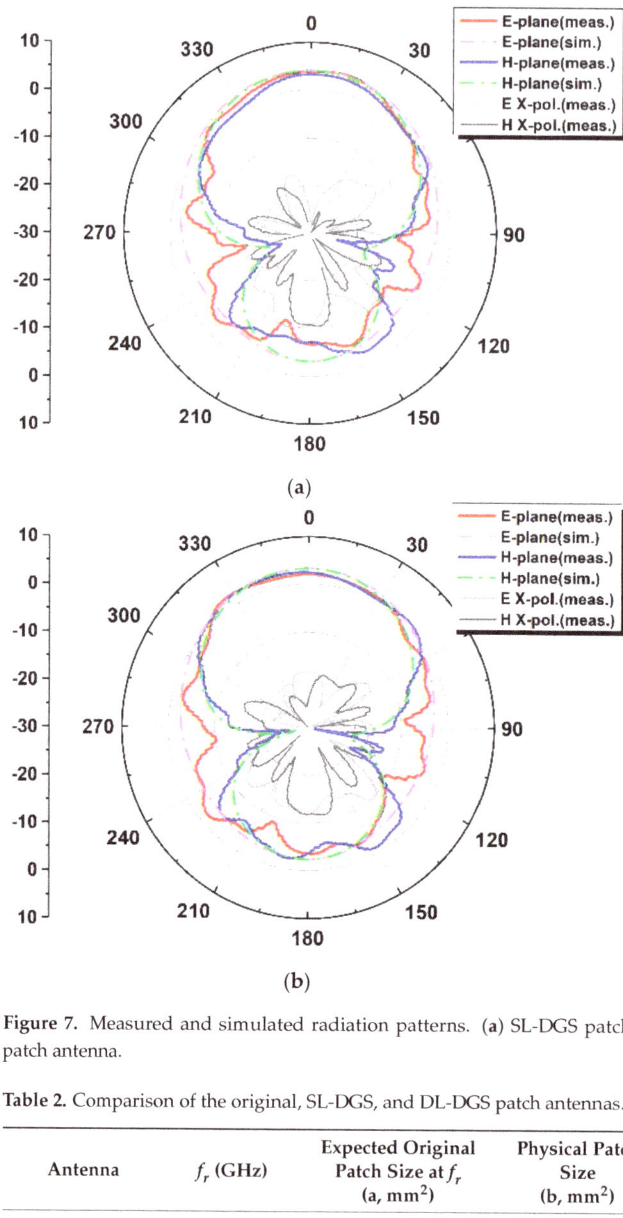

Figure 7. Measured and simulated radiation patterns. (**a**) SL-DGS patch antenna. (**b**) DL-DGS patch antenna.

Table 2. Comparison of the original, SL-DGS, and DL-DGS patch antennas.

Antenna	f_r (GHz)	Expected Original Patch Size at f_r (a, mm^2)	Physical Patch Size (b, mm^2)	Comparison (b/a × 100)
Original Patch	2.399	1681	1681	100
SL-DGS Patch	1.801	3025	1681	55.6
DL-DGS Patch	1.690	3481	1681	48.3

From the performance evaluation of the proposed size reduction methods for a planar microstrip patch antenna, it was verified that the proposed design can provide a new, additional method with which to minimize the patch with any ground plane slot. The proposed DL-DGS patch antenna maintains the original resonant element and radiation characteristics, as compared to the previous studies on the modification of patches [15,16,22–25]. In particular, as compared to the patch size reductions utilizing ground plane slots among the previous works [17–21], the patch size can be further minimized using the proposed method.

Even though it is difficult to compare size reductions, the previous works using ground plane slots were compared for the size reduction relative to a reference original patch antenna, as shown in Table 3. It is assumed that all the patch antennas have a patch area of $\lambda \times \lambda$ mm^2 that is the inverse of the square of resonant frequency, because each paper uses a different comparison. Because these design methods were compared for each design technique, the method in this work was compared for the SL-DGS patch, and the additional reduction method for the DL-DGS patch was appended. The paper [17] presents a coaxial-fed patch with a linear slot cross on a ground backside at 2.87 GHz. The position of the linear slot is a variable of the size reduction. The maximum rate of the size reduction is approximately 77.3%. The authors of [18] designed a planar microstrip-fed patch with meandering line slots operating at 5.989 GHz. The patch with an open-end slot type has an 83% maximum size reduction. The authors of [19] explored similar slot types with coaxial feeding at 2.387 GHz that has a 55.8% reduction rate, while the authors of [20] explored a cross-type array of four slots for a coaxial-fed patch operating at 2.84 GHz with a 73.2% reduction. The authors of [21] investigated a CSRR (Complementary Split-Ring Resonator)-type slot array for a microstrip patch operating at 5.0 GHz with a 42.2% reduction rate. The proposed structures of the SL-DGS and DL-DGS patch antennas were compared with those in the previous works for the reduction techniques, feeding types of the microstrip patches, resonant frequencies, and the size reduction rates. The SL-DGS patch antenna (the conventional DGS patch) is comparable to the previous works. However, the DL-DGS technique can provide additional size reductions compared to every comparable patch antenna design with defected slots on a ground plane.

Table 3. Comparison of the recent patch antenna reduction technologies with ground slots.

		Reduction Method	Feeding Type	f_r (GHz) of Ref. Antenna	Size Reduction (%)
	[17]	line slot position	coaxial	2.87	77.3
	[18]	meandering line type	planar	5.98	83.0
	[19]	meandering line offset	coaxial	2.387	55.8
	[20]	slot size	coaxial	2.84	73.2
	[21]	CSRR slot	planar	5.0	42.2
This work	SL-DGS	additional substrate	planar	2.4	44.4
	DL-DGS				51.7

5. Conclusions

A size reduction design method for microstrip patch antennas with defected ground structures was proposed using a double layered substrate. While the size of a normal patch antenna can be reduced by designing DGS patterns on a ground plane, an additional size reduction can be achieved by attaching a lower substrate. The operating mechanism was described based on an equivalent circuit model with each LC resonant circuit for the characteristics of the additional dielectric substrate. According to the experimental results, the proposed DL-DGS patch antenna can reduce the circuit size by up to 51.7%, while the conventional SL-DGS patch antenna reduces the size by 44.4%. The proposed size reduction method can maintain the resonator structure and resonance characteristics of the original microstrip patch antenna. Therefore, as compared to the original patch antenna, almost the same radiation performance can be achieved with the small resonant radiator. Furthermore, for all the resonant antenna configurations with various DGS patterns and sizes, the proposed method can be applied to reduce the size of the planar antenna resonators. The proposed new double-layered DGS antenna design technology can be used for various efficient designs of compact and novel microwave devices.

Author Contributions: Conceptualization, S.-M.H.; methodology, W.-S.Y. and D.A.; software, W.J.L validation, W.J.L. and S.-M.H.; formal analysis, D.A.; investigation, W.-S.Y.; resources, W.J.L.; data curation, W.J.L.; writing—original draft preparation, W.J.L.; writing—review and editing, S.-M.H. visualization, W.J.L.; supervision, S.-M.H.; project administration, S.-M.H.; funding acquisition S.-M.H. and D.A. All authors have read and agreed to the published version of the manuscript.

Funding: This work was supported by "Regional Innovation Strategy (RIS)" through the National Research Foundation of Korea funded by the Ministry of Education (2021RIS-004), the MSIT, Korea under the ICAN program (IITP-2023-2020-0-01832) supervised by the IITP, and the Soonchunhyang University research fund.

Data Availability Statement: The dataset is not applicable.

Conflicts of Interest: The authors declare no conflict of interest.

References

1. Salih, A.; Sahrawi, M.S. A dual-band highly miniaturized patch antenna. *IEEE Antennas Wirel. Propag. Lett.* **2016**, *15*, 1783–1786. [CrossRef]
2. Oulhaj, O.; Touhami, N.A.; Aghoutane, M.; Tazon, A. A miniature microstrip patch antenna array with defected ground structure. *Int. J. Microw. Opt. Technol.* **2016**, *11*, 32–39.
3. Kim, C.-S.; Park, J.-S.; Ahn, D.; Lim, J.-B. A novel 1-D periodic defected ground structure for planar circuits. *IEEE Microw. Guid. Wave Lett.* **2000**, *10*, 131–133.
4. Ting, S.-W.; Tam, K.-W.; Martins, R.P. Miniaturized microstrip lowpass filter with wide stopband using double equilateral U-shaped defected ground structure. *IEEE Microw. Wirel. Comp. Lett.* **2006**, *16*, 240–242. [CrossRef]
5. Choi, H.; Lim, J.; Jeong, Y. A new design of Doherty amplifiers using defected ground structure. *IEEE Microw. Wirel. Comp. Lett.* **2006**, *16*, 687–689. [CrossRef]
6. Jahan, N.; Rahim, S.A.E.A.; Mosalam, H.; Barakat, A.; Kaho, T.; Pokharel, R.K. 22-GHz-band oscillator using integrated H-shape defected ground structure resonator in 0.18-μm CMOS technology. *IEEE Microw. Wirel. Comp. Lett.* **2018**, *28*, 233–235. [CrossRef]
7. Woo, D.-J.; Lee, T.-K. Suppression of harmonics in Wilkinson power divider using dual-band rejection by asymmetric DGS. *IEEE Trans. Microw. Theory Tech.* **2005**, *53*, 2139–2144.
8. Han, S.-M.; Kim, J.O.; Yoon, W.-S.; Lim, J.; Ahn, D. Design and characterization of the double-layered defected ground structure transmission line with less radiation effect. *Microw. Opt. Technol. Lett.* **2019**, *61*, 903–906. [CrossRef]
9. Yoon, W.-S.; Baik, J.-W.; Lee, H.-S.; Pyo, S.; Han, S.-M.; Kim, Y.-S. A reconfigurable circularly polarized microstrip antenna with a slotted ground plane. *IEEE Antennas Wirel. Propag. Lett.* **2010**, *9*, 1161–1164. [CrossRef]
10. Thakur, J.P.; Park, J. An advance design approach for circular polarization of the microstrip antenna with unbalance DGS feedlines. *IEEE Antennas Wirel. Propag. Lett.* **2006**, *5*, 101–103. [CrossRef]
11. Lee, S.-J.; Lee, H.; Yoon, W.-S.; Park, S.-J.; Lim, J.; Ahn, D.; Han, S.-M. Circular polarized antenna with controlled current distribution by defected ground structures. In Proceedings of the 2010 International Symposium on Antennas and Propagation (ISAP 2010), Macau, China, 23–26 November 2010.
12. Luo, C.; Hong, J.; Zhong, L. Isolation enhancement of a very compact UWB-MIMO slot antenna with two defected ground structures. *IEEE Antennas Wirel. Propag. Lett.* **2015**, *14*, 1766–1769. [CrossRef]
13. Sharawi, M.S.; Numan, A.B.; Khan, M.U.; Aloi, D.N. A dual-element dual-band MIMO antenna system with enhanced isolation for mobile terminals. *IEEE Antennas Wirel. Propag. Lett.* **2012**, *11*, 1006–1009. [CrossRef]
14. Zeng, J.; Luk, K. A simple wideband magnetoelectric dipole antenna with a defected ground structure. *IEEE Antennas Wirel. Propag. Lett.* **2018**, *17*, 1497–1500. [CrossRef]
15. Chen, I.-F.; Peng, C.-M. A novel reduced-size edge-shorted patch antenna for UHF band applications. *IEEE Antennas Wirel. Propag. Lett.* **2009**, *8*, 475–477. [CrossRef]
16. Wu, D.-L.; Chen, J.H.; Yang, K.Y.; Zhu, W.J.; Ye, L.H. A compact dual-polarized patch antenna with L-shaped short pins. *IEEE Antennas Wirel. Propag. Lett.* **2023**, *22*, 689–693. [CrossRef]
17. Sarkar, S.; Majumdar, A.D.; Mondal, S.; Biswas, S.; Sarkar, D.; Sarkar, P.P. Miniaturization of rectangular microstrip patch antenna using optimized single-slotted ground plane. *Microw. Opt. Technol. Lett.* **2011**, *53*, 111–115. [CrossRef]
18. Rabhakar, H.V.; Kummuri, U.K.; Yadahalli, R.M.; Munnappa, V. Effect of various meandering slots in rectangular microstrip antenna ground plane for compact broadband operation. *Electron. Lett.* **2007**, *43*, 16–17.
19. Kuo, J.-S.; Wong, K.-I. A compact microstrip antenna with meandering slots in the ground plane. *Microw. Opt. Technol. Lett.* **2001**, *29*, 95–97. [CrossRef]
20. Lin, S.-Y.; Huang, K.-C. A compact microstrip antenna for GPS and SCS application. *IEEE Trans. Antennas Propag.* **2005**, *53*, 1227–1229.
21. Jang, H.A.; Kim, D.O.; Kim, C.Y. Size reduction of patch antenna array using CSRRs loaded ground plane. In Proceedings of the Progress in Electromagnetics Research Symposium, Kuala Lumpur, Malaysia, 27–30 March 2012; pp. 1487–1489.

22. Li, Y.; Podilchak, S.K.; Anagnostou, D.E.; Constantinides, C.; Walkinshaw, T. Compact antenna for picosatellites using a meandered folded-shorted patch array. *IEEE Antennas Wirel. Propag. Lett.* **2020**, *19*, 477–481. [CrossRef]
23. Lau, K.L.; Kong, K.C.; Luk, K.M. A miniature folded shorted patch antenna for dual-band operation. *IEEE Trans. Antennas Propag.* **2007**, *55*, 2391–2398. [CrossRef]
24. Chen, W.-L.; Wang, G.-M.; Zhang, C.-X. Small-size microstrip patch antennas combining Koch and Sierpinski fractal-shapes. *IEEE Antennas Wirel. Propag. Lett.* **2008**, *7*, 738–741. [CrossRef]
25. Amini, A.; Oraizi, H.; Chaychizadeh, M.A. Miniaturized UWB log-periodic square fractal antenna. *IEEE Antennas Wirel. Propag. Lett.* **2015**, *14*, 1322–1325. [CrossRef]
26. Melkeri, V.S.; Mallikarjun, S.L.; Hunagund, P.V. H-shape defected ground structure embedded square patch antenna. *Int. J. Adv. Res. Eng. Technol.* **2015**, *6*, 73–79.
27. Woo, D.-J.; Lee, T.-K.; Lee, J.W. Equivalent circuit model for a simple slot-shaped DGS microstrip line. *IEEE Microw. Wirel. Comp. Lett.* **2013**, *23*, 447–449. [CrossRef]

Disclaimer/Publisher's Note: The statements, opinions and data contained in all publications are solely those of the individual author(s) and contributor(s) and not of MDPI and/or the editor(s). MDPI and/or the editor(s) disclaim responsibility for any injury to people or property resulting from any ideas, methods, instructions or products referred to in the content.

electronics

MDPI

Article

Design of a Ka-Band Heterogeneous Integrated T/R Module of Phased Array Antenna

Qinghua Zeng [1], Zhengtian Chen [1], Mengyun He [1], Song Wang [1,2], Xiao Liu [1,*] and Haitao Xu [3]

1 Aerospace Information Research Institute, Chinese Academy of Sciences, Beijing 100101, China; zengqh@aircas.ac.cn (Q.Z.); chenzt@aircas.ac.cn (Z.C.); hemy@aircas.ac.cn (M.H.); wangsong@aircas.ac.cn (S.W.)
2 School of Electronic Science and Engineering, University of Electronic Science and Technology of China, Chengdu 611731, China
3 Shunde Innovation School, University of Science and Technology Beijing, Foshan 528399, China; alex_xuht@hotmail.com
* Correspondence: liuxiao@aircas.ac.cn

Abstract: The central element of a phased array antenna that performs beam electrical scanning, as well as signal transmission and reception, is the transceiver (T/R) module. Higher standards have been set for the integration, volume, power consumption, stability, and environmental adaptability of T/R modules due to the increased operating frequency of phased array antennas, the variability of application platforms, and the diversified development of system functions. Device-based multichannel T/R modules are the key to realizing low-profile Ka-band phased array antenna microsystem architecture. The design and implementation of a low-profile, high-performance, and highly integrated Ka-band phased array antenna T/R module are examined in this paper. Additionally, a dependable Ka-band four-channel T/R module based on Si/GaAs/Low Temperature Co-fired Ceramic (LTCC), applying multi-material heterogeneous integration architecture, is proposed and fabricated. The chip architecture, transceiver link, LTCC substrates, interconnect interface, and packaging are all taken into consideration when designing the T/R module. When compared to a standard phased array antenna, the module's profile shrunk from 40 mm to 8 mm, and its overall dimensions are only 10.8 mm × 10 mm × 3 mm. It weighs 1 g, and with the same specs, the single channel volume was reduced by 95%. The T/R module has an output power of ≥26 dBm for single-channel transmission, an efficiency of ≥25%, and a noise factor of ≤4.4 dB. When compared to T/R modules based on System-on-Chip (SOC) devices, the RF performance has significantly improved, as seen by an increase in single channel output power and a decrease in the receiving noise factor. This work lays a foundation for the devitalization and engineering application of T/R modules in highly reliable application scenarios.

Keywords: Ka-band; T/R module; phased array antenna; heterogeneous integration architecture

Citation: Zeng, Q.; Chen, Z.; He, M.; Wang, S.; Liu, X.; Xu, H. Design of a Ka-Band Heterogeneous Integrated T/R Module of Phased Array Antenna. *Electronics* **2024**, *13*, 204. https://doi.org/10.3390/electronics13010204

Academic Editors: Syed Muzahir Abbas, Yang Yang and Niamat Hussain

Received: 18 November 2023
Revised: 26 December 2023
Accepted: 29 December 2023
Published: 2 January 2024

Copyright: © 2024 by the authors. Licensee MDPI, Basel, Switzerland. This article is an open access article distributed under the terms and conditions of the Creative Commons Attribution (CC BY) license (https://creativecommons.org/licenses/by/4.0/).

1. Introduction

The phased array antenna is widely used in the field of communication [1–6]. Its flexible and agile beam scanning technology accomplishes directional communication between fast-moving and highly dynamic objects, while guaranteeing the tracking accuracy of high-speed mobile platform communication. Simultaneously, low sidelobe and bare nulls in directional diagrams can be achieved using array synthesis, enhancing the system's ability to withstand interference and producing more dependable and stable communication lines. Because of their outstanding performance and rapid development, phased array antennas are gaining importance and are extensively applied in the field of advanced military [7–10] and commercial applications [11–14].

The central element of a phased array antenna that performs beam electrical scanning, as well as signal transmission and reception, is the T/R module. Higher standards have been set for the integration, volume, power consumption, stability, and environmental adaptability of T/R modules due to the increased operating frequency of phased array antennas, the variability of application platforms, and the diversified development of system functions [15–18].

It is essential to improve the T/R components integration to meet the development trend of phased array antennas. Due to their complicated manufacturing and assembly processes, lengthy process flows, and multiple levels of interconnection ranging from chip circuits to system integration, traditional phased array antennas have trouble meeting the cost and profile requirements of modern information systems. In contrast, millimeter-wave phased array antenna microsystems with low profile and markedly improved functional density can effectively satisfy the requirements of modern information systems for the high integration and performance of phased array antennas, making them one of the research hotspots in the industry. Among them, the devitalization of the multichannel T/R modules is the key to achieving a low-profile phased array antenna microsystem architecture.

Radar's capacity to acquire a large instantaneous bandwidth and, consequently, improve imaging and detection accuracy can be facilitated by raising the operating frequency. The long wavelength and large bandwidth of millimeter-waves provide massive communication bandwidth attainment, enhance real-time transmission and anti-interference capabilities, and speed up communication system response [19–22]. In order to attain a broad scanning angle in a two-dimensional electric scanning array, it is imperative to reduce antenna size and enhance device integration, as the separation between array elements must be around half wavelength [23]. As demonstrated in Figure 1, millimeter-wave frequency antennas with the same aperture can attain greater SNR and EIRP values, allowing the terminal antenna to be smaller, while maintaining improved performance and energy accumulation [23].

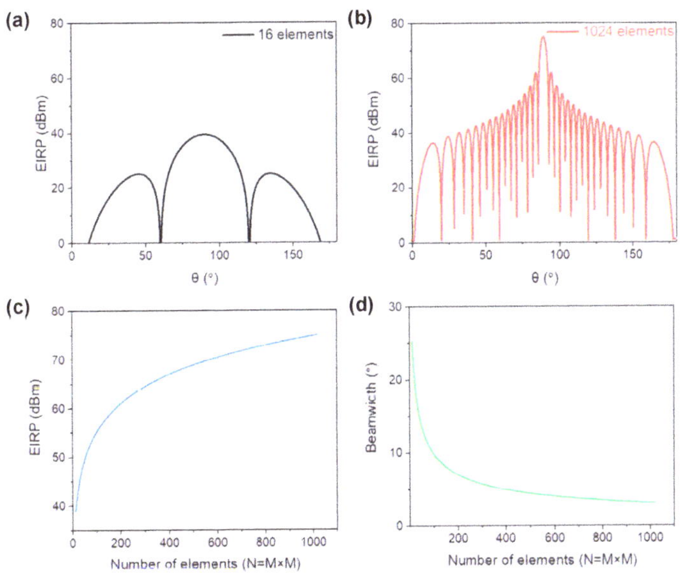

Figure 1. EIRP of phased array antenna with 16 (**a**) and 1024 (**b**) phased-array elements. (**c**) The EIRP grows as N × N, while (**d**) the beamwidth reduces as N versus the number of phased-array elements in a square array. Images reproduced from the literature [23].

Furthermore, the system must integrate more electronic devices with varying functions due to the devitalization and development trend of phased array antenna application scenarios. Higher requirements have been put forward to improve the integration and devitalization of T/R modules, and a low profile has become a very important factor to be considered in the design process of phased array antennas in order to avoid antenna equipment taking up too much volume on the flight platform, which affects the overall structural continuity and aerodynamic performance.

2. Related Works

Recently, both domestically and abroad, Ka-band phased array front-end microsystems were created. Table 1 lists the noteworthy accomplishments.

Table 1. Research achievements of Ka-band T/R modules domestically and internationally.

Time	Institution	Module	Technique	Size	Output Power (dBm)	Noise Factor (dB)	Other Parameter	Reference
2014	University of Electronic Science and Technology of China	Ka-band 6-channel switching delay line module	LTCC	84 mm × 47 mm × 15 mm	\	\	\	[24]
2015	The 13th Research Institute, CETC	Ka-band 16-channel transmission module	PCB multilayer wiring technique and multi-chip assembly technique	60 mm × 80 mm × 4.8 mm	>25	\	Linear gain > 25 dB	[25]
2015	Institute of Telecommunication Satellite, CAST	Ka-band T/R module	LTCC	36 mm × 20 mm × 1.1 mm	>24.6	<4.2	Receiving gain ~33 dB, output gain > 25 dB	[26]
2016	Beijing Institute of Technology	Ka-band T/R array	LTCC	Single channel cross-section 6 mm × 25 mm	>10.5	<4.5	\	[27]
2016	Xi'an Research Institute of Navigation Technology	Ku-band 3D miniaturized T/R module	LTCC, BGA	9.5 mm × 9.5 mm × 3.8 mm	>24.5	<3.5	Receiving gain > 25 dB	[28]
2016	Ching-Yun Chu	Ka-band 4-channel T/R chip	65 nm CMOS technique	4 mm × 2.5 mm	~18.5	~4.4	\	[29]
2017	Nanjing Research Institute of Electronics Technology	Ka-band 8-channel T/R module	LTCC	43 mm × 39.5 mm × 3.5 mm	\	\	Return loss > 15 dB	[30]
2020	Southwest China Institute of Electronic Technology	Ka-band circularly polarized phased array antenna	LTCC	46 mm × 44 mm × 2.8 mm	\	\	\	[31]
2020	Nanjing Research Institute of Electronics Technology	Ka-band 64-channel phased array antenna	Silicon based packaging integration, chip embedded packaging	\	\	\	\	[32]
2020	Chengdu radio wave technology Co., Ltd.	Ka-band 128-unit tile phased array antenna	Integrated packaging of Si-based multifunctional and GaAs transceiver chips	93 mm × 93 mm × 52 mm	>20	<6.5	\	[33]
2021	Aerospace Information Research Institute, Chinese Academy of Sciences	Ka-band 4-channel T/R module	LTCC, BGA, multi-material heterogeneous integration	10.8 mm × 10 mm × 3 mm	≥26	≤4.4	Efficiency ≥ 25%	This work
2023	The 13th Research Institute, CETC	Ka-band 4-channel T/R module	Silicon-based MEMs, TSV, 3D integration	18 mm × 19.5 mm × 3 mm	≥30	≤4.6		[34]

The comparison above makes clear how well the TR module developed in this work performs in terms of the noise coefficient, output power, and single channel volume. With a weight of 1 g and an overall dimension of only 10.8 mm × 10 mm × 3 mm, this device-based design greatly enhances T/R module integration when compared to conventional phased array antenna T/R modules. The volume of a single channel drops by 95% with the same specs. When compared to T/R modules based on SOC devices, the RF performance has significantly improved, manifested by an increase in single-channel output power and a decrease in the receiving noise factor. The single-channel output power is >26 dBm, the efficiency is ≥25%, and the noise factor is <4.4 dB thanks to the comprehensive optimization from aspects of chip architecture, transceiver link, LTCC substrate, interconnection interface, and packaging architecture design.

A majority of the modules in earlier studies were created and integrated using single materials like silicon, PCBs, and LTCC. Despite being the most widely used and developed solutions for compact sizing requirements, CMOS, SiGe, and multilayer PCB technologies still struggle to balance the low loss, low noise, large output power, and high transmission of RF signals, among other things. Numerous levels of interlayer interconnections, intricate manufacturing and assembly processes, considerable signal losses, and noise introduced at the interlayer interconnections are among the factors that typically limit the reduction of component volume and profile, leading to large single-channel volume. In this regard, this article designed and implemented a highly reliable Ka-band device-based four-channel T/R module based on multi-material heterogeneous integration architecture applying Si/GaAs/LTCC. With its high integration, straightforward interface, and high output power per unit volume, the device-based T/R module presented in this paper offers a practical and efficient design example for device-based multichannel T/R modules with dependable applications below Ka-band frequency.

3. Architecture Design

3.1. Architecture Scheme of 3D Heterogeneous Integration

At present, silicon-based circuit integration, three-dimensional heterogeneous integration, brick or tile type, and three-dimensional heterogeneous flat plate type are the most common phased array antenna microsystem architectures [32,35,36]. Brick and tile types are classic phased array antenna microsystem designs that accomplish hybrid circuit integration, chip mechanical support, and environmental protection by using metal shells as packaging carriers for different functional modules. Their advantages include inexpensive research and development investment cost and high single-channel output power; yet there are several tiers of connections, significant manufacturing complexity, and challenges in cutting costs and profiles. Silicon-based circuit integration and heterogeneous integration belong to chip-level and wafer-level integration. Enhancing qualification ratios (like stacking success ratios), resolving problems with thermal mechanical reliability (such mismatches in thermal expansion coefficients), and enhancing interconnection performance are the main challenges associated with these two approaches. They also call for sophisticated machinery, intricate procedures, and significant investments in research and development. Through a three-dimensional assembly interconnection, the three-dimensional heterogeneous flat panel scheme integrates circuits based on diverse functional materials that are made independently. This can combine the benefits of various high-integration techniques and simplify connectivity.

Table 2 summarizes the characteristic indicators of the antenna microsystem architecture mentioned above. The three-dimensional heterogeneous flat plate scheme may not have the highest integration when compared to other architectures, but it has a flexible design and a strong universal system architecture, satisfies real-world needs, and has comparatively low research and development costs. It may successfully address the needs of the application, while taking into account a number of limitations, including those related to performance, cost, space, and reliability. This results in high integration and small size, making it a workable option for systems and devices with superior stability and high efficiency.

Table 2. Comparison of phase array antenna microsystem architecture.

Integrated Architecture of Phased Array Antenna Microsystems	Requirements for Chip Integration	Technique Requirements	Single Channel Output Power	Testability	Research and Development Cost	Million Level Batch Production Cost
Brick and wafer type	Low	Metal shells with multi-chip planar packaging; high- and low-frequency cable interconnection	High	Good	Low	High
Three-dimensional heterogeneous flat plate type	Relatively high	Ceramic, silicon, and glass adapter plates; embedded packaging; hybrid multilayer boards; and surface assembly	Relatively high	Relatively good	Relatively low	Relatively low
Silicon based RF circuit integration	High	SOC chips and their packaging techniques based on Si Complementary Metal Oxide Semiconductor (CMOS) and SiGe BiCMOS	Low	Relatively good	High	Low
heterogeneous integration	High	Si CMOS and SiGe BiCMOS wafer-based technique; direct heterojunction of silicon compounds	Low	Bad	Very high	Low
2016	Xi'an Research Institute of Navigation Technology	Ku-band 3D miniaturized T/R module	LTCC, Ball Grid Array (BGA)	9.5 mm × 9.5 mm × 3.8 mm	Transmission output power > 24.5 dBm, receiver gain > 25 dB, and receiving noise factor < 3.5 dB	[17]

Based on this, this article adopts a three-dimensional heterogeneous flat panel integration scheme for T/R module to achieve devitalization.

3.2. Link Design of T/R Module

The link design of T/R module needs to meet the requirements of high integration, large output power, and low loss. The functional modules of the transceiver link are divided after carefully weighing the benefits and drawbacks of Si and GaAs-based technologies in order to further enhance the integration of the T/R module. The Ka-band 0.5 W four-channel five-chip architecture, or the 1.25 chips per channel method, is employed in this article. It consists of one Si-based four-channel amplitude-phase multifunctional chip and four 0.5 W GaAs transceiver multifunctional chips. A 6-bit phase shifter, a 6-bit attenuator, a driver amplifier, and a transceiver switching interface are integrated inside each channel of the four-channel Si-based amplitude-phase multifunctional chip, as seen in Figure 2. A power amplifier, a low-noise receiving amplifier, and a transceiver switch are all internally integrated within the GaAs transceiver multifunctional device.

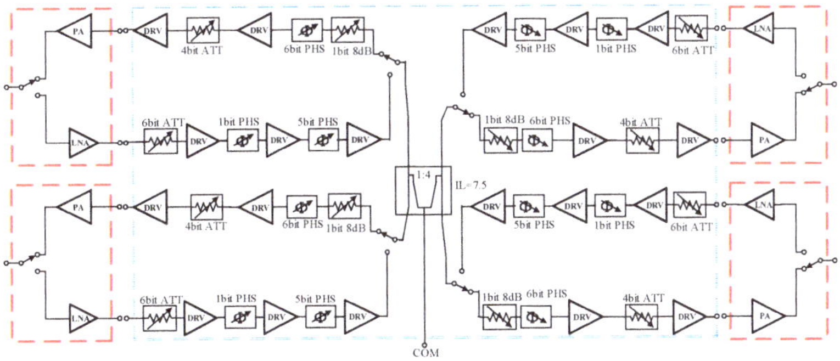

Figure 2. Schematic block diagram of Ka-band 0.5 W four-channel T/R module.

The GaAs transceiver multifunctional chip uses a two-stage amplification structure for its power amplifier, and a four-stage amplification and five-stage matching structure for its low-noise amplifier, which incorporates feedback networks. In order to increase the low-noise amplifier's stability, feedback networks are also added at the input and output ends of all stages of amplification. At the input stage, noise-coefficient matching is utilized to obtain good noise factors. The transceiver switch adopts a single pole double throw switch structure. The traditional Wilkinson power divider model is utilized by the four-channel Si-based amplitude-phase multifunctional power divider, which has an insertion loss of roughly 2 dB (not counting the 6 dB distribution loss). In order to improve channel isolation and make it easier to accurately calibrate the amplitude and phase of phased array antenna channels, the switch employs a single-pole three-throw switch, which has three states: transmission, reception, and load. Drive amplifiers 2 and 3 use an RC negative feedback structure to ensure a large output power, while amplifier 5 uses a two-stage amplification structure and negative feedback to achieve a high gain, while ensuring output power. Drive amplifiers 1 and 4 adopt a traditional amplifier structure due to their small input power. For the purpose of increasing the input P-1dB, the receiving phase shifter has a passive phase shifter structure, with a large displacement phase loss of roughly 3 dB and a small displacement phase loss of about 9.5 dB The transmission phase shifter adopts an active phase shifter structure to reduce the size, with a loss of approximately 2.5 dB. The attenuator adopts a traditional single-ended structure and adds parallel capacitors for phase compensation. The receiving large-bit attenuation of 16 dB is achieved by cascading two 8 dB attenuation circuits. Both chips are internally integrated with active bias, eliminating the need for additional power modulation circuits, which can save on the number of peripheral devices and simplify the assembly and production.

This design uses Si-based technology to achieve low-power and small-size amplitude-phase control and power management circuits, as well as GaAs technology to achieve excellent amplifier output power, efficiency, and noise factor.

Compared with single-chip integrated architecture, the discrete 1.25-chip architecture can fully exploit the advantages of various process technologies, effectively balancing integration, process feasibility, functionality, and performance. When it comes to large-scale and affordable production, the discrete 1.25-chip architecture is superior to the multi-chip discrete architecture because it can significantly improve integration, simplify the T/R module layout design, streamline the production and assembly process, and guarantee amplitude and phase consistency across channels. The discrete 1.25 chip architecture achieves a high performance, high integration, and good process implementation of the T/R module. The four-channel amplitude-phase multifunctional chip adopts the 55 nm Si CMOS technique, with a size of 3.50 × 5.00 × 0.10 mm, while the transceiver multifunctional chip uses a 0.25 μm GaAs p-type high-electron mobility transistor (pHEMT) technique measuring 2.42 mm × 1.85 mm × 0.10 mm.

Based on the above four-channel and five-chip T/R chip architecture, a device-based four-channel T/R module is constructed, consisting of one Si-based amplitude-phase multifunctional chip, four GaAs transceiver multifunctional chips, and a passive conversion structure. Under the guidance of wave control signals, each T/R channel is capable of independently achieving receiving phase shift and attenuation, transmission phase shift, and receiving/transmission/load state switching. Figure 3 displays the indicator allocation and link design. IL, NF, G, Pin, and Pout denote insertion loss, noise factor, gain, input power, and output power, respectively. It can be observed that the active gain of the receiving channel is greater than 30 dB, and the noise factor is less than 4.4 dB. The transmission channel's saturated output power is higher than 26 dBm.

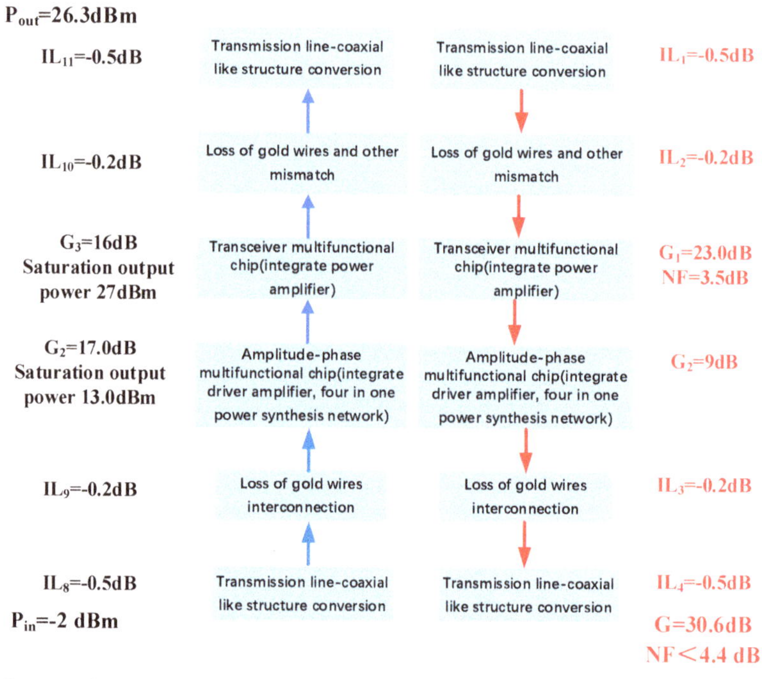

Figure 3. T/R module transmitting and receiving link design and indicator allocation.

The amplitude-phase multifunctional chip and transceiver multifunctional chip of the T/R module described in the paper are placed in the reserved slots of the LTCC multilayer substrate, and the LTCC substrate multilayer wiring provides electrical interconnection, heat dissipation channels, and mechanical support. The LTCC multilayer substrate material is FST07, with a relative dielectric constant of 6.6 and a loss angle tangent of 0.002. There are fifteen LTCC layers in total, and each layer is 96 μm thick, as shown in Figure 4a. Silver-based slurry is utilized on ceramic substrates to produce high-precision, high-density microwave and low-frequency circuits. The second layer of the LTCC substrate is a large-area grounding pad. Punch the substrate from the surface layer to the second layer and then stack and press them to form a blind cavity. Place the four-channel T/R sleeve on the large area grounding pad in the cavity, so that the chips and the surface pads are basically on the same horizontal plane, facilitating bonding operations and shortening the length of gold wires. The conductors from the third-to-fifth layers of the substrate form the power and control circuits, connected to the surface and bottom conductors through the conductor through holes, providing power and control signals for the four-channel T/R chips. The sixth-to-fifteenth layers of the LTCC substrate are Ka-band signal horizontal transmission layers, where the sixth layer is the upper ground layer for the stripline, isolating the microwave signal layer and the low-frequency signal layer to enhance shielding. The eleventh layer is the Ka-band signal horizontal transmission line layer. The bottom layer serves as the lower ground plane for the stripline, and the sixth ground plane is connected to the bottom ground plane through the conductor through holes arranged on both sides of the stripline, playing a shielding role. A Ka-band signal, power supply, and control signal pad are also located on the bottom layer. The BGA solder balls implanted on the bottom pad are interconnected with external circuit boards to achieve both high- and low-frequency signal input and output. The last surface coating layer is an electroless Ni/Pd/Au film that is utilized for bonding pads for bare chips, surface-mounted resistive and capacitive devices, wire-bonding connecting pads, and bonding pads for metal frame packaging. Figure 4b exhibits the printing and punching layout for each layer.

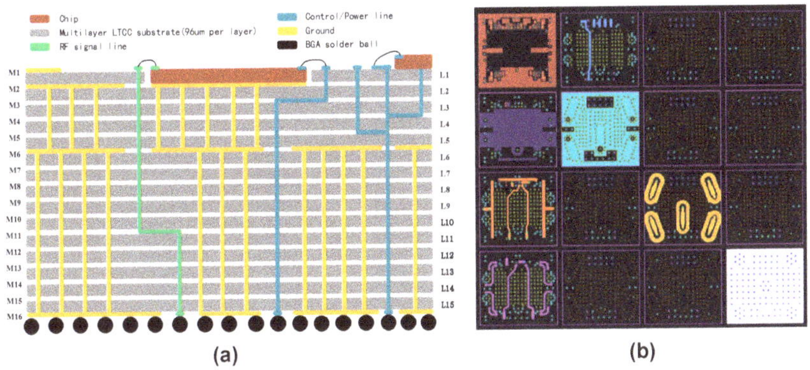

Figure 4. Schematic and layout of multilayer LTCC substrate: (**a**) schematic of multilayer LTCC substrate and (**b**) layout of LTCC substrate.

3.3. Device-Based Packaging Design

The four-channel T/R module adopts device-based packaging. As shown in Figure 5, the device-based T/R module uses an LTCC substrate as the packaging substrate, and the 4J29 Kovar alloy with a similar thermal expansion coefficient to the LTCC substrate within a wide temperature range (−70~500 °C) is selected as the enclosure material. The electrical interconnection interface adopts the BGA form, and the LTCC substrate is coated with Ni/Pd/Au film on the surface and then welded with gold tin alloy solder to form an airtight packaging for the metal enclosure and cover plate. The device-based T/R module achieves functional integration of mechanical support, 3D wiring, and heat dissipation by utilizing the dielectric characteristics, mechanical strength, thermal conductivity, and sealing qualities of the ceramic materials themselves. The components are interconnected with external circuits through a surface mount assembly, saving the size and weight of high- and low-frequency connectors and improving the integration of phased array antenna systems. The silver slurry is used as the conductor of 3D wiring, and the surface is coated with an electroless Ni/Pd/Au-plating process to meet the assembly requirements of various components and greatly reduce the process cost.

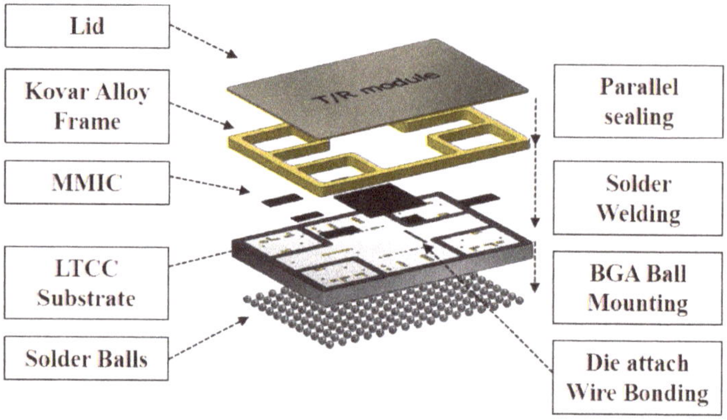

Figure 5. Schematic of tile four-channel T/R module.

4. Optimization of Interconnection Structure

4.1. Design of Device Based Interconnection Interface

When mounting and interconnecting components of this device-based T/R module, the interconnection is required to occupy a small area of the multifunctional board, have low millimeter-wave RF transmission loss, and have good isolation inside and among modules. In order to satisfy the high-performance and low-profile connecting requirements of phased array antenna microsystems, the T/R module referred to in this article uses a BGA as its interconnection interface for signal transmission, regardless of frequency.

BGA packaging can accomplish high-density RF and low-frequency connections by using solder balls in place of conventional metal leads or connectors. The balls are arranged in a two-dimensional array on the back of the packing substrate. Its short signal transmission path and minimal cable parasitic impact help to achieve a good transmission performance. BGA solder balls can be arranged in a coaxial-like structure, as seen in Figure 6, with a solder ball located in the center to replace the inner conductor of the coaxial line, and a circle of solder balls on the periphery to replace the grounding shielding layer of the standard coaxial line. The coaxial-structure BGA interface can achieve great shielding of electromagnetic wave signals in the Ka-band range, as shown in Figure 6, thereby achieving high isolation between RF ports. Figure 7 presents HFSS models comparing electric fields, as well as magnetic field distributions of the quasi-coaxial structure and coaxial structure, showing excellent and similar performance in confining the electromagnetic field inside for both structures.

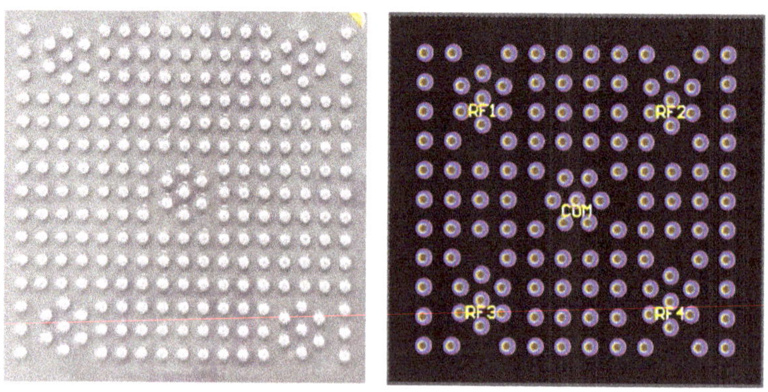

Figure 6. Device-based four-channel T/R module BGA interconnection interface, applying coaxial like structure.

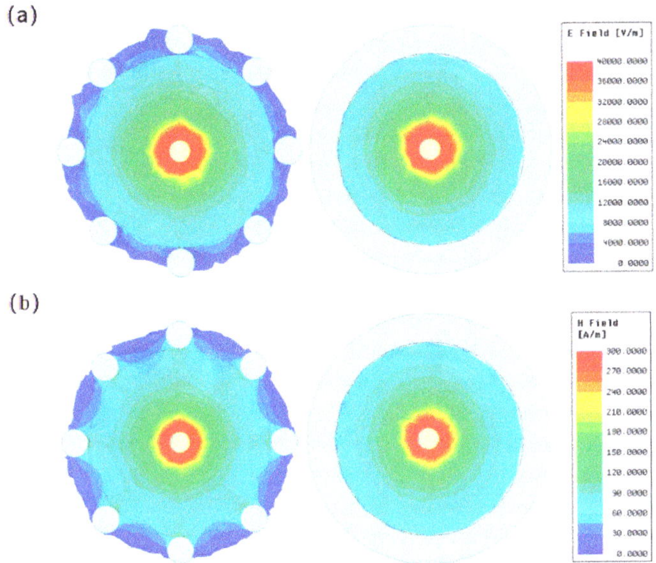

Figure 7. Comparison of electromagnetic field distributions between quasi-coaxial structure and coaxial structure: (**a**) electric fields and (**b**) magnetic fields.

4.2. Design of RF Full-Path Interconnection

In the article, the transceiver multifunctional chip is connected to LTCC through wire bonding after being pasted to a large-area solder pad, using conductive adhesive. The millimeter-wave signal is transmitted from the surface to the BGA solder ball output at the bottom of the LTCC substrate in three-dimensional form. It is necessary to optimize the design of the millimeter-wave signal full path interconnection structure of "chip–gold wire–LTCC surface–vertical interconnection–stripline–vertical interconnection–BGA solder ball".

The high-frequency structural simulator (HFSS) model's simulation results demonstrate that an impedance mismatch arises from the discontinuous structure between the bonding wire and the pad when chips are directly bonded through gold wire to the inner conductor of a vertical coaxial structure. This leads to a notable decline in the performance of signal transmission. Z_{in} is also in a state of impedance mismatch. Since this degradation is more noticeable at higher frequencies, the transmission performance must be improved on successive striplines by conducting impedance matching.

As illustrated in Figure 8a, a model of the RF path "chip–gold wire–coaxial structure–stripline–coaxial structure–BGA solder ball" was established for the device-based T/R module in order to optimize the interconnection structure. This was achieved by adding multiple matching structures with varying line widths and lengths to the stripline.

Figure 8b presents the simulation findings. It can be seen that, in the wide frequency range of 20–30 GH, the reflection coefficient is less than −20 dB, and the transmission loss is less than 0.4 dB. The electric field distribution results are shown in Figure 8c and show that the electromagnetic field is confined to the transmission path, indicating that the interconnection structure has an excellent transmission performance.

The difference between the RF input channel and output interface of the device-based T/R module is that the overall port of the amplitude-phase multifunctional chip on the RF input channel has larger wiring space, and the LTCC substrate's surface can accommodate the addition of a coplanar waveguide. The HFSS simulation model is shown in Figure 9a.

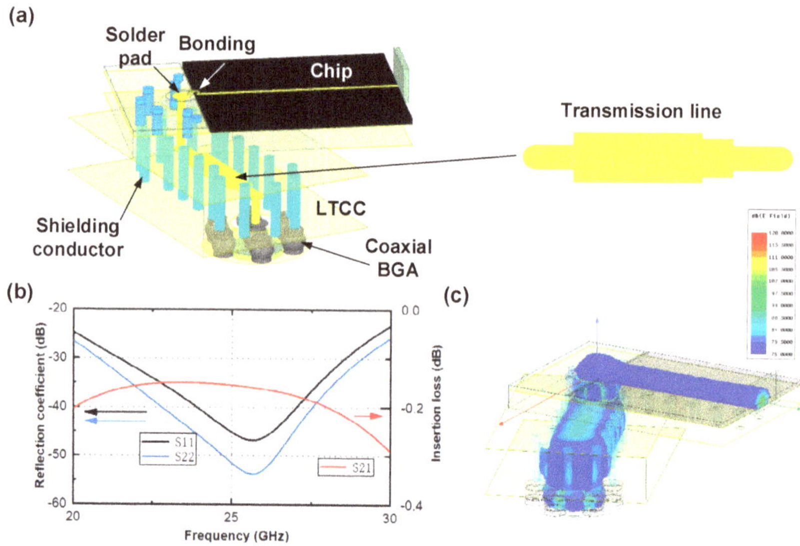

Figure 8. (**a**) HFSS model of radio frequency output interconnection structure. (**b**) Simulation results of radio frequency output interconnection structure. (**c**) Simulation results of electric field distribution in interconnection structure.

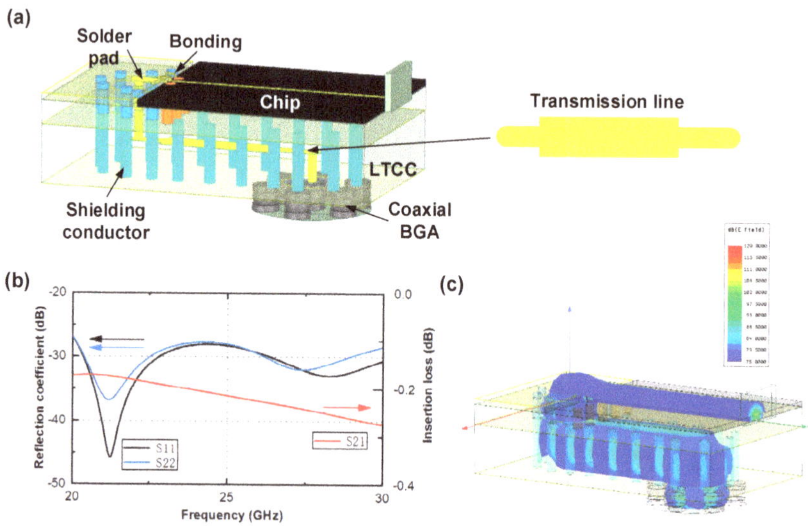

Figure 9. (**a**) HFSS model of radio frequency input interconnection structure. (**b**) Simulation results of tile T/R module radio frequency input interconnection structure. (**c**) Simulation results of electric field distribution in interconnection structure.

Simulation results are shown in Figure 9b, where the reflection coefficient is less than −20 dB and the transmission loss is less than 0.4 dB in the wide frequency range of 20–30 GHz. The electric field simulation model is shown in Figure 9c, presenting that the electromagnetic field is also well confined to the transmission path.

5. Testing the Key Parameters of the Device-Based T/R Module

Figure 10 displays the device-based T/R module's photo and testing environment. The module weighs 1 g and has overall dimensions of 10.8 × 10 × 3 mm. The testing results are summarized in Table 1.

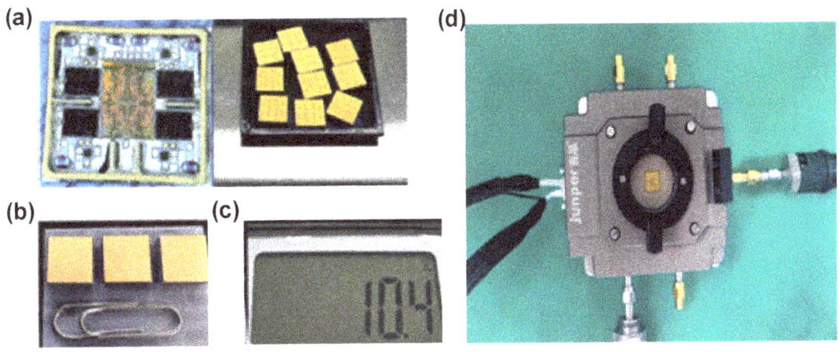

Figure 10. (**a**) Fabricated prototype of tile T/R module. (**b**) Size comparison between T/R modules and a clip. (**c**) A T/R module weight of 10.4 g. (**d**) Test environment of tile T/R module.

To measure the receiving performance of the module's single channel, put the other three channels into receiving standby mode, meaning that the receiving amplifier is turned off; and put the RF port into load mode.

The results of the single-channel ground-state receiving test are shown in Figure 11, and it can be seen that the in-band receiving gain exceeds 26 dB. Considering that the four-in-one power synthesis network introduces an additional insertion loss of about 6 dB, the in-band receiving gain for a single channel ground state is greater than 33 dB, which matches the predicted value 30.6 dB in Figure 3 well. The in-band noise factor is less than 4.4 dB, agreeing with the value 4.4 dB put forward in Figure 3. The test results meet our expectations.

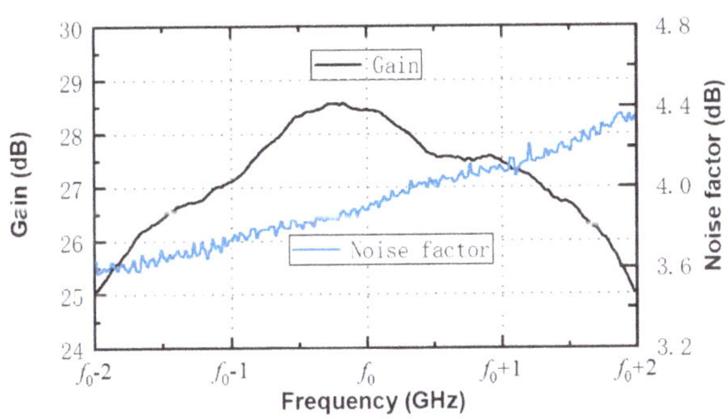

Figure 11. Measurement results of single-channel base-state reception.

The test results of the 6-bit attenuator and phaser are shown in Figure 12. The amplitude or phase measured following a particular amplitude modulation (a) or phase modulation (b) on the basis of a single-channel ground state is represented by each curve in Figure 12, with the particular amplitude modulation or frequency modulation value shown on the label of the curve. It is shown that the test results agree well with the set values within a wide band range (f_0 − 2~f_0 + 2 GHz, f_0 = 24), verifying that this T/R module has stable and accurate attenuation and phase-shifting effects.

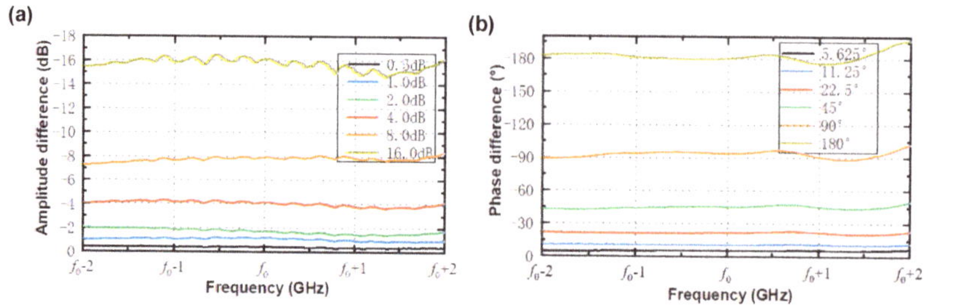

Figure 12. Measurement results of phase shift and attenuation. (**a**) amplitude modulation. (**b**) phase modulation.

Set the RF port to the load state and the other three channels to the transmitting standby state—that is, with the transmitting amplifier turned off—when evaluating the transmission performance of the module's single channel.

The test results of the single-channel transmission output power and current are shown in Figure 13. It can be seen that the output power of the device-based four-channel T/R module within a single channel 4 G bandwidth is greater than 26.2 dBm, which is superior, as seen from the comparison in Table 1. Additionally, the emission current, which has a maximum value of 300 mA and is related to efficiency at different frequencies, has an efficiency of more than 25%. The test results comply with the design values in Figure 3, considering a 6 dB insertion loss induced by the one-to-four power divider.

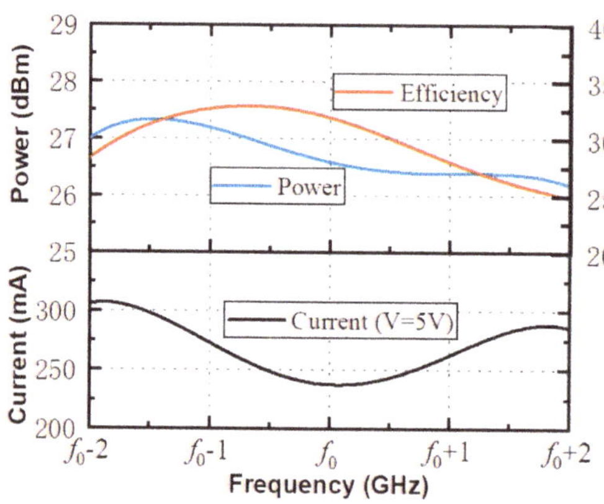

Figure 13. Measured results of output power, efficiency, current with single channel transmitting.

6. Conclusions

This work offers a device-based four-channel T/R module scheme based on a multi-material heterogeneous integration architecture incorporating Si/GaAs/LTCC to meet the demand for the devitalization of T/R modules in the millimeter-wave phased array antenna microsystem architecture design. By designing the chip architecture, connections, wiring, packaging, interfaces for interconnection, and full path interconnection, millimeter-wave four channel T/R modules are devitalized, and their critical parameters are tested. This module breaks through the device-based packaging of millimeter-wave signal full-path three-dimensional transmission interconnection, achieving a four-channel module overall size of 10.8 mm $\times$ 10 mm $\times$ 3 mm, weight of 1 g, single-channel transmission output power $\geq$ 26 dBm, efficiency $\geq$ 25%, and noise factor $\leq$ 4.4 dB, meeting the high-performance T/R module requirements of Ka-band phased array antenna microsystems for high dynamic platform communication. When compared to other T/R modules, this one has advantages like simple integration, a straightforward interface for connecting, and a high output power per unit volume.

With its low profile, high power, and low loss, this research offers a practical and workable design example for a device-based multichannel T/R module design. It has a wide range of applications and enormous potential, but more research and development are required.

Author Contributions: Methodology, Q.Z.; software, Z.C.; validation, M.H., S.W. and X.L.; formal analysis, Q.Z.; investigation, X.L.; resources, H.X.; data curation, X.L.; writing—original draft preparation, Q.Z. and X.L.; writing—review and editing, Q.Z.; project administration, H.X.; funding acquisition, H.X. All authors have read and agreed to the published version of the manuscript.

Funding: This research was funded the Aerospace Information Research Institute, Chinese Academy of Sciences, grant numbers E1Z221030F, and partly by the Scientific and Technological Innovation Foundation of Shunde Graduate School, USTB (BK22BF002).

Data Availability Statement: Data are contained within the article.

Conflicts of Interest: The authors declare no conflict of interest.

References

1. Han, G.; Du, B.; Wu, W.; Yang, B. A Novel Hybrid Phased Array Antenna for Satellite Communication on-the-Move in Ku-band. *IEEE Trans. Antennas Propag.* **2015**, *63*, 1375–1383. [CrossRef]
2. Alakija, C.; Stapleton, S.P. A mobile base station phased array antenna. In Proceedings of the 1992 IEEE International Conference on Selected Topics in Wireless Communications, Vancouver, BC, Canada, 25–26 June 1992.
3. Hu, R.; Li, X.; Deng, J.; Zhou, J.; Dong, S.; Liu, L.; MA, W. Research on the Architecture of Phased Array Antenna of LEO Communication Satellites Based on Hybrid Beamforming. *Space-Integr.-Ground Inf. Netw.* **2021**, *2*, 8.
4. Hong, W.; Jiang, Z.H.; Yu, C.; Zhou, J.; Chen, P.; Yu, Z.; Zhang, H.; Yang, B.; Pang, X.; Jiang, M.; et al. Multibeam Antenna Technologies for 5G Wireless Communications. *IEEE Trans. Antennas Propag.* **2017**, *65*, 6231–6249. [CrossRef]
5. Gan, Z.; Zhang, G. Current development of satellite communications technology. *J. Commun.* **2006**, *8*, 8.
6. Yang, W. *Array Antenna Technology in Mobile Communications*; Beijing Jiaotong University Press: Beijing, China, 2005.
7. Bil, R.; Holpp, W. Modern phased array radar systems in Germany. In Proceedings of the 2016 IEEE International Symposium on Phased Array Systems and Technology (PAST), Waltham, MA, USA, 18–21 October 2016; pp. 1–7.
8. Morrison, G.D.; McLachlan, A.D.; Kinghorn, A.M. "Tile"-based airborne phased array radar systems. In Proceedings of the International Conference on Radar Systems (Radar 2017), Belfast, UK, 23–26 October 2017; pp. 1–4.
9. Pirkl, M.; Holpp, W. From research to application: How phased array radars conquered the real world. In Proceedings of the 2013 14th International Radar Symposium (IRS), Dresden, Germany, 19–21 June 2013; pp. 17–22.
10. Brookner, E. Advances and breakthroughs in radars and phased-arrays. In Proceedings of the 2016 CIE International Conference on Radar (RADAR), Guangzhou, China, 10–13 October 2016; pp. 1–9.
11. Lambard, T.; Lafond, O.; Himdi, M.; Jeuland, H.; Bolioli, S.; Coq, L.L. Ka-Band Phased Array Antenna for High-Data-Rate SATCOM. *IEEE Antennas Wirel. Propag. Lett.* **2012**, *11*, 256–259. [CrossRef]
12. Lai, Q.; Li, P.; Jin, M.; Gao, C.; Yuan, T. A mm-wave multiple-beam phased array breadboard for GEO-LEO inter-satellite links. In Proceedings of the 2014 IEEE MTT-S International Microwave Symposium (IMS2014), Tampa, FL, USA, 1–6 June 2014; pp. 1–3.
13. Cooley, M. Phased Array Fed Reflector (PAFR) antenna architectures for space-based sensors. In Proceedings of the 2015 IEEE Aerospace Conference, Big Sky, MT, USA, 7–14 March 2015; pp. 1–11.
14. Ivanov, A.; Teplyakov, V.; Kalinin, V. Mobile communication system with a hybrid phased array antenna system. In Proceedings of the 2015 IEEE East-West Design & Test Symposium (EWDTS), Batumi, Georgia, 26–29 September 2015; pp. 1–4.
15. Lee, W.-B.; Im, C.-S.; Zhou, C.; Bhandari, B.; Choi, D.-Y.; Lee, S.S. Metasurface doublet-integrated bidirectional grating antenna enabling enhanced wavelength-tuned beam steering. *Photonics Res.* **2022**, *10*, 248–255. [CrossRef]
16. Li, Y.; Chen, B.; Na, Q.; Xie, Q.; Song, J. Wide-steering-angle high-resolution optical phased array. *Photonics Res.* **2021**, *9*, 2511–2518. [CrossRef]
17. Ginel-Moreno, P.; Hadij-ElHouati, A.; Sánchez-Postigo, A.; Wangüemert-Pérez, J.G.; Molina-Fernández, Í.; Schmid, J.H.; Cheben, P.; Ortega-Moñux, A. On-Chip Metamaterial Antenna Array with Distributed Bragg Deflector for Generation of Collimated Steerable Beams. *Laser Photonics Rev.* **2022**, *16*, 2200164. [CrossRef]
18. Bai, X.; Zhang, F.; Sun, L.; Cao, A.; Wang, X.; Kong, F.; Zhang, J.; He, C.; Jin, R.; Zhu, W.; et al. Cuiradiation-type programmable metasurface for direct manipulation of electromagnetic emission. *Laser Photonics Rev.* **2022**, *16*, 2200140. [CrossRef]
19. Xue, J. *Fundamentals of Millimeter-Wave Engineering*; Harbin Institute of Technology Press: Shenzhen, China, 2004.
20. Xiang, J.; Zhang, M. *Millimeter-Wave Radar and Its Applications*; National Defense Industry Press: Beijing, China, 2005.
21. Ruan, C. *Millimeter-Wave Theory and Technology*; University of Electronic Science and Technology of China Press: Chengdu, China, 2001.

22. Qi, S.; He, Z. Study of Millimeter Wave Phased Array Antenna. *Shipboard Electron. Countermeas.* **2004**, *27*, 14–16.
23. Sadhu, B.; Gu, X.; Valdes-Garcia, A. The More (Antennas), the Merrier: A Survey of Silicon-Based mm-Wave Phased Arrays Using Multi-IC Scaling. *IEEE Microw. Mag.* **2019**, *20*, 32–50. [CrossRef]
24. Li, S. *Ka Band Delay-Line Components Based on LTCC Technology*; University of Electronic Science and Technology of China: Chengdu, China, 2014.
25. Liu, X.; Hao, J. Design of Ka-band Multi-channel transmitting. *Shipboard Electron. Countermeas.* **2015**, *38*, 210–219.
26. Guo, F.; Jiang, W.; Wang, Z. A Ka-band transceiver based on LTCC technology. In Proceedings of the 2013 International Workshop on Microwave and Millimeter Wave Circuits and System Technology, Chengdu, China, 24–25 October 2013.
27. Wang, Y. *Research on a Ka Band T/R Module Array Based on LTCC Technolog*; Beijing Institute of Technology: Beijing, China, 2016.
28. Wan, T.; Wang, Y. Miniaturization and multilayer for 3-D T/R module. *J. Terahertz Sci. Electron. Inf. Technol.* **2016**, *14*, 758–762.
29. Chu, C.Y.; Tseng, S.T.; Gao, J.Z.; Chen, Y.P.; Chang, Y.C.; Tseng, C.W.; Huang, T.Y.; Hu, B.S.; Su, B.; Chu, T.S.; et al. A fully-integrated Ka-band 4TX/4RX phased-array transceiver IC in 65nm CMOS. In Proceedings of the 2016 IEEE International Symposium on Radio-Frequency Integration Technology (RFIT), Taipei, Taiwan, 24–26 August 2016.
30. Cui, L.; Zhang, Z.; Hu, Y.; Ji, L. Packaging Technology for Ka-band T/R Modules Based on LTCC Thick-thin Film Hybrid Substrate. *Mod. Radar* **2017**, *9*, 86–89.
31. He, Q.Q.; Ding, S.; Xing, C.; Chen, J.Q.; Yang, G.Q.; Wang, B.Z. Research on Structurally Integrated Phased Array for Wireless Communications. *IEEE Access* **2020**, *8*, 52359–52369. [CrossRef]
32. Sun, L. A Review of Antenna-in-package Technology for Millimeter-wave Phased Array Antennas. *Mod. Radar* **2020**, *42*.
33. Liu, X. A novel airproof miniaturized millimeter-wave TR module. *Electron. Test.* **2021**, *20*.
34. Yang, D.; Zhao, Y. A Ka Band Four-Channel Short Brick-Type 3D Integrated T/R Microsystem with Silicon-Based MEMS Technology. *IC Des. Appl.* **2023**, *48*, 506–511. [CrossRef]
35. Lu, J.G.; Wang, Y. From active phased array antenna to antenna array microsystem in post-Moore era. *Sci. Sin. Inf.* **2020**, *50*, 19.
36. Herd, J.S.; Conway, M.D. The Evolution to Modern Phased Array Architectures. *Proc. IEEE* **2016**, *104*, 519–529. [CrossRef]

Disclaimer/Publisher's Note: The statements, opinions and data contained in all publications are solely those of the individual author(s) and contributor(s) and not of MDPI and/or the editor(s). MDPI and/or the editor(s) disclaim responsibility for any injury to people or property resulting from any ideas, methods, instructions or products referred to in the content.

electronics

MDPI

Article

Ultracompact SIRC-Based Self-Triplexing Antenna with High Isolation

Rusan Kumar Barik [1] and Slawomir Koziel [1,2,*]

1 Engineering Optimization and Modeling Center, Reykjavik University, 102 Reykjavik, Iceland; rusanb@ru.is
2 Faculty of Electronics, Telecommunication and Informatics, Gdansk University of Technology, 80-233 Gdansk, Poland
* Correspondence: koziel@ru.is

Abstract: An ultracompact self-triplexing antenna realized on a substrate-integrated rectangular cavity (SIRC) is discussed in this study. The proposed structure employs two L-shaped slots and an inverted U-shaped slot to radiate at three independent operating frequency bands. Three 50-ohm microstrip feed lines were used to excite the radiation in these slots. The operating frequency was individually tuned using the slot size. The slot placement and size were designed having in mind considering obtaining one or more frequency bands below the SIRC cutoff frequency, which had the advantage of enabling an ultracompact size. High port isolation was achieved by applying one of the ports orthogonally to the two remaining ones, which created a weak cross-coupling channel. A lumped-circuit model was created to examine the antenna operation. The presented design has been prototyped and experimentally validated with the measured operating frequencies of 1.92 GHz, 4.43 GHz, and 5.25 GHz for GSM, 5G, and WLAN applications, respectively. The port isolations are better than 32.4 dB according to both EM simulations and measurements. Meanwhile, the measured realized gain of the antenna is better than 4.3 dBi at all bands.

Keywords: substrate-integrated rectangular cavity; self-triplexing; antenna; multiband; isolation

Citation: Barik, R.K.; Koziel, S. Ultracompact SIRC-Based Self-Triplexing Antenna with High Isolation. *Electronics* **2023**, *12*, 2112. https://doi.org/10.3390/electronics12092112

Academic Editors: Syed Muzahir Abbas and Yang Yang

Received: 13 March 2023
Revised: 20 April 2023
Accepted: 4 May 2023
Published: 5 May 2023

Copyright: © 2023 by the authors. Licensee MDPI, Basel, Switzerland. This article is an open access article distributed under the terms and conditions of the Creative Commons Attribution (CC BY) license (https://creativecommons.org/licenses/by/4.0/).

1. Introduction

Personal communication devices frequently employ multiband communication systems. Usually, these systems have two or more wireless transceivers for a variety of applications. Multiband antennas are typically utilized instead of multiple antennas to deliver better performance for transceivers. It is challenging to arrange antenna components tightly in a limited space while maintaining adequate isolation between them. One method for enhancing input port isolation is a decoupling network [1]. However, as a result, the size of the circuitry increases, limiting its applicability for handheld devices.

Therefore, current research has focused on creating multiband antennas, with monopole-, slot-, and SIW-based structures being the most prevalent architectures [2–11]. A rectangular monopole truncated patch loaded with a U-shaped slot was employed for the design of a tri-band millimeter-wave antenna in [2]. In [3], a monopole dual-band antenna was realized by applying a microstrip resonant-cell-based low-pass filter feed. A multiband antenna was designed by using the ground plane radiation mode in [4]. A triband dual-polarized antenna was constructed by using cross-shaped slots and L-shaped slots in [5]. A triband antenna was developed based on a substrate-integrated waveguide (SIW) loaded with a pair of open-loop slots in [7]. In [8], a triband antenna was realized on a half-mode SIW loaded with multiple slots. A quarter-mode SIW cavity loaded with multiple slots was employed to design a triband antenna [9]. In [10], a triangular SIW cavity loaded with slots was applied to build a wearable antenna. A rectangular loop and a rectangular slot were used to construct a planar antenna-triplexer in [11]. At each working frequency, the aforementioned designs provide strong in-band performance, and the structures are small. However, when used with several transceivers, a single-input arrangement becomes a deficiency. To facilitate connectivity between several transceivers and the multiband antenna, frequency-selective devices, such as a diplexer or triplexer, are needed because the differences in frequency interfere with other transceiver circuits. However, this new component will take up more space, and the design will become complex. Many researchers have suggested the construction of self-diplexing [12–24], self-triplexing [25–27], or self-multiplexing [28–30] antennas to reduce the required area and to improve isolation.

Substrate-integrated waveguide (SIW) technology has recently become an attractive alternative for constructing planar self-multiplexing antennas [12–30]. A self-diplexing antenna (SDA) was realized based on the SIW cavity-backed transverse slots in [12]. A self-diplexing antenna involving SIW cavity-backed bowtie-ring slots has been reported in [13]. In [14], the construction of a self-diplexing antenna using a shielded-quarter mode SIW loaded with a Y-shaped slot was described. By attaching two U-shaped slots to the top of the SIW cavity, a self-diplexing antenna was created in [15]. In [16], a rectangular SIW cavity-based asymmetric cross-shaped slot antenna was designed for circularly polarized applications. A self-diplexing antenna was constructed in [17] using square SIW cavity-backed stepped patches. In [18], a high-isolation self-diplexing antenna was developed using a cavity-backed U-shaped slot. A quarter-mode SIW was employed to design an ultracompact SDA [19]. In [20], a tunable SDA based on an SIW cavity loaded with open-ended slots was proposed. Rectangular and triangular radiating patches were employed on the SIW to realize a high isolation SDA in [21]. In [22], a shielded quarter-mode SIW cavity was employed to design an SDA with high isolation. An ultracompact SDA was developed by using a shielded quarter-mode SIW [23]. In [24], a tunable SIW-based SDA was constructed by employing microfluidic channels.

Although the aforementioned SDA structures work effectively, the design of self-triplexing or self-multiplexing antennas should be investigated further. Based on different SIW cavities, several self-triplexing antennas (STAs) [25–27] and self-multiplexing antennas (SMAs) [28–30] have been realized. In [25], an STA was built based on an SIW cavity loaded with two bowtie slots. Two SIW cavities, one of which is incorporated within the other, have been used to build a self-triplexing antenna [26]. In [27], a self-triplexing antenna was designed using a square SIW cavity loaded with a modified I-shaped slot. In [28], a substrate-integrated rectangular cavity was used to create a self-quadruplexing antenna. A self-hexaplexing antenna based on a substrate-integrated rectangular cavity loaded with two Pi-shaped slots was constructed [29]. In [30], a self-multiplexing antenna (SMA) was constructed using a rectangular SIW cavity loaded with modified radiating patches. Despite the satisfactory performance of the aforementioned antenna topologies, designing a highly miniaturized self-triplexing antenna featuring high port isolation remains a strenuous endeavor.

This paper presents an ultracompact self-triplexing antenna constructed on a substrate-integrated rectangular cavity (SIRC). Two L-shaped slots and an inverted U-shaped slot were used by the proposed structure to radiate at three separate operational frequency bands. The operating frequency can be independently altered using the slot size. For validation, a prototype was manufactured and tested at 1.94 GHz, 4.4 GHz, and 5.3 GHz. As demonstrated by both EM simulations and measurements, the isolations were better than 32.4 dB. The observed realized gains of the antenna are greater than 4.3 dBi. The unique technical contributions of the work are as follows:

(i) A systematic design approach for SIRC cavity-based slot antennas is presented.
(ii) We suggest, fabricate, and test an ultracompact single-layer tri-band antenna with high isolation and gain in all three bands.
(iii) The proposed STA configuration exhibits a highly miniaturized footprint area.
(iv) It is possible to tune all three bands independently without disturbing the other radiating frequency bands.

(v) For a better comprehension of the working principle of the proposed STA, an equivalent lumped-circuit model is developed.
(vi) Port one is orthogonally connected to ports two and three, resulting in a very weak cross-coupling path between ports and increased isolation. As a result, all three ports achieve excellent isolation of greater than 32.42 dB.

2. Development and Analysis of the Proposed SIRC-Based Self-Triplexing Antenna

2.1. Antenna Configuration

Figure 1 shows the layout of an ultracompact self-triplexing antenna (STA). The proposed STA uses a substrate-integrated rectangular cavity (SIRC), a U-shaped slot (USS), two L-shaped slots (LSS), and three microstrip feed lines. The design steps followed to realize the STA are shown in Figure 2. In Step 1, an SIRC with the dimensions of 26 mm × 34 mm is designed to operate at 4.65 GHz in the fundamental mode (TE_{110}). In Step 2, three microstrip 50 Ω lines are used to feed the cavity to obtain a self-triplexing property. The matching at each port is obtained by employing inline feeds. Since port 1 is orthogonal to ports 2 and port 3, a very weak cross-coupling path is formed that enables maximum port isolation. In Step 3, an inverted U-shaped slot (USS) is created on the top plane and excited by port 1, resulting in the first radiating frequency band and an isolation greater than 20 dB. 1. The resonant frequency of the fundamental mode is shifted towards a lower value as a result of the suggested U-shaped slot loading. This reduction in resonant frequency from the SIW cut-off frequency results in an ultracompact antenna footprint. This provides a size reduction factor of 80.4% for the suggested STA. Finally, two L-shaped slots (LSS) are attached to both sides of the USS, which results in the generation of two more radiating bands and provides freedom to control the isolation level. These slots create a very weak cross-coupling path between ports, which results in isolation greater than 32 dB. The suggested STA is optimized using CST Microwave Studio to operate at 1.92 GHz, 4.43 GHz, and 5.25 GHz. For a better understanding of the radiation mechanism, the surface current density at each port is demonstrated, as shown in Figure 3. It can be inferred from the figure that the maximum current flows through the port that has been excited, and almost negligible current passes through the other ports.

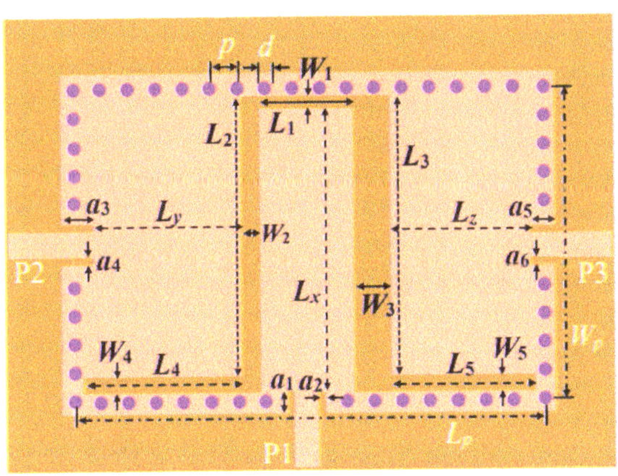

Figure 1. Parameterized geometry of the proposed SIRC-based self-triplexing antenna. The design parameter values are: L_p = 34, W_p = 26, L_1 = 7.0, L_2 = 24.6, L_3 = 24.6, L_4 = 10.95, L_5 = 8.65, W_1 = 0.8139, W_{12} = 0.8139, W_2 = 1.1168, W_3 = 4.1696, W_4 = 1.5, W_5 = 1.5, a_1 = 2.111, a_2 = 0.511, a_3 = 3.2191, a_4 = 0.5922, a_5 = 3.0945, a_6 = 0.5302, d = 1.0, p = 2.0; unit: mm.

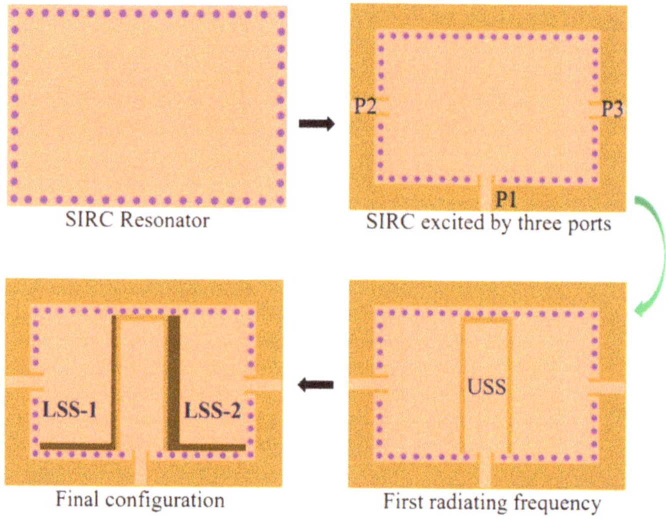

Figure 2. Development steps of the proposed STA.

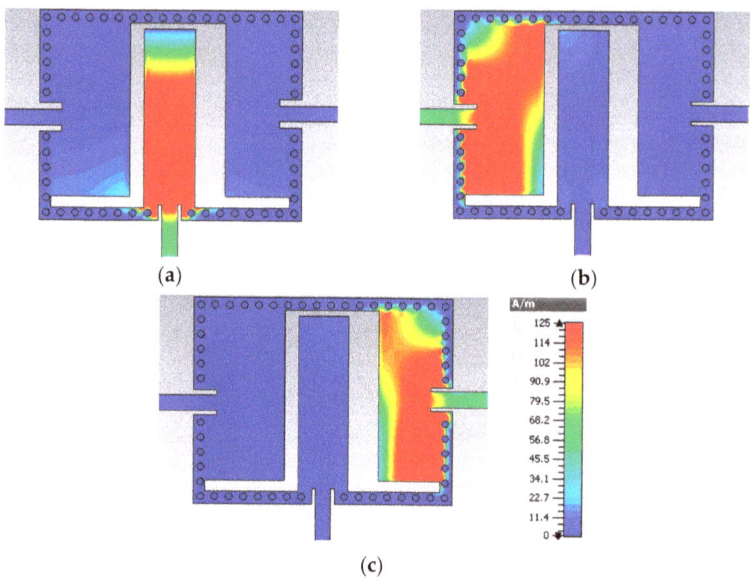

Figure 3. Surface current density by excitation at individual ports: (**a**) Port 1; (**b**) Port 2; (**c**) Port 3.

2.2. Equivalent Lumped-Circuit Model

The working principle of the STA was explained by developing an equivalent lumped-circuit model (LCM), as presented in Figure 4. The cavity-backed radiating aperture is expressed as shunt-connected resistance (R_{xj}), capacitance (C_{xj}), and inductance (L_{xj}), with j = 1, 2, and 3 standing for ports 1, 2, and 3, respectively. The U-shaped slot provides extra capacitive loading that is represented as shunt-connected capacitance (C_{y1}), which is responsible for producing the first radiation frequency. The L-shaped slots excited by ports 2 and 3 provide extra capacitive loadings, which are expressed as the shunt-connected capacitances C_{y2} and C_{y3}, respectively. These capacitive loadings are responsible for the second and third radiating frequencies. Since the capacitances C_{y1}, C_{y2}, and C_{y3} depend on the geometry of the USS and LSS, the radiating frequencies can be controlled by altering the corresponding slot dimensions. Series-connected inductance (L_k) and capacitance are used to describe the mutual connections between the ports (C_k). Impedance matching between cavity-backed patches and feed lines is accomplished using impedance transformers. The Keysight Advanced Design System simulates the proposed equivalent network model, and Table 1 lists the optimized component values. Figure 5 displays the circuit and simulated EM reflection coefficients.

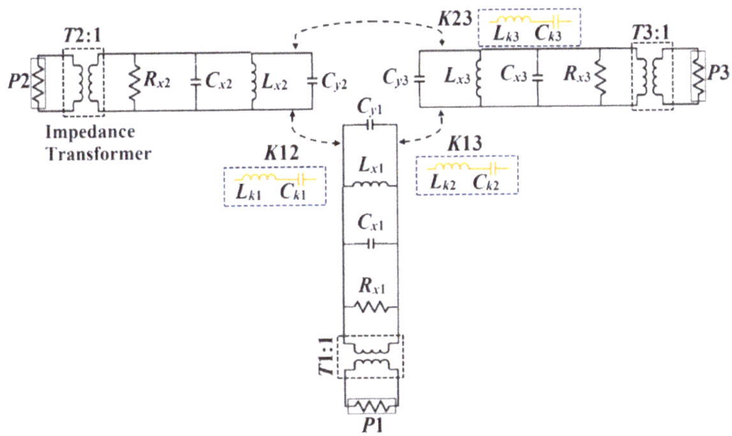

Figure 4. Equivalent lumped-circuit model of the proposed STA.

Table 1. Lumped-circuit model component values.

R_{x1} (Ω)	L_{x1} (nH)	C_{x1} (pF)	C_{y1} (pF)	L_{k1} (nH)	C_{k1} (pF)	T_1
239	0.373	18.6007	0.012	12.914	12.1519	0.4597
R_{x2} (Ω)	L_{x2} (nH)	C_{x2} (pF)	C_{y2} (pF)	L_{k2} (nH)	C_{k2} (pF)	T_2
281	0.189	6.4557	0.399	7.103	10.6529	0.429
R_{x3} (Ω)	L_{x3} (nH)	C_{x3} (pF)	C_{y3} (pF)	L_{k3} (nH)	C_{k3} (pF)	T_3
471	0.2495	1.9587	1.7988	13.808	0.4597	0.329

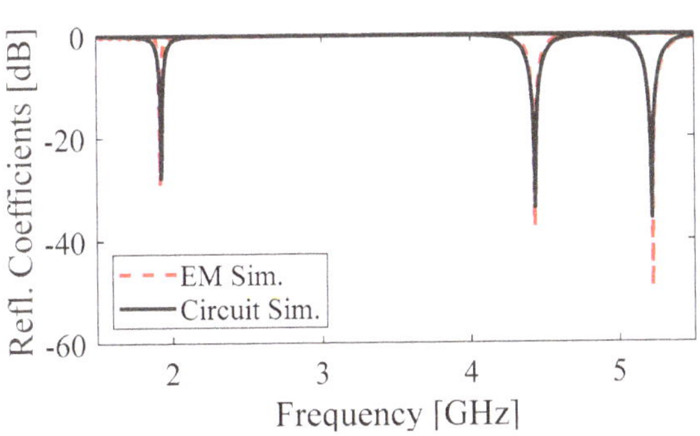

Figure 5. EM simulation and circuit model simulation of the suggested antenna.

2.3. Independent Frequency Tunability

The proposed STA operating at 1.92 GHz, 4.43 GHz, and 5.25 GHz is suitable for, but not limited, to Global System for Mobile Communications (GSM), 5G, and WLAN applications. However, the dimensions of inverted U-shaped and L-shaped slots can be varied to tune the operating frequencies f_{01} (1.93 GHz to 2.80 GHz), f_{02} (4.16 GHz to 5.60 GHz), and f_{03} (4.36 GHz to 5.73 GHz), which support various applications such as WiFi, Bluetooth, Extended PCS, 5G, WiMAX, and WLAN wireless standards. The frequency tunability range in relation to the design parameters and supporting application bands is shown in Table 2. It is clear that the suggested STA is appropriate for a variety of communication standards. Parametric studies were carried out with respect to the radiating patch height to establish a broad concept of independent control of the resonant frequencies, as shown in Figure 6. The heights of L_x, L_y, and L_z corresponding to USS and LSS are employed for the tunability of f_{01}, f_{02}, and f_{03}, respectively. From Figure 6, it can be inferred that the radiating frequency increases by decreasing the height of the radiating apertures. Referring to Figure 6a, the radiating frequency f_{o1} increases from 1.93 GHz to 2.80 GHz by decreasing the height L_x from 23.09 mm to 15.09 mm. When the height L_y decreases from 10.35 mm to 6.35 mm, the operating frequency f_{o2} increases from 4.16 GHz to 5.60 GHz, as depicted in Figure 6b. From Figure 6c, it is found that by decreasing the height from 10.13 mm to 6.63 mm, the radiating frequency increases from 4.36 GHz to 5.73 GHz. Moreover, it can be concluded that when one of the frequency bands is changed, the other bands remain unaffected.

Table 2. Tunable range of the proposed STA and possible application standards.

Parameter Range	Frequency Range	Application Standards
$15.09 \leq L_x \leq 23.09$ (port 1)	$1.93 \leq f_{01} \leq 2.80$	GSM/WiFi/ISM/Bluetooth/LTE/Extended PCS
$6.35 \leq L_y \leq 10.35$ (port 2)	$4.16 \leq f_{02} \leq 5.60$	5G/WLAN/WiFi/C-band
$6.63 \leq L_z \leq 10.13$ (port 3)	$4.63 \leq f_{03} \leq 5.73$	5G/WLAN/WiFi/Standard C-band/LTE

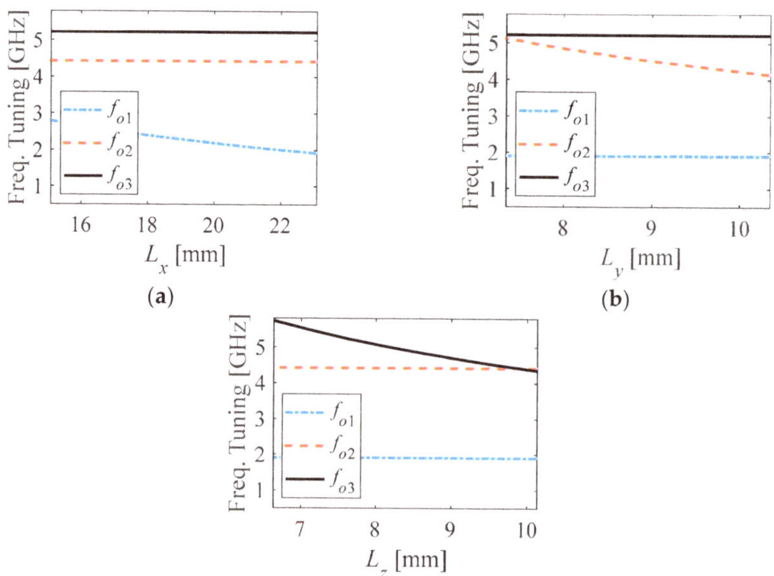

Figure 6. Independent frequency tunability of the proposed STA: (**a**) f_{o1} (port 1); (**b**) f_{o2} (port 2); (**c**) f_{o3} (port 3).

3. Fabrication, Measurement, and Results Discussion

As a proof of concept, a prototype of the self-triplexing antenna (STA) based on the SIRC was devised, manufactured, and experimentally validated. The prototype was fabricated on a low-loss 0.762 mm thick Arlon AD250 substrate with a permittivity of 2.5 and a loss tangent of 0.0014. The proposed STA is ultracompact with a size smaller than $0.086\lambda_g^2$. Photographs of the STA prototype and the experimental setup are displayed in Figures 7 and 8, respectively. The comparison between the EM-simulated and tested S-parameters is shown in Figure 9. The measured reflection coefficients were recorded as −24.11 dB, −28.76 dB, and −25.73 dB, whereas the EM simulated reflection coefficients were obtained as −29.03 dB, −37.6 dB, and −49.5 dB at 1.92 GHz, 4.43 GHz, and 5.25 GHz, respectively. The port isolations for the EM simulation and experiment were greater than 32.42 dB at each band, as shown in Figure 10. Considering a 10 dB reflection coefficient level, the measured impedance bandwidths were computed as 0.93%, 1.17%, and 1.88%, whereas the EM simulated impedance bandwidths were found to be 1.36%, 1.58%, and 1.89% at 1.92 GHz, 4.43 GHz, and 5.25 GHz, respectively. As the narrow band is coming, we can use it for pinpoint applications in the mentioned commercial wireless application band, such as Global Systems for Mobile Communications (GSM) (at 1.92 GHz), 5G (at 4.43 GHz), and WLAN (at 5.25 GHz). If a thicker substrate with the same dielectric constant is utilized, then the bandwidth and the radiation efficiency will increase [31–33]. Farfield radiation patterns were observed and recorded in an automatic anechoic chamber. A single port was excited at each measurement, with the remaining ports terminated using a matched 50-ohm load. The estimated realized gains and EM-simulated radiation efficiencies are shown in Figures 11 and 12, respectively. In Figure 11, the measured realized gains are recorded as 4.10 dBi, 5.59 dBi, and 4.94 dBi, whereas the EM simulated realized gains are calculated as 3.8 dBi, 5.3 dBi, and 4.86 dBi at 1.92 GHz, 4.43 GHz, and 5.25 GHz, respectively. A small discrepancy was found between the tested and EM-simulated realized gains. The deviation was attributed to the precision of the radiating aperture, imperfection of substrate properties, fabrication, and SMA connector. However, the deviation of the realized gains is less than 7.31%, 5.18%, and 1.62% at center frequencies of 1.92 GHz, 4.43 GHz, and 5.25 GHz, respectively. Referring to Figure 12, it was discovered that the proposed STA's radiation efficiency was higher than 82% across all bands. Table 3 provides an overview of the measured reflection coefficients, isolations, realized gains, and EM simulated radiation efficiencies. Figure 13 shows the normalized radiation patterns of the proposed STA. Linear polarization and unidirectional patterns were obtained using the STA prototype. As expected, an overall excellent agreement between the LCM model results, the EM simulation results, and the measurement results was observed. Nevertheless, minor differences were noticed, resulting from the connection loss and manufacturing tolerances.

(a)

(b)

Figure 7. Fabricated prototype: (**a**) Front view; (**b**) Back view.

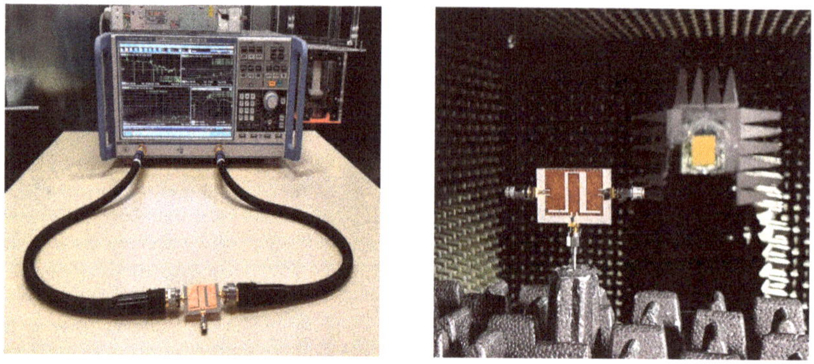

(**a**) (**b**)

Figure 8. Measurement setup for the suggested STA: (**a**) Reflection coefficients and isolations; (**b**) Radiation pattern and realized gain.

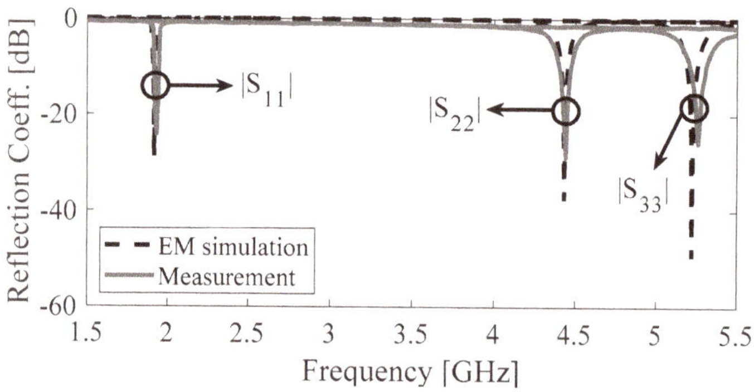

Figure 9. EM simulation and measurement reflection coefficients of the suggested STA.

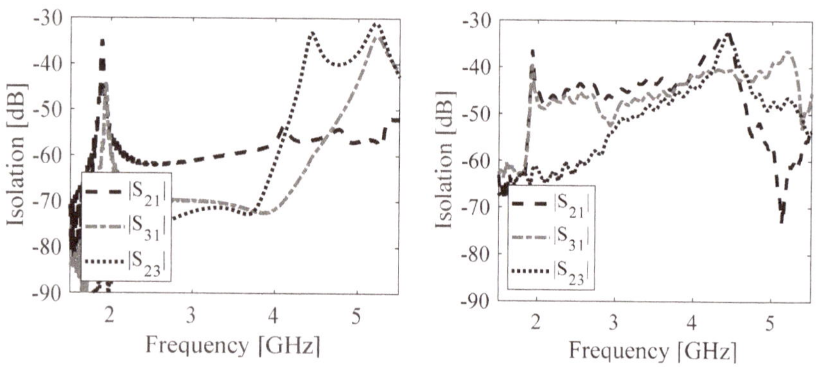

(**a**) (**b**)

Figure 10. Port isolation of the proposed STA: (**a**) EM simulation; (**b**) measurement.

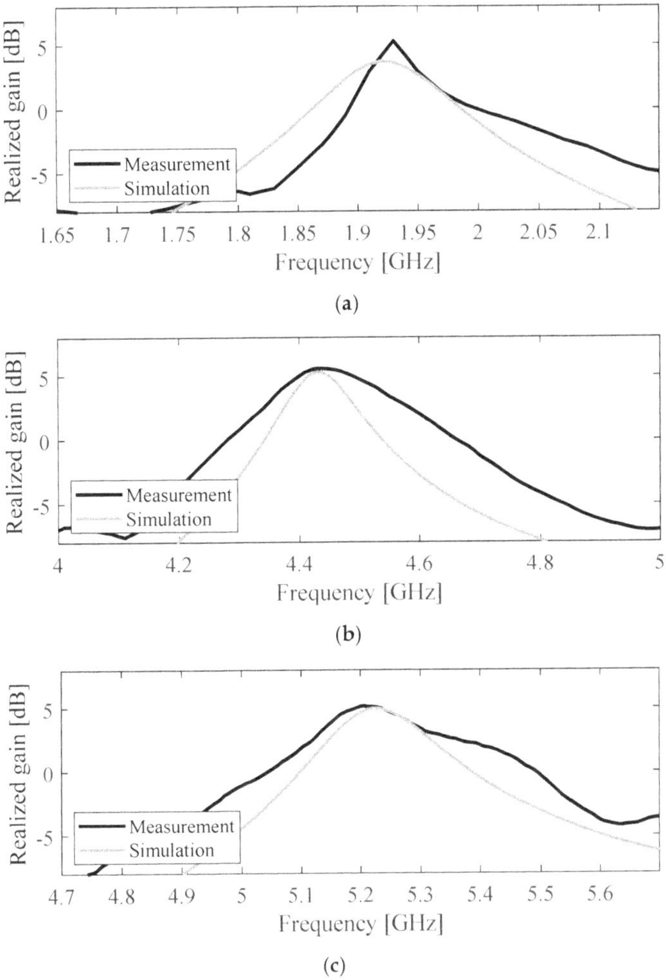

Figure 11. EM simulation and measured realized gains of the proposed STA: (**a**) Port 1; (**b**) Port 2; (**c**) Port 3.

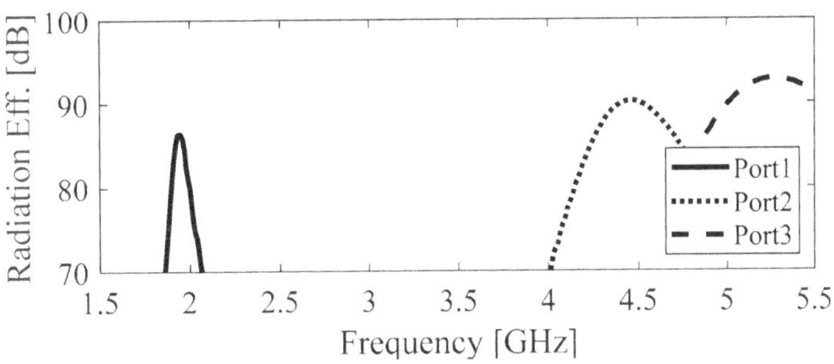

Figure 12. EM-simulated radiation efficiencies at different port excitation.

Table 3. Measured reflection coefficients, isolations, realized gains, and EM-simulated radiation efficiencies.

Ports	Freq. (GHz)	Reflection Coeff. (dB)	ISL (dB)	Realized Gain (dBi)	Radiation Efficiency (%)
1	1.92	24.11	32.42	4.10	86.4
2	4.43	28.76	32.74	5.59	90.24
3	5.25	25.73	39.45	4.94	92.84

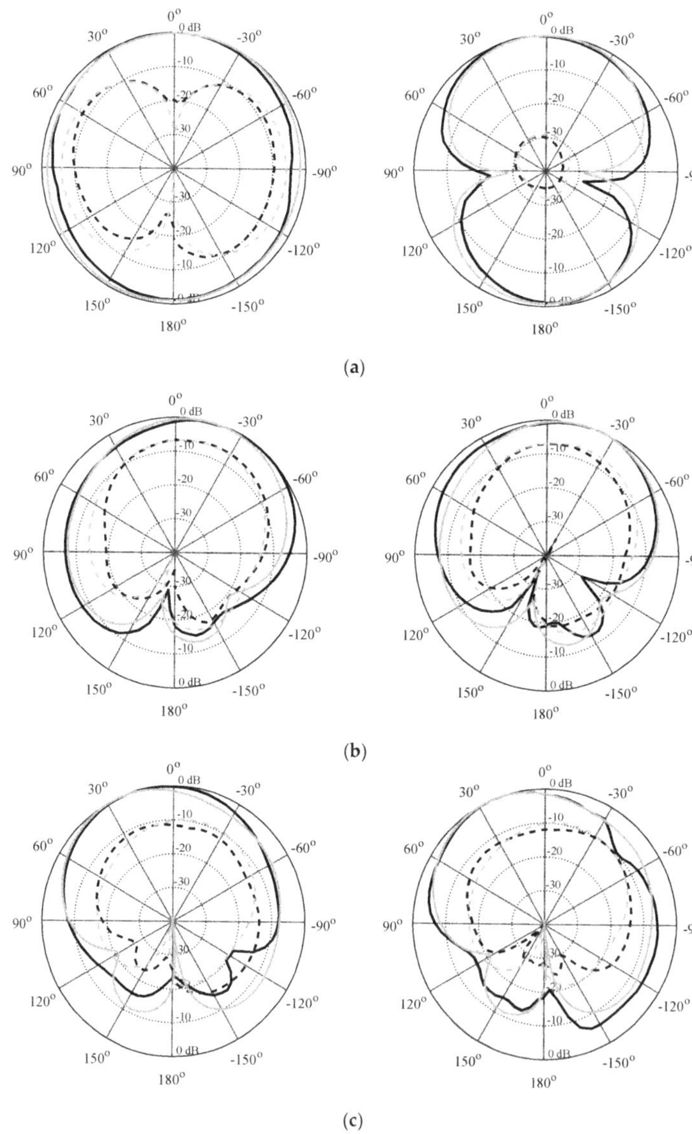

Figure 13. Normalized radiation pattern of the suggested STA associated with (**a**) port 1, (**b**) port 2, and (**c**) port 3. [Simulation—grey, measurement—black); H-plane (**left**), E-plane (**right**); copol—solid, crosspol—dashed].

Table 4 shows a performance comparison analysis of the proposed STA and previously published STAs or triband antennas. The proposed STA achieves an ultracompact size, high isolation, and excellent realized gain, according to the table. The proposed STA has a smaller footprint area than those in [2,5,7,10,11,25–27]. Furthermore, the suggested STA is 50% smaller than the most compact STA reported in [27], and there is no significant difference in the realized gains. The proposed STA has a minimum port isolation of 32.4 dB across all bands, which is greater than the isolation level obtained in previously documented antennas [2,5,7,10,11,25–27]. It should be mentioned that the proposed STA includes a lumped-circuit model to analyze the antenna's working principle, whereas, with the exception of [26], no lumped-circuit model was provided for antenna analysis in [2,5,7,10,11,25,27]. These findings indicate that the proposed STA is potentially more attractive for use in triband wireless communication networks.

Table 4. Performance of the suggested STA and previously reported STAs.

Ref.	Opertion	Self-Triplexing	Size (λ_g^2)	ISL (dB)	Gain (dBi)	LC Model
[2]	Tri-band	No	0.195	NA	2.1/1.9/5.1	No
[5]	Tri-band	No	0.462	>30	5.0/6.0/6.5	No
[7]	Tri-band	No	0.232	NA	5.05/4.58/3.17	No
[10]	Tri-band	No	0.912	NA	1.1/0.9/2.1	No
[11]	Tri-band	Yes	0.28	>19	0.85/4.0/4.23	No
[25]	Tri-band	Yes	0.42	>22.5	7.2/7.2/7.0	No
[26]	Tri-band	Yes	0.69	>26	4.5/5.9/6.0	Yes
[27]	Tri-band	Yes	0.17	>30.8	4.26/4.41/6.27	No
This work	Tri-band	Yes	0.086	>32.4	4.10/5.59/4.94	Yes

ISL: Isolation, λ_g: Guided wavelength at the lower frequency band, NA: Not available.

4. Conclusions

This work introduced an ultracompact self-triplexing antenna realized on a substrate-integrated rectangular cavity (SIRC). For transmission in three distinct operational frequency bands, the proposed antenna employs two L-shaped slots and an inverted U-shaped slot. To stimulate the radiation in these slots, three 50-ohm microstrip feed lines were utilized. The operating frequency can be individually controlled using the slot size. The slot arrangement and size were chosen with the purpose of obtaining one or more frequency bands below the SIRC cutoff frequency, which allowed for an ultracompact size. High port isolation was achieved by connecting one port orthogonally to two other ports, resulting in a weak cross-coupling channel. To explain the operating principles of the proposed antenna, a lumped-circuit model was developed and analyzed. The proposed antenna was prototyped and experimentally validated at 1.92 GHz, 4.43 GHz, and 5.25 GHz for GSM, 5G, and WLAN applications, respectively. The prototype was manufactured using a low-loss 0.762 mm thick Arlon AD250 substrate with a relative permittivity of 2.5. In the EM simulations and tests, the isolations were more than 30 dB. The estimated realized gains for the suggested antenna are more than 4.1 dBi. The agreement between the full-wave EM analysis results and experimental data is satisfactory with respect to all antenna characteristics. The proposed antenna structure is suitable for a number of applications, which include the Global System for Mobile Communications (GSM), 5G mobile technology, and WLAN applications, among others. Upon appropriate changes in antenna dimensions, it can also support a broad range of other applications, such as WiFi, Bluetooth, Extended PCS, and WiMAX.

Author Contributions: Conceptualization, R.K.B. methodology, R.K.B. and S.K.; software, R.K.I and S.K.; validation, R.K.B. and S.K.; formal analysis, R.K.B.; investigation, R.K.B.; resources, S.K data curation, R.K.B. and S.K.; writing—original draft preparation, R.K.B.; writing—review anc editing, S.K.; visualization, R.K.B. and S.K.; supervision, S.K.; project administration, S.K.; fundin acquisition, S.K. All authors have read and agreed to the published version of the manuscript.

Funding: The research leading to these results has received funding from the Norway Grants 2014-202 via the National Centre for Research and Development, grant NOR/POLNOR/HAPADS/0049/2019-0(This work was also supported in part by the Icelandic Centre for Research (RANNIS) Grant 217771

Data Availability Statement: Not applicable.

Acknowledgments: The authors would like to thank Dassault Systemes, France, for making CST Microwave Studio available.

Conflicts of Interest: The authors declare no conflict of interest.

References

1. Chen, S.C.; Wang, Y.S.; Chung, S.J. A decoupling technique for increasing the port isolation between two strongly coupled antennas. *IEEE Trans. Antennas Propag.* **2008**, *56*, 3650–3658. [CrossRef]
2. Ahmad, S.; Boubakar, H.; Naseer, S.; Alim, M.E.; Sheikh, Y.A.; Ghaffar, A.; Al-Gburi, A.J.A.; Parchin, N.O. Design of a Tri-Band Wearable Antenna for Millimeter-Wave 5G Applications. *Sensors* **2022**, *22*, 8012. [CrossRef]
3. Wang, D.; Chan, C.H. Multiband antenna for WiFi and WiGig communications. *IEEE Antennas Wirel. Propag. Lett.* **2015**, *15* 309–312. [CrossRef]
4. Anguera, J.; Andújar, A.; García, C. Multiband and small coplanar antenna system for wireless handheld devices. *IEEE Trans Antennas Propag.* **2013**, *61*, 3782–3789. [CrossRef]
5. Cui, Y.; Wang, X.; Shen, G.; Li, R. A triband SIW cavity-backed differentially fed dual-polarized slot antenna for WiFi/5G applications. *IEEE Trans. Antennas Propag.* **2020**, *68*, 8209–8214. [CrossRef]
6. Bozzi, M.; Georgiadis, A.; Wu, K. Review of substrate-integrated waveguide circuits and antennas. *IET Microw. Antennas Propag* **2011**, *5*, 909–920. [CrossRef]
7. Zheng, Z.; Li, D.; Tan, X.; Chen, Q. Single-Layer Dual-/Tri-band SIW Filtenna Based on Multifunctional Cavity-Backed Slots *IEEE Trans. Antennas Propag.* **2023**, *71*, 4498–4503. [CrossRef]
8. Niu, B.J.; Tan, J.H. Half-mode SIW cavity antenna for tri-band MIMO applications. *Microw. Opt. Technol. Lett.* **2020**, *62*, 1697–1701 [CrossRef]
9. Niu, B.J.; Tan, J.H. Compact tri-band MIMO antenna based on quarter-mode slotted substrate-integrated-waveguide cavity. *Int. J. RF Microw. Comput.-Aided Eng.* **2020**, *30*, e22101. [CrossRef]
10. Mandal, B.; Parui, S.K. Wearable tri-band SIW based antenna on leather substrate. *Electron. Lett.* **2015**, *51*, 1563–1564. [CrossRef]
11. Cheong, P.; Chang, K.F.; Choi, W.W.; Tam, K.W. A highly integrated antenna-triplexer with simultaneous three-port isolations based on multi-mode excitation. *IEEE Trans. Antennas Propag.* **2014**, *63*, 363–368. [CrossRef]
12. Nandi, S.; Mohan, A. An SIW cavity-backed self-diplexing antenna. *IEEE Antennas Wirel. Propag. Lett.* **2017**, *16*, 2708–2711. [CrossRef]
13. Barik, R.K.; Cheng, Q.S.; Dash, S.K.; Pradhan, N.C.; Subramanian, K.S. Design of a compact orthogonal fed self-diplexing bowtie-ring slot antenna based on substrate integrated waveguide. *Int. J. RF Microw. Comput.-Aided Eng.* **2020**, *30*, e22422. [CrossRef]
14. Pradhan, N.C.; Subramanian, K.S.; Barik, R.K.; Cheng, Q.S. A shielded-QMSIW-based self-diplexing antenna for closely spaced bands and high isolation. *IEEE Antennas Wirel. Propag. Lett.* **2021**, *20*, 2382–2386. [CrossRef]
15. Barik, R.K.; Koziel, S.; Cheng, Q.S.; Szczepanski, S. Highly Miniaturized Self-Diplexed U-Shaped Slot Antenna Based on Shielded QMSIW. *IEEE Access* **2021**, *9*, 158926–158935. [CrossRef]
16. Priya, S.; Kumar, K.; Dwari, S.; Mandal, M.K. Circularly polarized self-diplexing SIW cavity backed slot antennas. *IEEE Trans. Antennas Propag.* **2019**, *68*, 2387–2392. [CrossRef]
17. Dash, S.K.K.; Cheng, Q.S.; Barik, R.K.; Pradhan, N.C.; Subramanian, K.S. A compact substrate integrated self-diplexing antenna for WiFi and ISM band applications. In Proceedings of the 50th European Microwave Conference (EuMC), Utrecht, The Netherlands, 12–14 January 2021; pp. 232–235.
18. Barik, R.K.; Cheng, Q.S.; Dash, S.K.K.; Pradhan, N.C.; Karthikeyan, S.S. Compact high-isolation self-diplexing antenna based on SIW for C-band applications. *J. Electromagn. Waves Appl.* **2020**, *34*, 960–974. [CrossRef]
19. Iqbal, A.; Al-Hasan, M.; Mabrouk, I.B.; Nedil, M. Ultracompact quarter-mode substrate integrated waveguide self-diplexing antenna. *IEEE Antennas Wirel. Propag. Lett.* **2021**, *20*, 1269–1273. [CrossRef]
20. Iqbal, A.; Tiang, J.J.; Lee, C.K.; Mallat, N.K. SIW cavity backed self-diplexing tunable antenna. *IEEE Trans. Antennas Propag.* **2021**, *69*, 5021–5025. [CrossRef]
21. Dash, S.K.K.; Cheng, Q.S.; Barik, R.K.; Khan, T.; Subramanian, K.S. A compact dual-fed highly isolated SIW based self-diplexing antenna. *AEU-Int. J. Electron. Commun.* **2021**, *132*, 153613. [CrossRef]

22. Pradhan, N.C.; Subramanian, K.S.; Barik, R.K.; Dalal, P.; Cheng, Q.S. Design of Compact Shielded QMSIW Based Self-Diplexing Antenna for High-Isolation. In Proceedings of the IEEE Wireless Antenna and Microwave Symposium (WAMS), Rourkela, India, 5–8 June 2022; pp. 1–4.
23. Pradhan, N.C.; Sholampettai, S.K.; Barik, R.K.; Cheng, Q.S. A shielded QMSIW ultra-compact self-diplexing antenna for WiMAX/WLAN applications. *J. Electromagn. Waves Appl.* **2022**, *36*, 1869–1881. [CrossRef]
24. Pradhan, N.C.; Reddy, M.G.; Subramanian, K.S.; Barik, R.K.; Koziel, S.; Cheng, Q.S. Microfluidic SIW-Based Tunable Self-Diplexing Antenna for Sub-6 GHz Band Applications. *IEEE Trans. Circuits Syst. II Express Briefs* **2022**, *70*, 1435–1439. [CrossRef]
25. Kumar, K.; Dwari, S. Substrate integrated waveguide cavity-backed self-triplexing slot antenna. *IEEE Antennas Wirel. Propag. Lett.* **2017**, *16*, 3249–3252. [CrossRef]
26. Iqbal, A.; Selmi, M.A.; Abdulrazak, L.F.; Saraereh, O.A.; Mallat, N.K.; Smida, A. A compact substrate integrated waveguide cavity-backed self-triplexing antenna. *IEEE Trans. Circuits Syst. II Express Briefs* **2020**, *67*, 2362–2366. [CrossRef]
27. Dash, S.K.K.; Cheng, Q.S.; Barik, R.K.; Pradhan, N.C.; Subramanian, K.S. A compact triple-fed high-isolation SIW-based self-triplexing antenna. *IEEE Antennas Wirel. Propag. Lett.* **2020**, *19*, 766–770. [CrossRef]
28. Barik, R.K.; Koziel, S. Highly-Miniaturized Self-Quadruplexing Antenna Based on Substrate-Integrated Rectangular Cavity. *IEEE Antennas Wirel. Propag. Lett.* **2023**, *22*, 482–486. [CrossRef]
29. Barik, R.K.; Koziel, S. A Compact Self-Hexaplexing Antenna Implemented on Substrate-Integrated Rectangular Cavity for Hexa-Band Applications. *IEEE Trans. Circuits Syst. II Express Briefs* **2023**, *70*, 506–510. [CrossRef]
30. Dash, S.K.K.; Cheng, Q.S.; Barik, R.K.; Jiang, F.; Pradhan, N.C.; Subramanian, K.S. A compact SIW cavity-backed self-multiplexing antenna for hexa-band operation. *IEEE Trans. Antennas Propag.* **2021**, *70*, 2283–2288. [CrossRef]
31. Harrington, R.F. Effect of antenna size on gain, bandwidth, and efficiency. *J. Res. Natl. Bur. Stand. Sect. D Radio Propag.* **1960**, *64*, 1–12. [CrossRef]
32. Luo, G.Q.; Hu, Z.F.; Dong, L.X.; Sun, L.L. Planar slot antenna backed by substrate integrated waveguide cavity. *IEEE Antennas Wirel. Propag. Lett.* **2008**, *7*, 236–239.
33. Jin, C.; Li, R.; Alphones, A.; Bao, B. Quarter-mode substrate integrated waveguide and its application to antennas design. *IEEE Trans. Antennas Propag.* **2013**, *61*, 2921–2928. [CrossRef]

Disclaimer/Publisher's Note: The statements, opinions and data contained in all publications are solely those of the individual author(s) and contributor(s) and not of MDPI and/or the editor(s). MDPI and/or the editor(s) disclaim responsibility for any injury to people or property resulting from any ideas, methods, instructions or products referred to in the content.

electronics

MDPI

Article

A THz Slot Antenna Design Using Analytical Techniques

Shay Rozenberg and Asher Yahalom *

Department of Electrical & Electronic Engineering, Ariel University, Ariel 40700, Israel
* Correspondence: asya@ariel.ac.il

Abstract: Slot antennas are very popular microwave antennas, and slotted waveguides are used for high frequency radar systems. A thin slot in an infinite ground plane is the complement to a dipole in free space. This was described by H.G. Booker, who extended Babinet's principle from optics to show that the slot will have the same radiation pattern as a dipole such that the E and H fields are swapped. As a result, the polarization is rotated, so that radiation from the vertical slot is polarized horizontally. In this work, we show how this straightforward analytical technique can be used for the design of high-frequency THz slot antennas. The analysis is then corroborated by using a numerical simulation which validates the performance parameters predicted by the analytical technique. We show by simulation that, despite the simplicity of our classical approach, we obtained useful results, even at THz frequencies. We show that gradually moving the slot position from the centerline improves the antenna's performance.

Keywords: THz; slot antenna; Babinet's principle

1. Introduction

Slot antennas are very popular microwave antennas, and slotted waveguides are used for high frequency radar systems [1–16]. A thin slot in an infinite ground plane is the complement to a dipole in free space. This was described by H.G. Booker [2], who extended Babinet's principle from optics to show that the slot will have the same radiation pattern as a dipole such that the E and H fields are swapped. As a result, the polarization is rotated, so that radiation from the vertical slot is polarized horizontally. For instance, a vertical slot has the same pattern as a horizontal dipole with the same dimensions, and we are able to calculate the radiation pattern of a dipole. Thus, a longitudinal slot in the broad wall of a waveguide radiates just like a dipole perpendicular to the slot. By using this principle, it is easier to analyze slot antennas using the theory of dipole antennas. The slots are typically thin and 0.5 wavelengths long. The position of the slots affects the intensity of the transmitted power by directly causing the impedance matching of the antenna; good impedance matching produces the maximum efficiency of power transmission. In [15], a waveguide-fed transverse slot was analyzed using the Method of Moments (MoM) formulation, wherein the internal admittance matrix elements were evaluated from image theory and using the images of the slot from waveguide walls. The contribution from all the images not in the same row as the slot was evaluated using a closed-form expression and a rapidly converging series with negligible contribution. The contribution from the row containing the reference slot was evaluated by adding the mutual admittance between a few images and the reference slot using some recently developed simple and accurate approximations. The formulation was extended to transverse slots radiating between the baffles. The results calculated using this simple, efficient, and intuitive method were in excellent agreement with other theoretical as well as measured results. The technique can be extended to other slot configurations, such as an inclined slot, and for other slot aperture distributions in configurations permitting image formation, such as with an edge condition. Despite the fact that we do not consider images, our results are satisfactory in that the simulation results show that the antenna gain is 18.4 dBi and the azimuth beam width is

Citation: Rozenberg, S.; Yahalom, A. A THz Slot Antenna Design Using Analytical Techniques. *Electronics* **2023**, *12*, 2233. https://doi.org/10.3390/electronics12102233

Academic Editors: Syed Muzahir Abbas, Yang Yang and Liming Si

Received: 9 March 2023
Revised: 1 May 2023
Accepted: 10 May 2023
Published: 14 May 2023

Copyright: © 2023 by the authors. Licensee MDPI, Basel, Switzerland. This article is an open access article distributed under the terms and conditions of the Creative Commons Attribution (CC BY) license (https://creativecommons.org/licenses/by/4.0/).

1.7°, compared to 21.8 dBi and 0.7° in the analytical design. It may be that adding images may reduce the discrepancy further; however, this is beyond the scope of the current work.

In [14], a straightforward design procedure for a slotted antenna fed by a single-ridge waveguide is introduced. The proposed method does not use the analytical method to derive the general relation between the backward- and forward-scattering dominant mode coefficients and slot voltage, as is done in the current paper. Rather, in [14], the general relation connecting the backward- and forward-scattering dominant mode coefficients to the slot electric field distribution is derived using an available electromagnetic field simulator and by fitting some proper polynomials to the derived data. Then, the well-known Elliott's design procedure for arrays of longitudinal slots in the broad wall of an air-filled rectangular waveguide is developed for the slot array feed by a single-ridge waveguide. A four-by-five array is designed and simulated which verifies the design specifications and the proposed design procedure. Compared with that approach, our approach is simpler, more analytical, and straightforward.

In this paper, an antenna based on a slotted waveguide is analyzed analytically and designed for the THz band (frequencies of about 330 GHz).

Such antennas may be useful for multiple applications. Medical applications include tooth imaging, cancer treatment and detection, fungus treatment [17–25], and surgery.

The existence of unique spectral lines in the THz band allows for the characterization and analysis of materials. Thus, a THz antenna can be useful to monitor the composition of the atmosphere, including gases and particles which cause air pollution.

Electromagnetic energy transfer to drones and balloons using THz radiation may elongate the duration of their mission without the need to land for refueling. For this application, the shorter wavelength of THz is an advantage as this allows a higher gain for a smaller antenna.

There are also security applications of the suggested THz antenna in places such as airports to discover concealed weapons or illegal drugs. In this case, passengers are irradiated by the non-ionizing and harmless THz radiation, and the reflected radiation is used to create an image which delineates the hidden object. Furthermore, spectroscopical analysis of the THz radiation reveals the nature of the hidden object, determining if it is a harmless candy or a dangerous illegal drug.

Communication applications may include point-to-point or point-to-multipoint and "last mile" connections.

The structure of the paper is as follows: Section 2 will be devoted to the THz spectral range and its applications. In Section 3, the antenna design is presented, and the waveguide dimensions and the operation frequency are briefly discussed. First, we describe a single slot on a rectangular waveguide and show that it can be described using a model of a transmission line. By using this model, equations that relate the slot position to the transmitted power are developed. This model was expanded for an array of slots on a rectangular waveguide in order to design the slot antenna. We show that gradually moving the slot position from the centerline improves the antenna's performance. In Section 4, an antenna based on a slotted rectangular waveguide in the THz band was designed and simulated. The conclusions are given in Section 5.

2. The Importance of the THz Frequency Band

Terahertz radiation occupies a large portion of the electromagnetic spectrum between infrared and millimeter waves. The common definition of the terahertz range is the spectral range of 100 GHz −10 THz (see Figure 1). This corresponds to wavelengths from 30 μm to 3 mm or approximately 3 cm^{-1} to 300 cm^{-1} on the wave number scale.

The THz range is important as it allows the identification of molecules based on their total structure, thus making chemical identification of closely resembling molecules possible. Moreover, THz radiation is 'soft' and non-ionizing. The high intensity of the thermal background radiation and a lack of robust sources and sensitive detectors restricted for many years both exploration and applications in the THz range. For a long time, the submillimeter radiation could only be obtained from weak blackbody radiators and a few gas lasers which could only be produced in laboratory environments. This situation changed around 1990, when advances in the coherent generation and detection of short pulses of THz radiation initiated an intensive development of basic and applied studies. At present, extensive interest and activity exists in fundamental as well as application-oriented research in the THz spectral region [26–28]. Thousands of publications directly addressing the subject have been published in the previous decades in scientific periodicals.

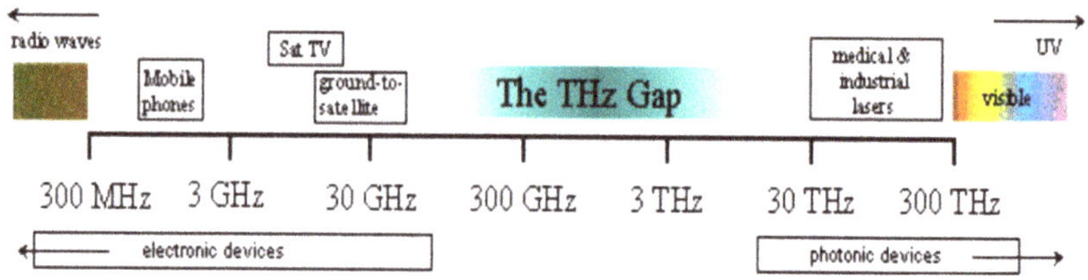

Figure 1. The THz gap in the electromagnetic spectrum.

2.1. THz Radiation and Safety Issues

Over the past years, there has been significant interest in employing terahertz (THz) technology, spectroscopy, and imaging for security applications. There are three prime motivations for this interest:

a. THz radiation can detect concealed weapons since many non-metallic, non-polar materials such as clothing, envelopes etc., are transparent to THz radiation (but are not transparent to visible radiation).
b. Target compounds such as explosives and illicit drugs have characteristic THz spectra that can be used to identify these compounds.
c. THz radiation poses no health risk for scanning people (as opposed to X-rays).

Different materials 'look' different in THz radiation:

a. Typical clothing items and paper and plastic packaging appear transparent in the THz regime.
b. Metals absorb or reflect completely.
c. Ceramic guns and knives partially reflect.
d. Skin, because of its high-water content, absorbs THz radiation. Its energy is harmlessly dissipated as heat in the first 100 μm of skin tissue.

A THz reflection image of a person would show the outline of clothing and the reflection of objects beneath (such as weapons or key chains), but the person's skin would appear substantially darker. False colors in a THz image taken from [29] is shown in Figure 2 below. A millimeter-wave image produced in the Ariel laboratory is shown in Figure 3.

Many drugs and explosives have unique spectral lines in the THz regime, which facilitate their identification even when they are hidden or otherwise invisible to the naked eye. A list of such lines is given in Table 1 below.

The absorption spectra of some typical explosives are shown in Figure 4. The spectra show features that help to identify the explosives. One must bear in mind that the way that the explosives are prepared and measured can affect their spectrum; nevertheless, some spectral features are robust and do not change despite the preparation method [30]. A comparison of the spectra from different publications [30] shows the similarities and differences of measurements described in different references.

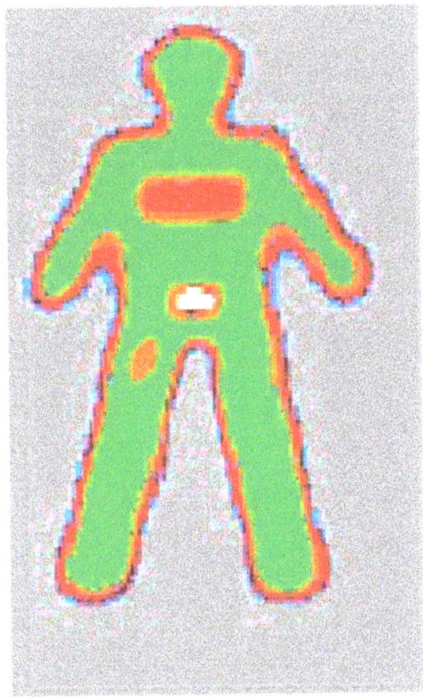

Figure 2. THz Image—False Colors. white = metal, red = RDX, orange = candy, green = skin, gray = background, blue = unknown [29].

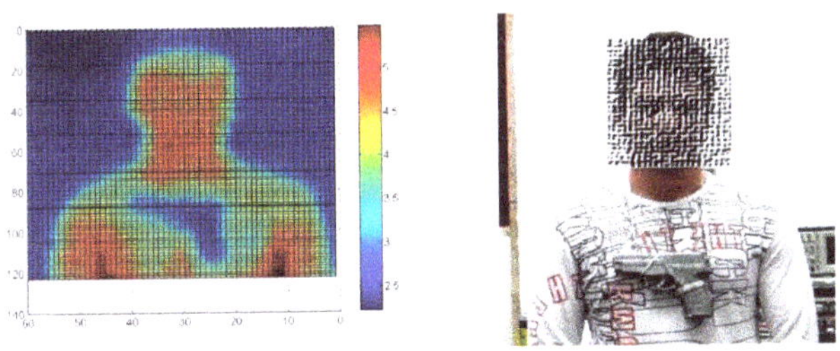

Figure 3. Millimeter-Wave Image—produced in the Ariel laboratory.

Table 1. Unique Spectral Lines of Explosives and Drugs (from [30]). Most individual features are robust against the different preparation techniques of these materials (see below).

Material	Feature Band Center Position Frequency (THz)
Explosives	
Semtex-II	0.72, 1.29, 1.73, 1.88, 2.15, 2.45, 2.57
PE4	0.72, 1.29, 1.73, 1.94, 2.21, 2.48, 2.69
RDX/C4	0.72, 1.26, 1.73
PETN	1.73, 2.01, 2.51
HMX	1.58, 1.84, 1.91, 2.21, 2.57
TNT	1.44, 1.7, 1.91, 5.6, 8.2, 9.1, 9.9
NH_4NO_3	4, 7
Drugs	
Methamphetamine	1.2, 1.7–1.8
MDMA	1.4, 1.8
Lactose α-monohydrate	0.54, 1.20, 1.38, 1.82, 2.54, 2.87, 3.29

Table 1. *Cont.*

Material	Feature Band Center Position Frequency (THz)
Icing sugar	1.44, 1.61, 1.82, 2.24, 257, 2.84, 3.44
Co-codamol	1.85, 2.09, 2.93
Aspirin, soluble	1.38, 3.26
Aspirin, caplets	1.4, 2.24
Acetaminophen	6.5
Terfenadine	3.2
Naproxen sodium	5.2, 6.5

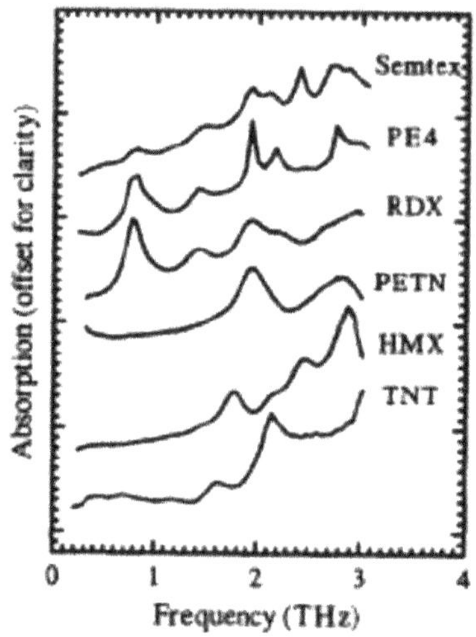

Figure 4. Examples of frequency-resolved absorption of explosives and drugs [30].

The target of remote-sensing systems may also be hazardous gases or aerosols spread in the atmosphere. These agents also have typical spectroscopic signatures, which make it possible to recognize the threat by remote sensing in situations when it is not accessible or too risky to approach.

The NATO was involved in "Chemical Agents Detection and Atmosphere Monitoring (SFP-981415)", in which investigators from Denmark, the Netherlands, the Russian Federation, and Ukraine cooperated in the design of a superconducting integrated submillimeter spectrometer. This spectrometer comprises a Superconductor–Insulator– Superconductor (SIS) tunnel junction with a planar superconducting antenna and a superconducting Flux Flow Oscillator (FFO). Such an integrated spectrometer can be used in the laboratory and for distance monitoring of the atmosphere to detect chemical warfare agents. The same detectors can also serve as part of the proposed stand-off detection imaging system.

2.2. *Commercial THz Systems*

2.2.1. Complete Systems

THz imaging systems are available today for short distances using a single moving detector. Figure 5 describes such a system available from TeraView (UK); Figures 6 and 7 describe such a system from Picometrix (US). Existing commercial systems enable imaging and spectroscopy using beams of ultra-short THz pulses. A whole technology of detectors, detection schemes, etc., has already been developed for the detection of THz radiation frequency in short-pulse waveform. We do not need nor intend to develop these sensing technologies, but we will make it possible to extend its use for stand-off range by using a THz slot antenna. State-of-the-art terahertz-imaging security systems are capable of raster scanning at a rate of 100 pixels per second, certainly not fast enough for video and only marginally sufficient for scanning a bag on a conveyor belt. A briefcase containing a gun, a glass bottle, and a knife would take half an hour to scan at a resolution of 1.5 mm per pixel using a THz pulse-based imager from Picometrix (see Figure 7). For fast scanning, a THz array of detectors is needed.

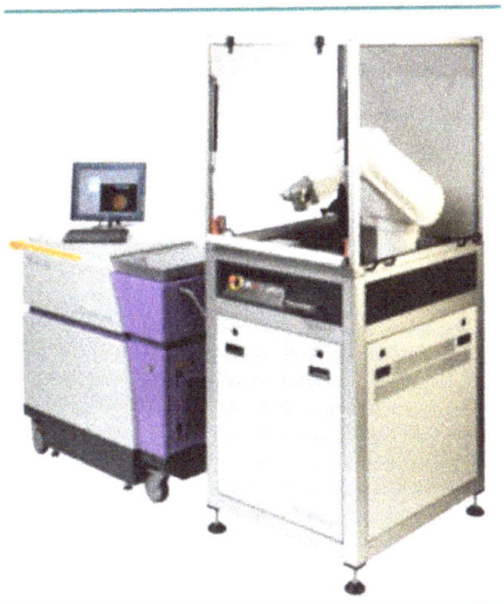

Figure 5. Imaging system from Tera View (UK).

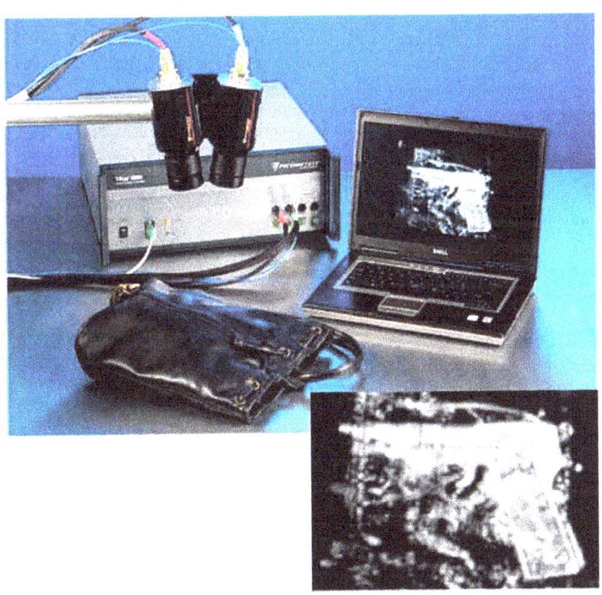

Figure 6. Imaging system from Picometrix (US).

Figure 7. An image produced by Picometrix.

2.2.2. THz Array Detectors

In order to produce images, fast THz array detectors are needed. Although there are no terahertz camera chips, there are infrared camera chips, which can be tweaked to pick up THz rays. Infrared radiation is detected at each pixel as the radiation affects the resistance or other electric properties of the semiconductor elements. However, a decent image requires a bright source with more than 100 milliwatts of power. Therefore, a high-gain antenna such as the one we suggest in the current paper along with a powerful THz source that we developed [31] are needed not only for remote detection but also for short distances if one is interested in rapid video-pace imaging. Spiricon's Pyrocam™ III pyroelectric camera (Figure 8) is an excellent tool for measuring THz lasers and sources. The coating of the crystal absorbs all wavelengths, including from 1 μm to over 3000 μm (0.1 THz to 300 THz). For THz sources, the sensitivity of the Pyrocam™ III is relatively low, being about 300 mW/cm^2 at full output.

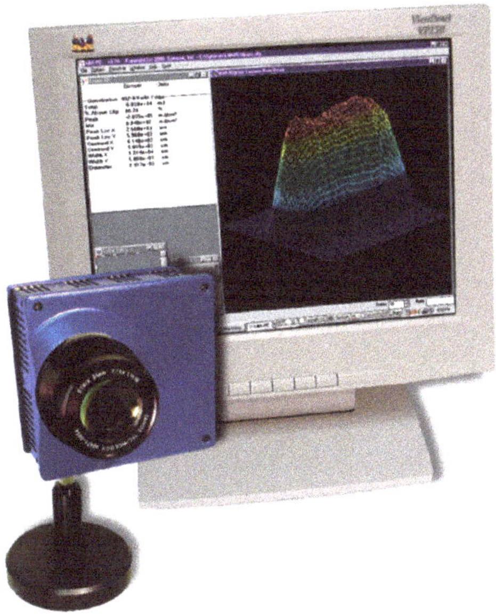

Figure 8. Spiricon's Pyrocam™ III pyroelectric camera.

The Pyrocam™ III sensor array is shown in Figure 9. It consists of a rugged LiTa03 pyroelectric crystal mounted with indium bumps to a solid-state readout multiplexer. This system would benefit from a high-gain antenna and a powerful THz radiation source, as we suggest in the current paper.

Figure 9. Pyrocam™ III sensor array and window assembly.

2.2.3. Remote Sensing of Dangerous Compounds

The problem of remote detection is that it requires the generation of THz radiation, its propagation through the atmosphere up to a barrier, the penetration of this barrier and the explosive material, which is hidden behind the barrier, and its reflection from the skin explosive interface back through a detector. Figure 7 shows examples of absorption of various explosive compounds. Most explosive spectral characteristics are robust and are not affected by modification of the explosive material's preparation method. This indicates that such characteristics are useful for detection schemes. Figure 10 shows the principle of detection of hidden explosives.

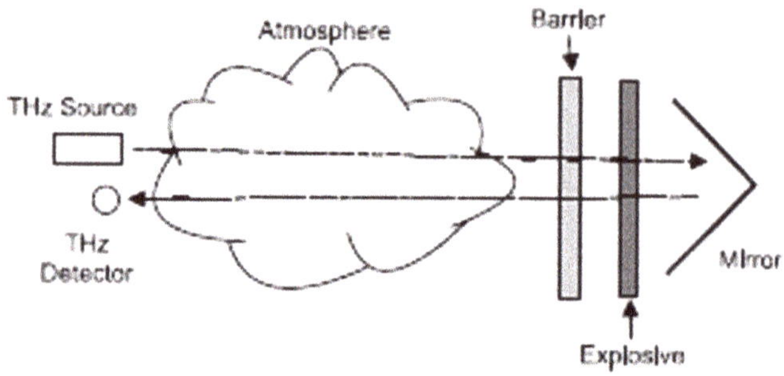

Figure 10. Remote detection scheme. Notice that the "mirror" in real-life scenarios is the terrorist's skin.

Propagation of THz radiation can be calculated using the TRANSM 1.2 software, which is described in [32]. A typical screen of this software is given in Figure 11. Note that the atmospheric absorption is high in the THz regime compared with the millimeter-wave regime. THz remote sensing and imaging requires strong sources and high-gain antennas.

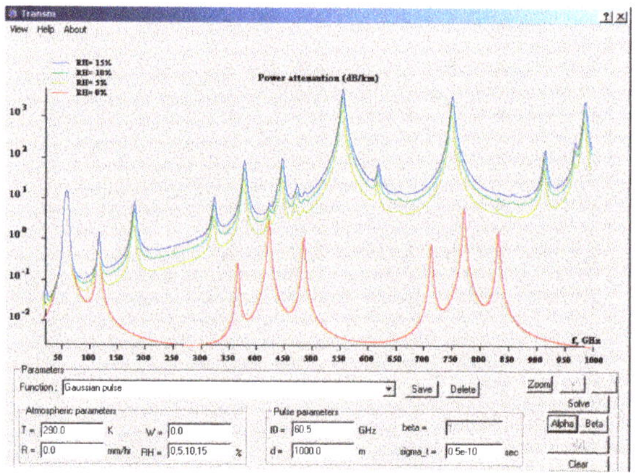

Figure 11. Atmospheric attenuation and phase shift using TRANSM. RH represents the relative humidity, the strongest absorber in normal air. The temperature was assumed to be 17 degrees Celsius. Absorption is given in dB/km.

The power calculation can be performed using the following formula:

$$P_r = A_r \left[\frac{P_0}{\Theta(2D)^2} \right] e^{-2\alpha_b L_b} e^{-2\alpha_a D} e^{-2\alpha_e L_e}$$

in which P_0 is the transmitter power; Θ is the solid angle in which the transmitted power is directed; A_r is the effective area of the receiver; L_b and L_e are the thicknesses of the barrier and explosive layers, respectively; and α_b, α_a, and α_e are the attenuation coefficients of the barrier material, atmosphere, and explosive, respectively. The calculation results are given in Figure 12 for various clothes and, in most cases, demonstrate the possibility of detection. These power levels represent an advance in the current state-of-the-art compact/portable THz source technology and will force us to develop more powerful sources as well as a high-gain antenna to be discussed below.

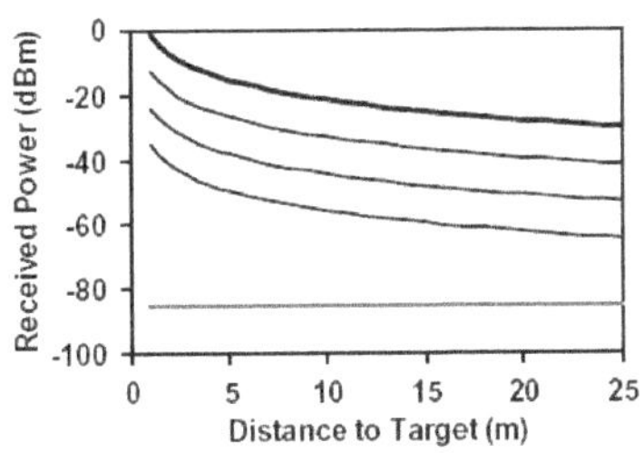

(a)

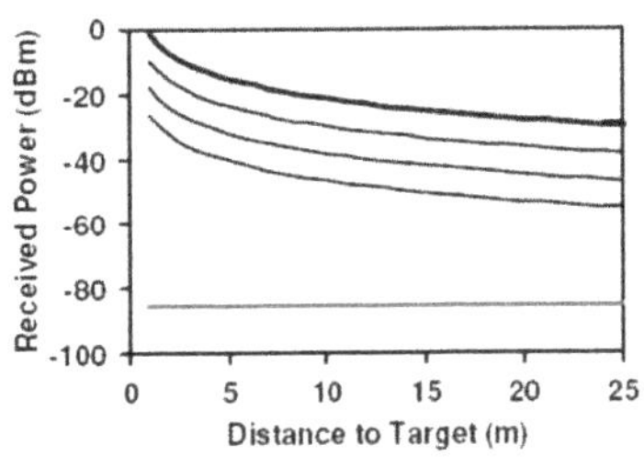

(b)

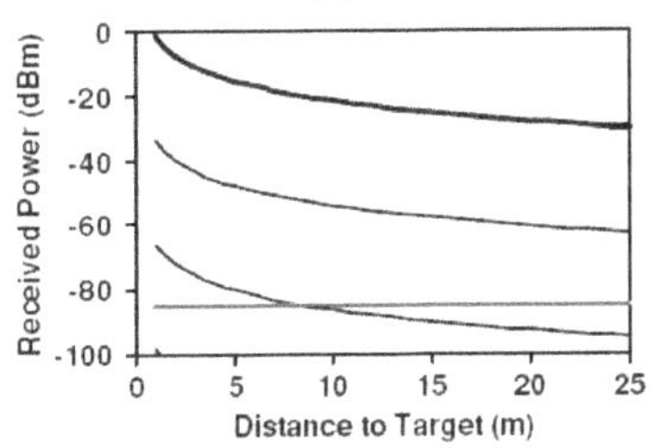

(c)

Figure 12. Received power through various clothing items. The blue line designates the receiver noise limit. Minimum detectable power $P_{min} = 3 \cdot 10^{-11}$ W as the receiver noise limit. The example is taken at 0.8 THz frequency within the ~0.8 THz transmission band in the atmosphere, equal to a peak in the RDX THz spectrum (40 dB mm^{-1} at 0.8 THz). $\Theta = 1.6 \cdot 10^{-3}$ steradians. The RDX thickness is 0.1 mm. P_0 = 100 mW, allowing fast image acquisition. (**a**) Ski jacket—width 0, 3, 6, 9 mm. (**b**) Corrugated cardboard—width 0, 3, 6, 9 mm. (**c**) Wool sweater—width 0, 3, 6 mm.

Of course, a low initial power level shifts the received power levels downwards in the figures. For example, a 1 mW backward wave oscillator (BWO) source shifts the calculated received power curves downwards 20 dB. This implies that current lower power sources might be used for shorter stand-off distances, while the detection of explosives and hazardous gases and aerosols at significant stand-off distances would require substantial development in compact/high power THz sources.

The tunable radiation sources available today in the relevant spectral range are either solid-state or various low-power electron tubes and, in particular, backward wave oscillator tubes [29]. The power, which is about 10^{-2} Watts, is a million times weaker than the peak power generated by the Israeli Free Electron Laser (FEL) and the peak power expected at the NIJMEGEN THz-FEL. As a consequence of the low power, conventional sources are not fit for applications that require long-range propagation through the atmosphere (which in large parts of the THz regime is quite absorptive) or penetration through enclosures for the detection of concealed weapons. The subject of the development of powerful radiation sources will not be discussed here any further; rather, we turn our attention to the design of a high-gain antenna.

3. Antenna Design and Operation

In order to understand the slotted waveguide antenna, we first need to understand the fields within the waveguides. In a waveguide, we are looking for solutions to Maxell's equations that are propagating along the guiding direction (the z direction). Thus, the electric and magnetic fields are assumed to have the form

$$\begin{aligned} \tilde{\vec{E}}(x,y,z,t) &= \tilde{\vec{E}}(x,y)\, e^{i(\omega t-\beta z)} \\ \tilde{\vec{H}}(x,y,z,t) &= \tilde{\vec{H}}(x,y)\, e^{i(\omega t-\beta z)} \end{aligned} \tag{1}$$

where x, y, z are the spatial coordinates, t is time, $\vec{E}$ is the electric field, $\vec{H}$ is the magnetic field, ω is the angular frequency, and β is the propagation wavenumber along the guide direction. The corresponding guide wavelength is denoted by $\lambda_g = 2\pi/\beta$. It is assumed that the reader is familiar with waveguide mode theory [1], and thus it will be stated without proof that the TE field components in a rectangular guide have the form

$$\begin{aligned} \tilde{\vec{H}}_z &= \tilde{\vec{H}}_{az} e^{-i\beta z} \\ \tilde{\vec{E}}_t &= \tilde{\vec{E}}_{at} e^{-i\beta z} \\ \tilde{\vec{H}}_t &= \tilde{\vec{H}}_{at} e^{-i\beta z} \end{aligned} \quad . \tag{2}$$

where a is shorthand for the double index mn (m and n are integers) and

$$\begin{aligned} \tilde{\vec{H}}_{az} &= \cos\left(\tfrac{m\pi x}{a}\right) \cdot \cos\left(\tfrac{m\pi y}{b}\right)\hat{z} \\ \tilde{\vec{E}}_{at} &= \tfrac{-i\omega\mu}{k_c^2}\left(\hat{x}\tfrac{\partial \tilde{\vec{H}}_{az}}{\partial y} - \hat{y}\tfrac{\partial \tilde{\vec{H}}_{az}}{\partial x}\right) \\ \tilde{\vec{H}}_{at} &= \tfrac{-i\beta}{k_c^2}\left(\hat{x}\tfrac{\partial \tilde{\vec{H}}_{az}}{\partial x} + \hat{y}\tfrac{\partial \tilde{\vec{H}}_{az}}{\partial y}\right) \\ \beta &= \sqrt{k^2 - k_c^2} = \sqrt{k^2 - \left(\tfrac{m\pi}{a}\right)^2 - \left(\tfrac{n\pi}{b}\right)^2} \end{aligned} \tag{3}$$

where a and b are the dimensions of the waveguide (see Figure 1 below), μ is the magnetic permeability of the material filling the wave guide, k is the free space wave number, and

$$k_c \equiv \pi\sqrt{\left(\frac{m}{a}\right)^2 + \left(\frac{n}{b}\right)^2}$$

The cutoff frequency of the TE_{mn} mode is expressed by the form

$$f_{mn} = \frac{1}{2\pi\sqrt{\mu\varepsilon}}\sqrt{\left(\frac{m\pi}{a}\right)^2 + \left(\frac{n\pi}{b}\right)^2}. \tag{4}$$

where ε is the dielectric susceptibility. The dominant mode fields, TE_{10}, are given by

$$\begin{aligned} \tilde{\vec{E}}_{10,y} &= \tfrac{-i\omega\mu}{k_c^2}\tfrac{\pi}{a}\sin\tfrac{\pi x}{a} e^{-i\beta_{10} z} \\ \tilde{\vec{H}}_{10,x} &= \tfrac{i\beta}{k_c^2}\tfrac{\pi}{a}\sin\tfrac{\pi x}{a} e^{-i\beta_{10} z} \\ \tilde{\vec{H}}_{10,z} &= \cos\tfrac{\pi x}{a} e^{-i\beta_{10} z} \end{aligned} \quad . \tag{5}$$

Figure 13 shows the waveguide description; the waveguide walls are perfectly conducting, and the dimensions are chosen so that all modes except TE_{10} are cut off.

This information needs to be applied to scattering off a slot cut in one of the walls of the waveguide. If the waveguide is assumed to be infinitely long, and a TE_{10} mode is launched from $z = -\infty$, traveling in the positive z-direction, the incidence of this mode on the slot will cause backwards and forwards scattering of this mode and radiation into outer space is possible.

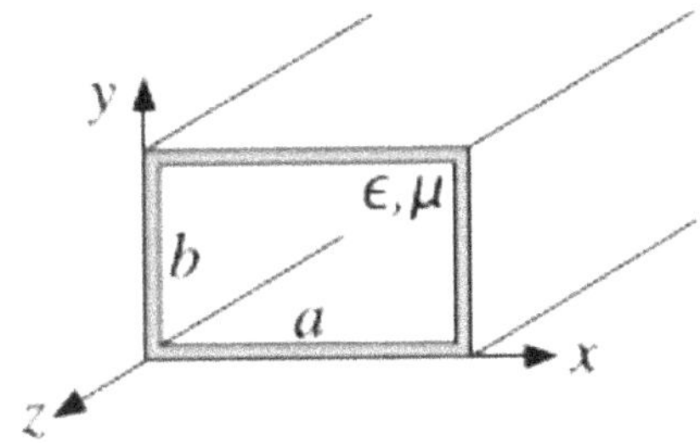

Figure 13. The waveguide description. (https://www.brainkart.com/article/TM-and-TE-waves-in-Rectangular-wave-guides_12511/ (accessed on 1 May 2023)).

Figure 14 shows the geometry of the slot antenna. It is assumed that the waveguide walls have negligible thickness and are composed of a perfect conductor. The slot is rectangular with length $2l$ and width w, where $2l >> w$.

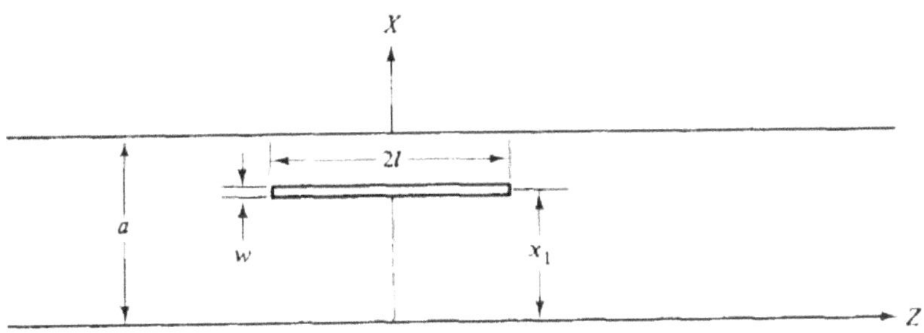

Figure 14. An offset longitudinal slot in the upper broad wall of a rectangular waveguide [1].

The forward and backward scattering off this slot in the TE_{10} mode will be [1]

$$B_{10} = \frac{-\int\limits_{x1-w/2}^{x1+w/2} \cos(\frac{\pi x}{a})dx \int\limits_{-l}^{l} E_{1x}(z)e^{-i\beta_{10}z}dz}{\omega\mu\beta_{10}ab/(\pi/a)^2} \quad . \quad (6)$$
$$C_{10} = \frac{-(\pi/a)^2 \cos(\frac{\pi x_1}{a}) \int\limits_{-l}^{l} V(z)e^{i\beta_{10}z}dz}{\omega\mu\beta_{10}ab}$$

in which $V(z) = w \cdot E_{1x}(z)$ is the voltage distribution in the slot. A slot in a large ground plane can be described as a two-wire transmission line that is fed at the central points; these "wires" are shorted at $z = \pm l$, as shown in Figure 15.

Detailed analysis shows that the voltage distribution in the slot will be a symmetrical standing wave of the form

$$V(z) = V_m \sin[k(l - |z|)]. \quad (7)$$

When inserting (7) in (6), we will obtain

$$B_{10} = C_{10} = \frac{2V_m}{\omega\mu(\beta_{10}/k)ab}(\cos\beta_{10}l - \cos kl)\cos(\frac{\pi x_1}{a}). \quad (8)$$

It is important to observe that the scattering off the slot is symmetrical, that is, $B_{10} = C_{10}$. This implies that the slot is equivalent to a shunt obstacle on a two-wire transmission line. To see this, consider the situation suggested by Figure 16. A transmission line of characteristic admittance G_0 is shunted at $z = 0$ by a lumped admittance Y.

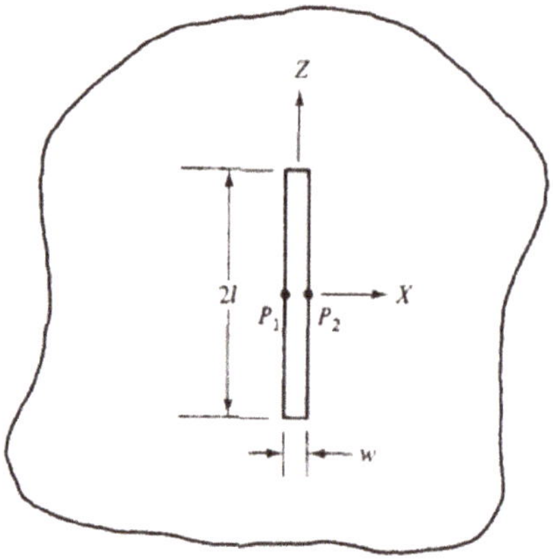

Figure 15. Center-fed slot in a large ground plane [1].

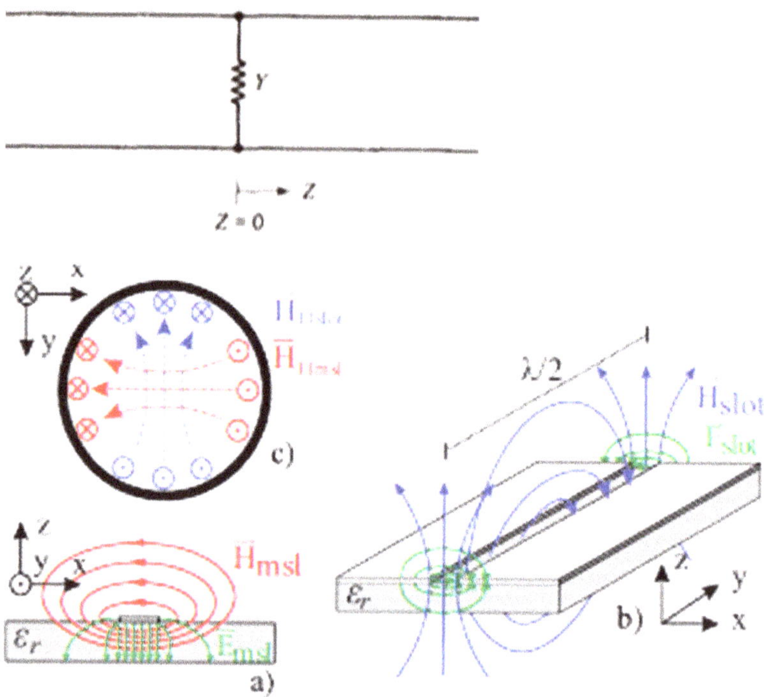

Figure 16. A shunt obstacle on a two-wire transmission line. Below we show how the electric and magnetic fields behave at the slot (**a**) x-z cross section (**b**) a view of the slot from a side perspective (**c**) magnetic field lines from above. The figure is taken from [16].

The voltage and current on the line are given by

$$\begin{aligned} V(z) &= Ae^{-i\beta z} + Be^{i\beta z} \\ & \qquad\qquad z<0 \\ I(z) &= AG_0e^{-i\beta z} - BG_0e^{i\beta z} \\ V(z) &= (A+C)e^{-i\beta z} \\ & \qquad\qquad z>0 \\ I(z) &= (A+C)G_0e^{-i\beta z} \end{aligned} . \tag{9}$$

The boundary conditions are

$$\begin{array}{l} V(0^-) = V(0) = V(0^+) \\ I(0^-) = V(0)Y + I(0^+) \end{array} . \tag{10}$$

Which, when inserted in (9), give

$$\begin{array}{l} B = C \\ \frac{Y}{G_0} = -\frac{2B}{A+B} \end{array} . \tag{11}$$

The usefulness of (11) lies in the fact that, by analogy, if one can find the ratio $-2B_{10}/(A_{10}+B_{10})$ for the slot, one can then say that the slot has equivalent normalized shunt admittance equal to that ratio.

The slot is said to be resonant if Y/G_0 is pure real. If we take A_{10} to be pure real, it follows that the resonant conductance of the slot is given by

$$\frac{G}{G_0} = -\frac{2B_{10}}{A_{10}+B_{10}}. \tag{12}$$

where B_{10} is pure real also. What this implies is that, for a given displacement x_1 of the slot, it is assumed in (12) that the length $2l$ of the slot has been adjusted so that B_{10} is either in phase with, or out of phase with, A_{10}.

The incident power is given by

$$\begin{aligned} p_{inc} &= \tfrac{1}{2}\mathrm{Re}\int_{S1} (A_{10}\vec{\tilde{E}}_{10,t} \times A_{10}{}^{*}\vec{\tilde{H}}_{10,t}{}^{*}) \cdot \hat{z}dS_1 \\ &= \frac{\omega\mu\beta_{10}ab}{4(\pi/a)^2} A_{10}A_{10}{}^{*} \end{aligned} . \tag{13}$$

In a similar manner, one finds that the reflected and transmitted powers are

$$\begin{aligned} p_{ref} &= \frac{\omega\mu\beta_{10}ab}{4(\pi/a)^2} B_{10}B_{10}{}^{*} \\ p_{tr} &= \frac{\omega\mu\beta_{10}ab}{4(\pi/a)^2} (A_{10}+C_{10})(A_{10}+C_{10})^{*} \end{aligned} . \tag{14}$$

If use is made of the information that $B_{10} = C_{10}$, and that all three amplitudes are pure real, then we can find the expression of the power radiation:

$$\begin{aligned} power\ radiation &= P_{inc} - P_{ref} - P_{tr} \\ power\ radiation &= -\frac{\omega\mu\beta_{10}ab}{2(\pi/a)^2} B_{10}(A_{10}+B_{10}) \end{aligned} . \tag{15}$$

In [2], it is shown that the impedance relationship between the slot and the complementary dipole is

$$R_{rad}{}^{dipole} \cdot R_{rad}{}^{slot} = \frac{\eta^2}{4}. \tag{16}$$

where $\eta = 377\Omega$ and $R_{rad}{}^{dipole} = 73\Omega$. By using (16), we can find that $R_{rad}{}^{slot} = 486\Omega$. If we express the radiation resistance of the slot by $R_{rad}{}^{slot} = 486\Omega = \frac{\pi\eta/4}{0.609}$, we can find that

$$P_{rad} = \frac{1}{2}\frac{V_m{}^2}{R_{rad}} = 0.609\frac{V_m{}^2 \cdot 2}{\pi\eta}. \quad (17)$$

Thus,

$$-\frac{\omega\mu\beta_{10}ab}{2(\pi/a)^2}B_{10}(A_{10}+B_{10}) = 0.609\frac{V_m{}^2}{\pi\eta}. \quad (18)$$

If V_m is eliminated from (18) and (8), the result is

$$\frac{G}{G_0} = [2.09\frac{(a/b)}{(\beta_{10}/k)}(\cos\beta_{10}l - \cos kl)^2\cos^2(\frac{\pi x_1}{a})]. \quad (19)$$

When the substitution $x = x_1 - (a/2)$ is made and $kl = \pi/2$ is used, one obtains

$$\frac{G}{G_0} = [2.09\frac{(a/b)}{(\beta_{10}/k)}\cos^2\left(\frac{\beta_{10}}{k}\frac{\pi}{2}\right)\sin^2\left(\frac{\pi x}{a}\right)]. \quad (20)$$

in which x is the offset from the center line of the broad wall. Equation (20) indicates that the normalized conductance of a resonant longitudinal slot in the broad wall of a rectangular waveguide is approximately equal to a constant times the square of the sine of an angle proportional to its offset. From the point of view of the waveguide, the slot is a shunt impedance across the transmission line or an equivalent admittance loading the transmission line. When the admittance of the slot (or combined admittance of all slots) equals the admittance of the guide, then we have matched the transmission line and the maximum power radiated.

In a similar manner, the circuit model of slotted waveguide antenna is depicted in Figure 17.

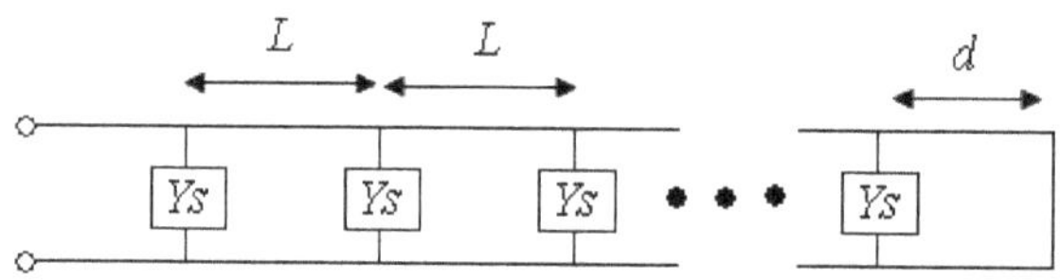

Figure 17. Circuit model of slotted waveguide antenna.

The last slot is a distance d from the end (which is shorted-circuited, as seen in Figure 17), and the slot elements are spaced a distance L from each other. The distance between the last slot and the end, d, is chosen to be a quarter-wavelength. Transmission line theory [1] states that impedance of a short circuit a quarter-wavelength down a transmission line is an open circuit, hence, Figure 17 then reduces to Figure 18.

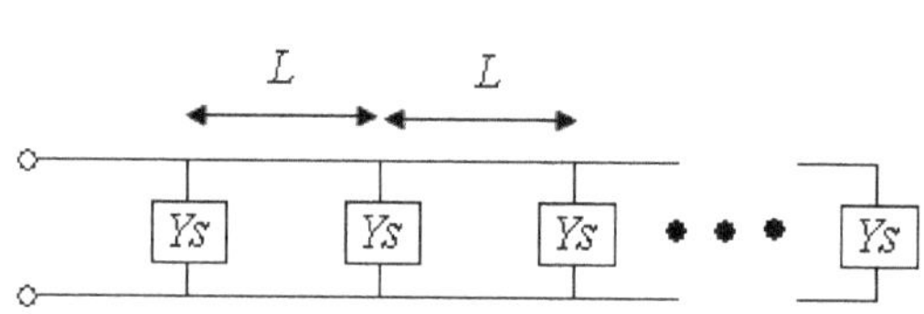

Figure 18. Circuit model of slotted waveguide using quarter-wavelength transformation.

The input admittance for an N element slotted array is:

$$Y = NY_S. \quad (21)$$

Sometimes the closed end is spaced $\frac{3}{4}\lambda_g$ for mechanical reasons; the additional half-wavelength is transparent. Spacing the slots at $\frac{1}{2}\lambda_g$ intervals in the waveguide represents an electrical spacing of 180°: each slot is exactly out of phase with its neighbors, so their radiation will cancel each other. However, slots on opposite sides of the centerline of the guide will be out of phase, so we can alternate the slot displacement around the centerline and have a total phase difference of 360° between slots, putting them back in phase.

A picture of a complete waveguide slot antenna is shown in Figure 19 This example has six slots on each side for a total of twelve slots. The slots have identical length and spacing along the waveguide. Note how the slot position alternates about the centerline of the guide. The far wall of the waveguide has an identical slot pattern, so you can see through the slots.

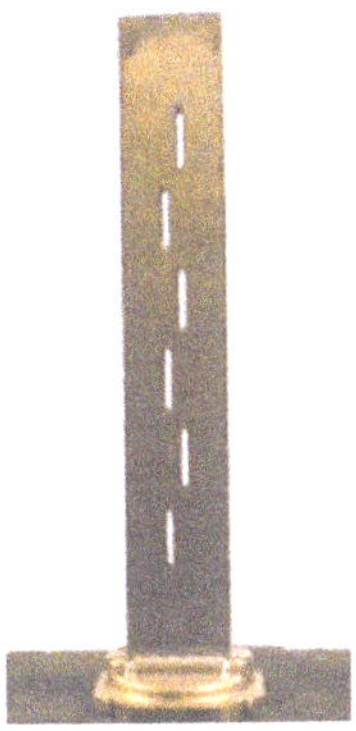

Figure 19. Example of waveguide slot antenna.

A simple way to estimate the gain of a slot antenna is to remember that is an array of dipoles. Each time we double the number of dipoles, we double the gain, or add 3 dB. The approximate gain formula is thus $Gain = 10\log(N)\ dBi$ for N total slots. Ref. [3] gives better formulas for the gain and beam width:

$$\begin{aligned} GAIN &= 10\log\left(\frac{N\cdot\frac{\lambda_g}{2}}{\lambda}\right)\ dBi \\ Beamwidth &= 50.7\left(\frac{\lambda}{\frac{N}{2}\cdot\frac{\lambda_g}{2}}\right)\ [^\circ] \end{aligned} \quad (22)$$

It is known from [4] that the slot width should be taken as $w = \frac{\lambda_g}{20}$. However, measurements taken by Stegen in [5] were based on a slot width of 1.5875 mm in a WR-90 waveguide. For other waveguide sizes, the slot width should be scaled accordingly.

4. Slot Antenna Design and Simulation

We can summarize the design procedure for waveguide slot antenna as follows:

1. Choose the number of slots required for the desired gain and beam width. ($Gain = 10\log(N)\ dBi$ so, for a desired gain one can calculate N, and later improve the estimate using Equation (22)).
2. Choose a waveguide size appropriate for the operating frequency (see Equations (3), (4) and (23) below for a specific example).
3. Calculate the wavelength in the waveguide at the operating frequency (see Equations (3), (4) and (23) below for a specific example).
4. Determine the slot dimensions, length, and width appropriate for the operating frequency (see Equation (25) below).
5. Determine the slot position from centerline for normalized admittance of 1/N, where N is the number of slots in both walls of the waveguide (see Equation (26) below).

6. Space $\frac{1}{4}\lambda_g$ or $\frac{3}{4}\lambda_g$ between the center of the last slot and the end of the waveguide.

For a THz slot antenna design at operation frequency of 330 GHz, the waveguide dimensions are chosen so that all modes except TE_{10} are at cut off. The waveguide is WR3 with the dimensions of $0.864 \times 0.432\ \text{mm}^2$.

The wavelengths are

$$\begin{array}{l} \lambda = \frac{c}{f} = 0.909[mm] \\ \lambda_c = \frac{2\pi}{k_c} = \frac{2\pi}{\frac{m\pi}{a}+\frac{n\pi}{b}} = 2a = 1.728[mm] \\ \lambda_g = \frac{1}{\sqrt{\frac{1}{\lambda^2}-\frac{1}{\lambda_c{}^2}}} = 1.07[mm] \end{array}. \quad (23)$$

By choosing 256 slots, we will obtain:

$$\begin{array}{l} GAIN = 10\log\left(\frac{N\cdot\frac{\lambda_g}{2}}{\lambda}\right) = 10\log\left(\frac{256\cdot\frac{1.07}{2}}{0.909}\right) = 21.8\ dBi \\ Beamwidth = 50.7\left(\frac{\lambda}{\frac{N}{2}\cdot\frac{\lambda_g}{2}}\right) = 50.7\left(\frac{0.909}{\frac{256}{2}\cdot\frac{1.07}{2}}\right) = 0.7\ [^\circ] \end{array}. \quad (24)$$

Thus, the resolution at 10 m is 11.74 [cm].

The slot dimensions are

$$\begin{array}{l} slot\_length = 2l = \frac{\lambda}{2} = 0.454[mm] \\ slot\_width = w = \frac{\lambda_g}{2} = 0.0535[mm] \end{array}. \quad (25)$$

The offset from the center line of the broad wall is:

$$\begin{array}{l} \frac{G}{G_0} = [2.09\frac{(a/b)}{(\beta_{10}/k)}\cos^2\left(\frac{\beta_{10}}{k}\frac{\pi}{2}\right)\sin^2(\frac{\pi x}{a})] \\ \frac{1}{N} = [2.09\frac{(a/b)}{(\beta_{10}/k)}\cos^2\left(\frac{\beta_{10}}{k}\frac{\pi}{2}\right)\sin^2(\frac{\pi x}{a})] \\ N\cdot[2.09\frac{(a/b)}{(\beta_{10}/k)}\cos^2\left(\frac{\beta_{10}}{k}\frac{\pi}{2}\right)\sin^2(\frac{\pi x}{a})] = 1 \\ \sin(\frac{\pi x}{a}) = \frac{1}{\sqrt{N[2.09\frac{(a/b)}{(\beta_{10}/k)}\cos^2\left(\frac{\beta_{10}}{k}\frac{\pi}{2}\right)]}} \\ x = \frac{a}{\pi}\arcsin\frac{1}{\sqrt{N[2.09\frac{(a/b)}{(\beta_{10}/k)}\cos^2\left(\frac{\beta_{10}}{k}\frac{\pi}{2}\right)]}} \\ x = 0.033\ [mm] \end{array}. \quad (26)$$

Figures 20 and 21 show a model performed using CST Microwave Studio according to the analytical design, the simulation result for the radiation pattern, gain and beam width depicted in Figures 22–26.

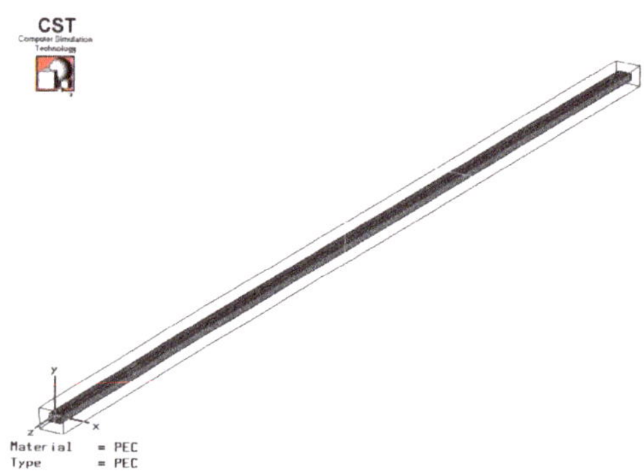

Figure 20. Slot antenna model using CST Microwave Studio.

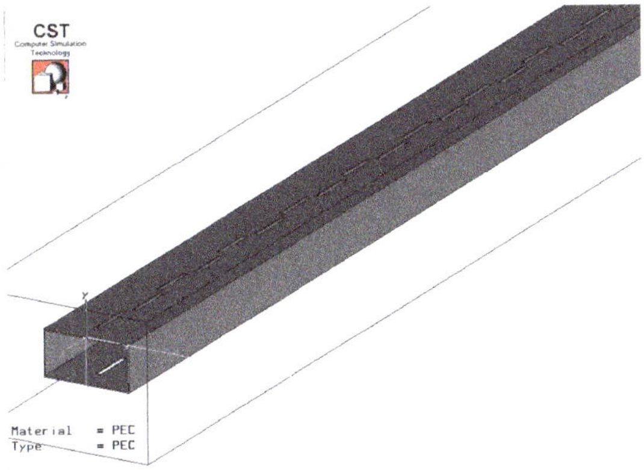

Figure 21. Slot antenna model using CST Microwave Studio (zoomed in).

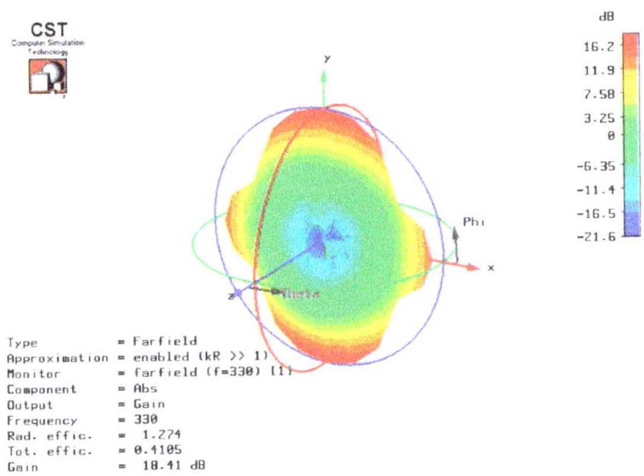

Figure 22. Radiation pattern simulation result.

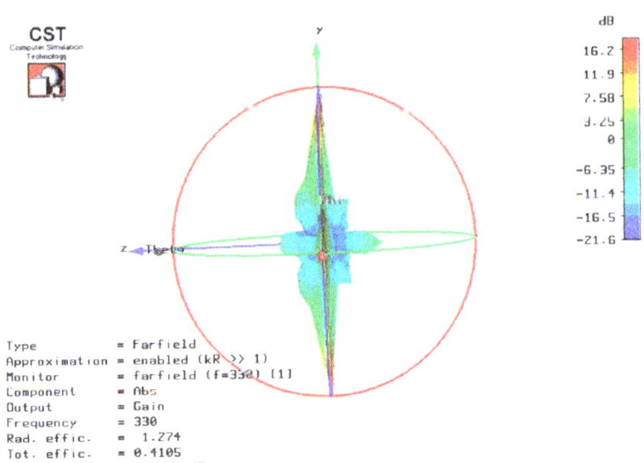

Figure 23. Radiation pattern simulation result (x-axis view).

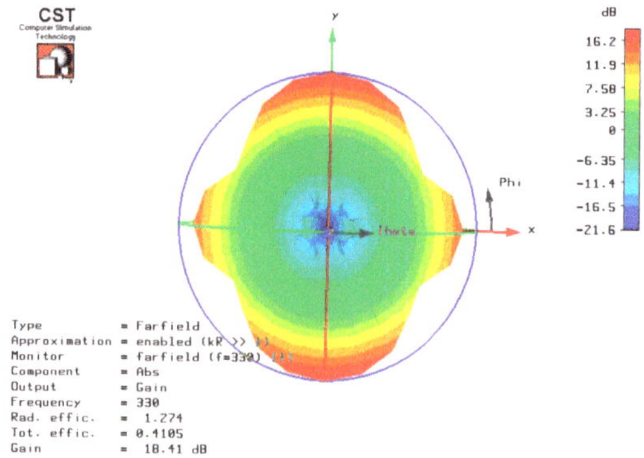

Figure 24. Radiation pattern simulation result (z-axis view).

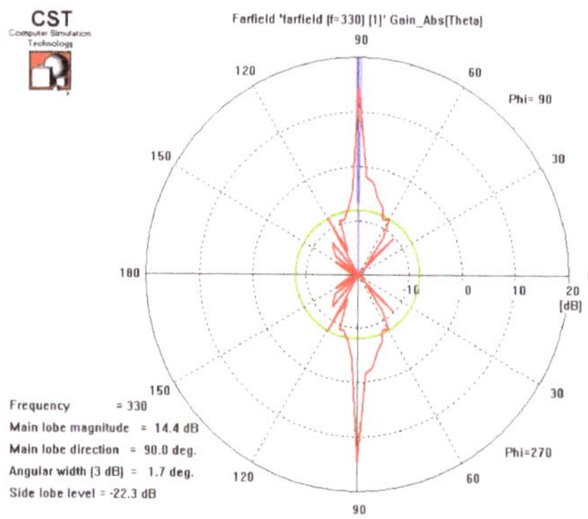

Figure 25. Azimuth radiation pattern simulation result.

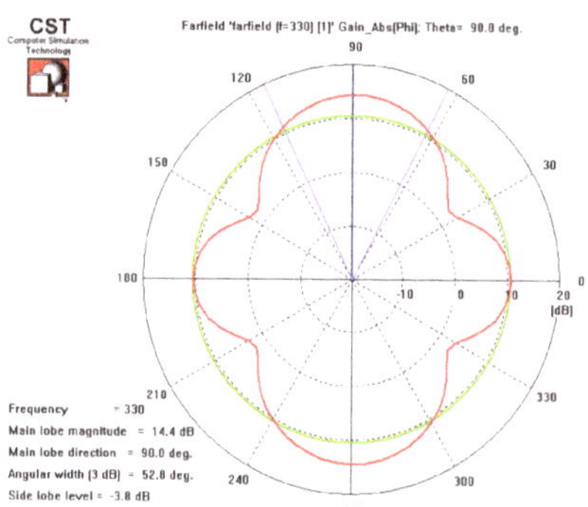

Figure 26. Elevation radiation pattern simulation result.

The simulation results show that the antenna gain is 18.4 dBi and the azimuth beam width is 1.7° compared to 21.8 dBi and 0.7° in the analytical design.

There is a small difference between the analytical and simulation results. The difference is due to the simplifying assumptions that were made in the analytical design (wall thickness was ignored, waveguide length was taken as infinity, etc.). Despite the difference between the analytical and simulation results, a good approximation of the antenna's performance, pattern, beamwidth, and gain can be made. To conclude the voltage standing wave ratio is given by

$$VSWR = \frac{|V|_{max}}{|V|_{min}} = \frac{1+|\Gamma|}{1-|\Gamma|} \tag{27}$$

where $|\Gamma|$ is the reflection coefficient which is given in Equation (6) through the coefficient B_{10}. S_{11} is the reflection coefficient if the other side of the wave guide is terminated by a matched load. A plot of B_{10} as a function of frequency is given in Figure 27 below.

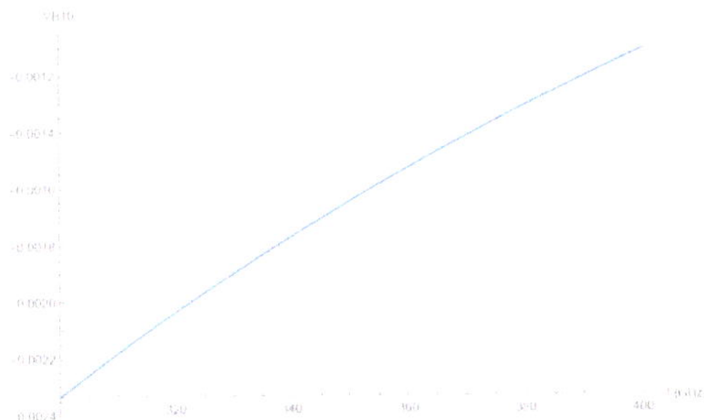

Figure 27. $|B_{10}|$ reflection coefficient as a function of frequency.

In addition, the radiated power as a function of frequency is shown in Figure 28 below.

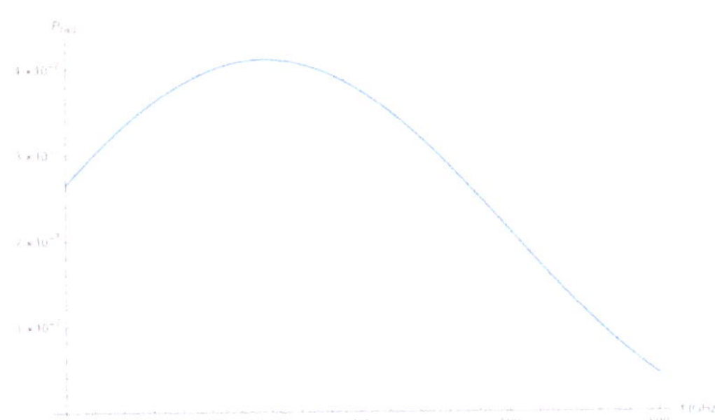

Figure 28. Radiated power as a function of frequency.

An efficiency of an antenna is defined as

$$e = \frac{R_r}{R_r + R_l} \tag{28}$$

in which R_r is the radiation resistance and R_l is the loss (ohmic resistance). Obviously, for a perfect conductor, $R_l = 0$, and thus $e = 1$.

5. Conclusions and Discussion

In this paper, a slot antenna based on a rectangular waveguide has been presented and investigated. It has been shown that a slot antenna in the high THz frequency regime can be analyzed and designed by an analytical model. We show that gradually moving the slot position from the centerline improves the antenna's performance. This was verified using simulation software. To obtain the required antenna's pattern and performance, it is first recommended to use an analytical model of the antenna. Using the classical analytical design parameter information, one can perform a simulation. Then, according to the simulation results, the parameters can be modified until optimum results are obtained. We emphasize that, despite the simplicity of our classical approach, the simulation shows that we obtain useful results even at THz frequencies. The relation between radiated power and coefficients is complicated because the parameter V_m is obtained by solving Equation (18) and is different for each frequency. Thus, the relation between reflection coefficient and radiated power is not trivial, as can be seen from Figures 27 and 28. The high gain and directivity of this antenna is well described in Figures 22–26.

A discussion of other related findings, but at different frequencies, is given in the introductory section.

Author Contributions: Conceptualization, A.Y.; methodology, A.Y.; software, S.R. and A.Y.; validation, S.R.; formal analysis, S.R.; investigation S.R.; resources, A.Y.; data curation, S.R. and A.Y.; writing—original draft preparation, S.R.; writing—review and editing, A.Y.; visualization, S.R.; supervision, A.Y.; project administration, A.Y. All authors have read and agreed to the published version of the manuscript.

Funding: This research received no external funding.

Data Availability Statement: No new data were created or analyzed in this study. Data sharing is not applicable to this article.

Conflicts of Interest: The authors declare no conflict of interest.

References

1. Collin, R.E. *Foundations for Microwave Engineering*, 2nd ed.; Mcgraw-Hill College: Columbus, OH, USA, 1992; ISBN 10: 0070118116.
2. Booker, H.G. Slot Aerials and Their Relation to Complementary Wire Aerials. *J. IEE* **1946**, *93*, 620–626. [CrossRef]
3. Wade, P. The W1GHZ Online Microwave Antenna Book, Chapter 7. Available online: http://www.w1ghz.org/antbook/preface.htm (accessed on 1 May 2023).
4. Bell, S. KB7TRZ, Slot Antenna Design. In Proceedings of the 40th Annual West Coast VHF/UHF Conference, Cerritos, CA, USA, 5–7 May 1995; Available online: http://www.qsl.net/w6by/techtips/slot_ant.html (accessed on 1 May 2023).
5. Stegen, R.J. Longitudinal Shunt Slot Characteristics, Hughes Technical Memorandum No. 261, Culver City, CA. November 1951. Available online: http://www.dtic.mil/dtic/tr/fulltext/u2/a800305.pdf (accessed on 1 May 2023).
6. Kraus, J. *Antennas*; McGraw-Hill: Columbus, OH, USA, 1950; pp. 624–642, ISBN 10: 007123201X.
7. Pozar, D.M. *Microwave Engineering*, 4th ed.; Wiley: New York City, NY, USA, 2011; pp. 65–73, ISBN 10: 0470631554.
8. Chang, K. *Encyclopedia of RF and Microwave Engineering*; John Wiley & Sons: New York City, NY, USA, 2005; Volume 1–6, pp. 4696–4702, ISBN 978-0-471-27053-9.
9. Elliott, R.S. *Antenna Theory and Design*; Prentice-Hall: Hoboken, NJ, USA, 1981; pp. 39–41, 86–99, 336–344, 407–414, ISBN 10: 0471449962.
10. Nell, O.; Solbach, K.; Dreier, J. Omnidirectional Waveguide Slot Antenna for Horizontal Polarization. VHF Communications, Part 1: 1991/4, pp. 200–205 and Part 2: 1992/1, pp. 11–17. Available online: http://www.vhfcomm.co.uk/index.html (accessed on 1 May 2023).
11. Stevenson, A.F. Theory of slots in rectangular waveguides. *J. Appl. Phys.* **1948**, *19*, 24–38. [CrossRef]
12. Croswell, W. Slots antenna, Chapter 8. In *Antenna Engineering Handbook*, 3rd ed.; Johnson, R., Jasik, H., Eds.; McGraw Hill, Inc.: New York, NY, USA, 2007; ISBN 10: 0071475745.
13. Gilbert, R.A. Slot antenna arrays, Chapter 9. In *Antenna Engineering Handbook*, 3rd ed.; Johnson, R.C., Jasik, H., Eds.; McGraw Hill, Inc.: New York, NY, USA, 2007; ISBN 10: 0071475745.
14. Moradian, M.; Khalaj-Amirhosseini, M.; Tayarani, M. Design of planar slotted array antenna fed by single ridge waveguide. *Int. J. RF Microw. Comp. Aid Eng.* **2010**, *20*, 593–602. [CrossRef]
15. Kakatkar, S.S.; Deshmukh, A.A.; Ray, K.P. Analysis of a rectangular waveguide transverse slot using MoM and image method. *IET Microw. Antennas Propag.* **2020**, *14*, 1006–1011. [CrossRef]
16. Stein, W.; Deckelmann, M.; Oborovski, A.; Vossiek, M. Planar cross-polarized transition to a circular waveguide feed for a 61 GHz dielectric antenna. In Proceedings of the 2015 German Microwave Conference, Nuremberg, Germany, 16–18 March 2015; pp. 162–165. [CrossRef]
17. Komoshvili, K.; Levitan, J.; Aronov, S.; Yahalom, A.; Kapilevich, B. Millimeter waves non-thermal effect on human lung cancer cells. In Proceedings of the 2011 IEEE International Conference on Microwaves, Communications, Antennas and Electronic Systems (COMCAS 2011), Tel Aviv, Israel, 7–9 November 2011. [CrossRef]
18. Aronov, S.; Einat, M.; Furman, O.; Pilossof, M.; Komoshvili, K.; Ben-Moshe, R.; Yahalom, A.; Levitan, J. Millimeter-wave insertion loss mice skin. *J. Electromagn. Waves Appl.* **2017**, *32*, 758–767. [CrossRef]
19. Rajput, S.; Komoshvili, K.; Aronov, S.; Kumar Patnaik, P.; Levitan, J.; Yahalom, A. Optimization of transmitted power of horn antenna for biomedical applications. In Proceedings of the 2019 IEEE International Conference on Microwaves, Antennas, Communications and Electronic Systems (COMCAS), Tel-Aviv, Israel, 4–6 November 2019; pp. 1–4. [CrossRef]

0. Fruman, O.; Komoshvili, K.; Levitan, J.; Yahalom, A.; Liberman-Aronov, S. The Lack of Toxic Effect of High-Power Short-Pulse 101 GHz Millimeter Waves on Healthy Mice. *Bioelectromagnetics* **2020**, *41*, 188–199. [CrossRef] [PubMed]
:1. Komoshvili, K.; Becker, T.; Levitan, J.; Yahalom, A.; Barbora, A. Liberman-Aronov. S. Morphological Changes in H1299 Human Lung Cancer Cells Following Millimeter-Wave, Irradiation. *Appl. Sci.* **2020**, *10*, 3187. [CrossRef]
:2. Komoshvili, K.; Levitan, J.; Israely, K.; Yahalom, A.; Barbora, A.; Liberman-Aronov, S. W-Band Millimeter Waves Targeted Mortality of H1299 Human Lung Cancer Cells without Affecting Non-tumorigenic MCF-10A Human Epithelial Cells In Vitro. *Appl. Sci.* **2020**, *10*, 4813. [CrossRef]
:3. Rajput, S.; Barbora, A.; Komoshvili, K.; Levitan, J.; Yahalom, A.; Liberman-Aronov, S. Scrutinizing Effects of 75 GHz MMW Irradiation on Biological Functions of Yeast. In Proceedings of the 2020 IEEE MTT-S International Microwave Biomedical Conference (IMBioC), Toulouse, France, 25–28 May 2020; pp. 1–4. [CrossRef]
:4. Barbora, A.; Rajput, S.; Komoshvili, K.; Levitan, J.; Yahalom, A.; Liberman-Aronov, S. Non-Ionizing Millimeter Waves Non-Thermal Radiation of Saccharomyces Cerevisiae-Insights and Interactions. *Appl. Sci.* **2021**, *11*, 6635. [CrossRef]
:5. Marks, H.S.; Kleinman, H.; Gover, A.; Einat, M.; Kanter, M.; Borodin, D.; Lasser, Y.; Lurie, Y.; Kapilevich, B.; Litvak, B.; et al. High-Power Long-Pulse Operation of a Millimeter-Wave FEL. In Proceedings of the IPSTA-2014, The 16th Israeli Plasma Science and Applications Conference, Tel Aviv University, Tel Aviv, Israel, 5 February 2014; p. 20.
:6. Siegel, P.H. THz Technology: An Overview. *Int. J. High Speed Electron. Syst.* **2003**, *13*, 351. [CrossRef]
:7. Mickan, S.P.; Zhang, X.-C. T-ray sensing and imaging. *Int. J. High-Speed Electron. Sys.* **2003**, *13*, 601. [CrossRef]
:8. Siegel, P.H. Terahertz technology. *IEEE Trans. Microw. Theory Tech.* **2002**, *50*, 910. [CrossRef]
:9. Medvedev, I.; Winnewisser, M.; De Lucia, F.C.; Herbst, E.; Białkowska-Jaworska, E.; Pszczółkowski, L.; Kisiel, Z. The millimeter- and submillimeter-wave spectrum of the trans–gauche conformer of diethyl ether. *J. Mol. Spectrosc.* **2004**, *228*, 314–328. [CrossRef]
30. Kemp, M.C.; Taday, P.F.; Cole, B.E.; Cluff, J.A.; Fitzgerald, A.J.; Tribe, W.R. Security applications of terahertz technology. *Proc. SPIE* **2003**, *5070*, 44.
31. Balal, N.; Magori, E.; Yahalom, A. Design of a Permanent Magnet Wiggler for a THz Free Electron Laser. *Acta Phys. Pol. A* **2015**, *128*, 259–263. [CrossRef]
32. Pinhasi, Y.; Yahalom, A.; Harpaz, O.; Vilner, G. Study of an Ultra Wideband Transmission in the Extremely High Frequency (EHF) Band. *IEEE Trans. Antennas Propag.* **2004**, *52*, 2833–2842. [CrossRef]

Disclaimer/Publisher's Note: The statements, opinions and data contained in all publications are solely those of the individual ıuthor(s) and contributor(s) and not of MDPI and/or the editor(s). MDPI and/or the editor(s) disclaim responsibility for any injury to ɔeople or property resulting from any ideas, methods, instructions or products referred to in the content.

electronics

Article

SIW Leaky Wave Antenna for THz Applications

Vivek Arya [1,*], Tanuj Garg [1] and Hamza Mohammed Ridha Al-Khafaji [2,*]

1 Department of ECE, FET, Gurukua Kangri (Deemed to Be University), Haridwar 249404, Uttarakhand, India
2 Biomedical Engineering Department, Al-Mustaqbal University College, Hillah 51001, Babil, Iraq
* Correspondence: ichvivekmalik@gmail.com (V.A.); hamza.alkhafaji@uomus.edu.iq (H.M.R.A.-K.)

Abstract: This paper proposes a new design of leaky wave antenna (LWA) based on substrate integrated waveguide (SIW) technology for THz applications. The suggested LWA structure has a combination of longitudinal and transverse slots and makes a 10-element linear array of radiating elements. To address the problem of open-stop-band (OSB), four additional smaller slots were etched on the corners of longitudinal and transversal slots. At the broadside, this LWA provided a gain of 12.33 dBi, and a continuous wide beam scanning range from +78° to −6° via the broadside while exhibiting efficient radiation performance over the operating frequency bands of 105 GHz to 109 GHz with a peak gain of 16.02 dBi.

Keywords: leaky wave antenna (LWA); THz applications; SIW; CBS

1. Introduction

In 1940, a leaky wave antenna (LWA) was investigated as a rectangular type of closed waveguide structure [1]. The presence of slots at regular intervals on an electromagnetic wave-carrying waveguide results in the emission of a plane wave from these slots. Later, in 1956, N. Marcuvitz named this plane wave the leaky wave [2]. LWAs are waveguides with periodic slots that are responsible for the production of leaky waves [3]. In general, LWAs are made up of waveguides with a single/double radiator in a unit cell. Owing to its high gain, elementary structure, as well as its inherent beam scanning feature w.r.t. frequency, the type of traveling wave antenna is a planar LWA. LWA has a large range of applications in the fields of broadband free-space communications as well as collision-avoidance radars. LWAs are a sort of frequency beam scanning antenna that may be loosely categorized into three kinds, specifically uniform [4,5], periodic [6–8], and quasi-periodic [9,10], depending on the modulations applied to the waveguide. Although it emits from the main space harmonics ($n = 0$), uniform and quasi-periodic antenna structures radiate only in an advance forward direction. Composite left/right-handed LWAs, on the other hand, are quasi-periodic structures that radiate with fundamental harmonic both in the forward and backward directions.

In the past, different variants of LWAs have been proposed by various researchers for beam scanning. The periodic and uniform LWAs scan only in one direction, that is, only in the forward direction. The traditional LWAs suffered from broadside radiation problems. Traveling wave LWAs transform into standing wave antennas at broadside frequencies. Continuous beam scanning (CBS) [11] is now possible with the introduction of composite left/right transmission lines metamaterial [12], although at the cost of difficulties in production, as a unit cell period is shorter than $\lambda_g < 4$. Various experts have obtained CBS for periodic LWAs by different methods, such as matching stubs, taking a quarter distance between pairs of identical elements, and using substrate integrated waveguide (SIW) [13]. SIW is the most significant breakthrough of the previous decade, bridging the gap between metallic waveguides and planar circuits [14]. When compared to rectangular waveguides, SIW offers the capabilities of a low profile, lightweightedness, and the ability to be simply constructed using conventional PCB technology. As a result, SIW has sparked

Citation: Arya, V.; Garg, T.; Al-Khafaji, H.M.R. SIW Leaky Wave Antenna for THz Applications. *Electronics* **2023**, *12*, 1839. https://doi.org/10.3390/electronics12081839

Academic Editors: Syed Muzahir Abbas and Yang Yang

Received: 20 February 2023
Revised: 24 March 2023
Accepted: 10 April 2023
Published: 12 April 2023

Copyright: © 2023 by the authors. Licensee MDPI, Basel, Switzerland. This article is an open access article distributed under the terms and conditions of the Creative Commons Attribution (CC BY) license (https://creativecommons.org/licenses/by/4.0/).

a lot of interest, and a variety of components employing this technology have been created. Half and full-mode substrate integrated waveguides efficiently work for CBS. Due to the simplicity of the SIW structure, many researchers have developed SIW-based LWAs for CBS that provide good scanning capability [14]. Ultimately, the SIW platform will be essential for producing larger scanning range LWAs.

LWAs have been extensively researched in the microwave spectrum and have recently been developed at THz frequencies. The LWA, in its most basic form, is a metallic waveguide that leaks radiation via a slot formed beside its length, while more complicated designs exist that use metamaterials. The radiating beam arises from the hole or slot at a frequency-dependent angle to contest the phases of both the propagating and free-space modes. Several of the issues connected with THz transmission, such as frequency division multiplexing, beam steering, radar-object tracking, and connection discovery, have been addressed using this one-to-one correlation between emission angle and frequency.

Over the years, there has been considerable interest in terahertz (THz) technologies owing to their potential for access to wideband wireless links. To address numerous difficulties related to THz wireless communications, LWAs that scale at THz frequencies (over 100 GHz) have been considered suitable candidates due to their authentic qualities, such as narrow beam, high gain, high efficiency, and wider CBS capabilities [15,16]. The SIW-based LWAs that work in the THz range have many important growing applications, such as high transmission rate communications [17], voice-recognizing radar systems [18], terrestrial remote sensing [19], as well as molecular spectroscopy [20].

A frequency scanning radiating device, i.e., an LWA, is a viable and efficient choice in this respect because of its straightforward feeding structure, excellent gain, and wide band nature [21,22]. The terms uniform and periodic refer to two types of LWAs, depending on the geometrical construction. For the uniform LWA, the waveguide must experience periodic disturbance in order to acquire leakage by leaky wave radiation. While in the periodic LWA structure, there is a constant perturbation along the waveguide structure. The primary beams of traditional uniform LWAs cannot be scanned in the backward quadrant space; however, LWAs with periodic structures have backward and forward scanning capabilities.

Nevertheless, the scanning area of a uniform LWA is limited to the front quadrant. The literature has described a number of perturbed guiding structures to date [23–25] in order to produce periodic LWAs. For their use in microwave and millimeter wave (mm-wave) applications, transverse metallic slotted waveguides, such as rectangular [23], substrate integrated [25,26], microstrip line [27,28], or coplanar [29], have been explored and achieved. However, LWA based on metallic waveguide is not a common option since there is a huge conductor loss and the creation of surface waves at higher frequencies exceeding 100 GHz [30]. Furthermore, their manufacturing is difficult since their sectional measurements must be 1/10th of the wavelength for a single-mode operation [31].The majority of past studies, which built on previous work at minimum frequencies, used typical leaky-wave waveguide structures that may not be optimal for usage at THz frequency ranges. In the microwave range, for example, the leaky structural measurements are invariably deeply sub-wavelength in the span. This keeps the guided wave consistent through the slot width. Additionally, the radiated field pattern may be estimated in this limit, utilizing a simple mathematical model based on leakage rate as well as a sinc function. In contrast, the majority of modern THz implementations employ a leaky structure with a size close to the wavelength. At the wavelength scale, the wave interaction with the leaky structure can no longer be ignored to generate an output beam that follows the frequency-angle relationship; one must employ very tiny rectangular slots. This reduces device capability since the slot length radiated energy is quite short. This is a severe disadvantage, as the power production at THz frequencies remains a considerable issue. To address this issue, an alternative slot geometry, whereby the slot width rises linearly with its length, has been provided by various experts and has demonstrated that the design allows for improved coupling efficiencies at the output while maintaining the phase-matching connection. Furthermore, because the trapezoidal slots have the larger effective aperture, the emitted beam is limited in both angular directions, which is notably not the case with a traditional rectangular slot aperture [18].

LWAs have been successfully developed utilizing a dielectric waveguide with enhanced loss characteristics to address these shortcomings at THz frequencies [32–41]. Different types of transmission lines called "dielectric waveguides" use one or additional insulating dielectrics to direct an electromagnetic wave. For more significant size reduction, different layers of dielectric material are typically employed. The waveguides that are based on dielectric have an inherent lack of conductor loss since their transmission mechanisms rely only on the reflections at the air and dielectric interfaces [42–44]. For high degrees of integration, a metal ground plane is a must when the waveguides are made from dielectric. One such transmission line is the dielectric image line (DIL), which has been employed in mm-wave as well as sub-THz circuits. It is perfectly integrated with the planar circuit. By focusing the electromagnetic energy at the interface of the dielectric, it considerably decreases the electric current on the ground plane and enables transmission with a significantly lower conductor loss [45].

Power leakages will occur when the DIL is perturbed on a regular basis as a semi-open structure as well as radiation, and can be generated deliberately to fulfill the needs of particular services. Numerous LWAs have been considered in recent literature to date on the basics of periodic DIL disturbances caused by metallic loadings or dielectric gratings. Due to the internal resonance, known as the open-stop-band (OSB) phenomenon, dielectric grating-based LWAs typically do not provide radiation in the precise broadside. Due to the various forms of dielectric grating, these LWAs also have manufacturing difficulties [32–40]. To circumvent the OSB problem, A.O. Salman [36] investigated the connection of sinusoidal metallic gratings having a dielectric line concave face. A second technique for producing leaky waves in a DIL environment is metallic perturbation on the surface of a DIL. The intrinsic ease of production and the audible radiation make this technology appealing. In [39], a leaky wave antenna was created using a regularly loaded DIL with rectangular metal strips. Beam scanning merely occurs in a forward direction in these LWAs. Five metal strips in a rectangular shape were utilized in a single unit cell in [40] to reduce the OSB phenomenon. This displays primary beam scanning between 80 and 120. Beam-scanning techniques other than LWAs, such as SIW edge-radiation, near-field transformation, and end-fire antennas, may be employed [40–50].

The design and optimization of such SIW-based leaky wave structures at THz frequencies need high computer computations due to complex calculations. A new SIW LWA in the THz frequency region is suggested in this paper, which is the foundation for the wider angle of scanning at THz. An array of periodic nature composed of metallic circular discs that overlapped each other was used in the suggested antenna between different dielectric layers to create a radiator of high efficiency having broadside OSB mitigation properties. The open-stop-band is the biggest challenge for leaky wave antennas. Generally, open-stop-band indicates the range of frequencies in the EM spectrum where radiation is stopped in a particular direction. As we know that LWA is a kind of directional antenna that produces radiation in a specific direction by allowing the propagation of waves in the antenna structure. The open-stop-band in an LWA is created by adding a periodic perturbation in the antenna structure. This perturbation results in a periodic variation of the guided wave's effective refractive index along the antenna. When the wavelength of the guided wave matches the antenna's periodicity, a strong reflection takes place that creates a stop band. In this stop band, the radiation waves will experience strong reflection and will not be allowed to propagate. As a result, the antenna beam pattern will exhibit a null in this direction and enhance the antenna's directional properties. Various techniques are available for removing OSB issue, such as impedance matching, and the asymmetry, ridged, reflection cancellation, RC, and composite right/left-handed (CRLH) techniques. Generally, researchers used all these techniques to remove the OSB problem completely for the leaky wave antennas. The ridged technique is highly complex, but the RC and CRLH techniques are less complex compared to others. From the perspective of performance, all the OSB removal techniques work effectively, while the impedance matching and CRLH techniques are highly efficient and show excellent performance. The CRLH transmission line (TL) is normally used for realizing backfire to end-fire LWA [13]. A special feature of this type of LWA, which cannot be obtained in conventional LWA, is the backfire-to-end-fire ability of LWA (uniform or periodic), which was first signified experimentally in [13,14]. SIW is a planar platform that demonstrates itself as an excellent possibility for the realization of CRLH LWAs. In OSB, the CRHL technique is most effective and gives qualitative results in broadside and perfect impedance matching at the broadside to remove OSB. Most of the techniques can remove the problem of OSB but do not achieve the best impedance bandwidth in the broadside for a restricted beam scanning range. Again, in the impedance matching technique, a SIW-based unit cell accommodates a longitudinal slot and inductive post, whose positions are opposite from the center of the unit cell. To achieve perfect impedance matching, a combination of longitudinal slots and inductive posts gives inductive and capacitive effects. In the top metallic plate, the longitudinal slots are located away from the midline for surface current radiation. The main lobe direction changes with frequency at the broadside direction [13,14]. No such complex techniques are used in the proposed antenna for the removal of OSB, while inherited slots solve this issue efficiently and effectively. The suggested antenna does not need elaborate prototyping and is light weight. The suggested design preserves most of the benefits of the aforementioned SIW-based LWAs, which have been successfully used at low mm-wave frequencies (<100 GHz), incorporating continuous broad beam-scanning abilities with OSB suppression employing a low-profile construction and cheap cost.

2. Design Methodology and Antenna Array

The proposed SIW-based LWA configuration for THz applications is depicted in Figure 1. In this, the longitudinal and transverse slots are integrated with SIW along with a feed network. First, we designed the SIW platform for the design of a SIW leaky wave antenna for THz applications.

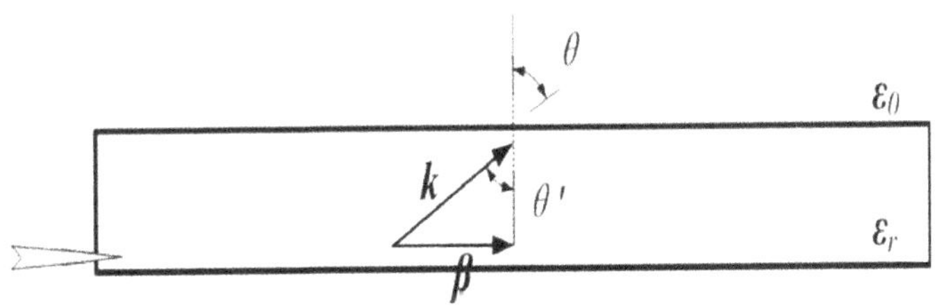

Figure 1. Side elevation of a leaky wave antenna.

2.1. Design of SIW

The side elevation for LWA along with its notations is shown in Figure 1, in which k denotes the propagation constant and β represents the phase constant for SIW, and they can be defined as [51–55]:

$$\sin\theta' = \frac{\beta}{k} \tag{1}$$

For simplifying the analysis, assume that when there is no leakage taking place, then β represents phase constant. The interface of dielectric and air medium refraction takes place following Snell's law as:

$$\frac{\sin\theta'}{\sin\theta} = \frac{1}{\sqrt{\varepsilon_r}} \tag{2}$$

where ε_r represents the permittivity of the substrate material and with minimal attenuation the pattern of the primary beam indicates an angle:

$$\theta = \arcsin\frac{\sqrt{\varepsilon_r}\beta}{k} = \arcsin\frac{\beta}{k_0} = \arcsin\frac{\lambda_0}{\lambda_g} \quad (3)$$

In expression (3), k_0 represents the propagation constant, λ_0 represents the wavelength and the guided wavelength of SIW is represented by λ_g.When leakage takes place, the value of β become less, hence β in SIW without any slot can be considered as the upper limiting case. Consider that the operating frequency is f and the required beam direction is θ, then:

$$\sin\theta = \sqrt{\varepsilon_r - \left(\frac{c}{2wf}\right)} \quad (4)$$

where w means the width of SIW, and c represents light speed in free space. The dimension of SIW is the width (w) and it can be computed as [53]:

$$w = a - \frac{d^2}{0.95s} \quad (5)$$

where the diameter of via is (d) and the space between the two neighboring vias is called pitch and is denotedby (s). With the optimization procedure, the measured length is 6.25 mm and the width of the tapered line is 7.6 mm. In addition, Rogers substrate was utilized for the designing of the SIW-based LWA having 0.009 loss tangent, 0.787 mm thickness, and 2.2 dielectric permittivity.

2.2. Design Concept

Periodic LWAs for THz were created by adding periodic perturbations in the waveguide's top surface. Two slots were added for the perturbations, which was a combination of transverse and longitudinal slots. These transverse and longitudinal slots were used in our case for periodic modulation. Apart from these longitudinal and transverse slots, four additional smaller slots were also added, which removed the open-stop-band problem effectively without using any OSB removal techniques. These disturbances produced infinity of space harmonics. In periodic LWAs, just a single rapid harmonic should be emitted for only one beam creation. This rapid wave space harmonic (i.e., $n = -1$ and $n = 0$) generates a complicated leaky propagation constant that is steered down the line. There is an important relationship between the phase constant and period (P) of a unit cell [33]. Apart from these longitudinal and transverse slots, four additional smaller slots were also added, which removed the open-stop-band problem effectively without using any OSB removal techniques:

$$\beta_n = \beta_0 + \frac{2\pi n}{P} \quad (6)$$

where P denotes the period, and β_0 represents the fundamental phase constant.

The phase constant value varied from a −ve to +ve zero crossing. As a direct consequence of this backward or negative, side-to-forward or positive side scanning takes place. However, the radiation from such antenna designs at the broadside is constrained via an OSB, which results from the standing wave production at broadside frequency. This results in a poor radiation pattern as well as a significant return loss [34]. Again, the attenuation (α) and the phase constant (β) are the two key factors that are responsible for the design of the LWA. These factors may be utilized to determine the angle of the main beam from the broadside as:

$$\sin\theta = \frac{\beta}{k_0} \quad (7)$$

where θ denotes the main angle of the beam from the broadside, and k_0 represents the free space wave number. The length of the leaky wave antenna is thereby chosen ~90% of its energy and gets transmitted while reaching to load and is expressed by:

$$\frac{L}{\lambda_0} = \frac{0.18}{\alpha/k_0} \tag{8}$$

For the proposed design, $n = -1$ harmonic was produced by creating periodic modulations. For the broadside $\beta_{-1} = 0$, as per Equation (6) we then get:

$$\beta_0 = \frac{2\pi n}{P} = \frac{2\pi}{\lambda_g} \tag{9}$$

where λ_g denotes the guided wavelength. By Equation (9), we can say that to achieve the CBS unit cell period must be equivalent to the guided wavelength.

2.3. Geometry

The SIW LWA unit cell design and array design are illustrated in Figure 2a,b. The LWA was made up of 10unit cells that resulted in 147 mm length. The unit cell of SIW design had dual artificial electric walls utilizing the vias. Also, the diameter 'd' as well as the spacing 's' (i.e., $\leq$2d) between these vias was assumed to be 1 mm as well as 2 mm, correspondingly. Otherwise, the SIW width can be determined by Equation (5). The proposed LWA was ingested from the left side and a matched impedance via a 50 Ω termination was achieved. The unit cell dimensions are shown in Table 1. Furthermore, a unit cell dispersion diagram is represented by Figure 2c, which indicates that scanning took place from the back side to the forward side through the broadside.

Table 1. Parameters of a unit cell design.

Parameters	Values (mm)
P	14.7
W	16
A_L	5.5
A_T	3.6
D	1
B	0.8
C	1.5
S	2
W_{eff}	14

For the SIW design to work as an LWA, the rectangular slots were etched on the SIW's top surface. Four smaller additional slots were also made along with the longitudinal and transverse slots, which provide leaky wave radiation. The selected longitudinal and transverse slot was 90°, since the total power transmitted by these slots was equal to $\sin^2 \varphi$ where φ represents the angle of the intersection of the longitudinal and transverse slots.

To determine the angle between the longitudinal and transverse slots for the SIW LWA, a parametric analysis was carried out and the scattering parameter (S_{11}) is depicted in Figure 3. In the parametric analysis, to get the optimum scattering parameter, different angles between the longitudinal and transversal slots were applied, such as 70°, 80°, 90°, 100°, and 110°, as shown in Figure 3. As can be seen from Figure 3, the best scattering parameter plot was obtained when the angle between the longitudinal and transversal slots was 90°. For matching the antenna, the angle between the longitudinal and transverse slots was very important. The angle between the longitudinal and transverse slots must be 90 degrees to produce the optimum impedance matching and maximize the radiation from these slots. The nature of the electric field distribution for the proposed design is represented by Figure 4, which shows the optimum distribution that is necessary for leaky waves.

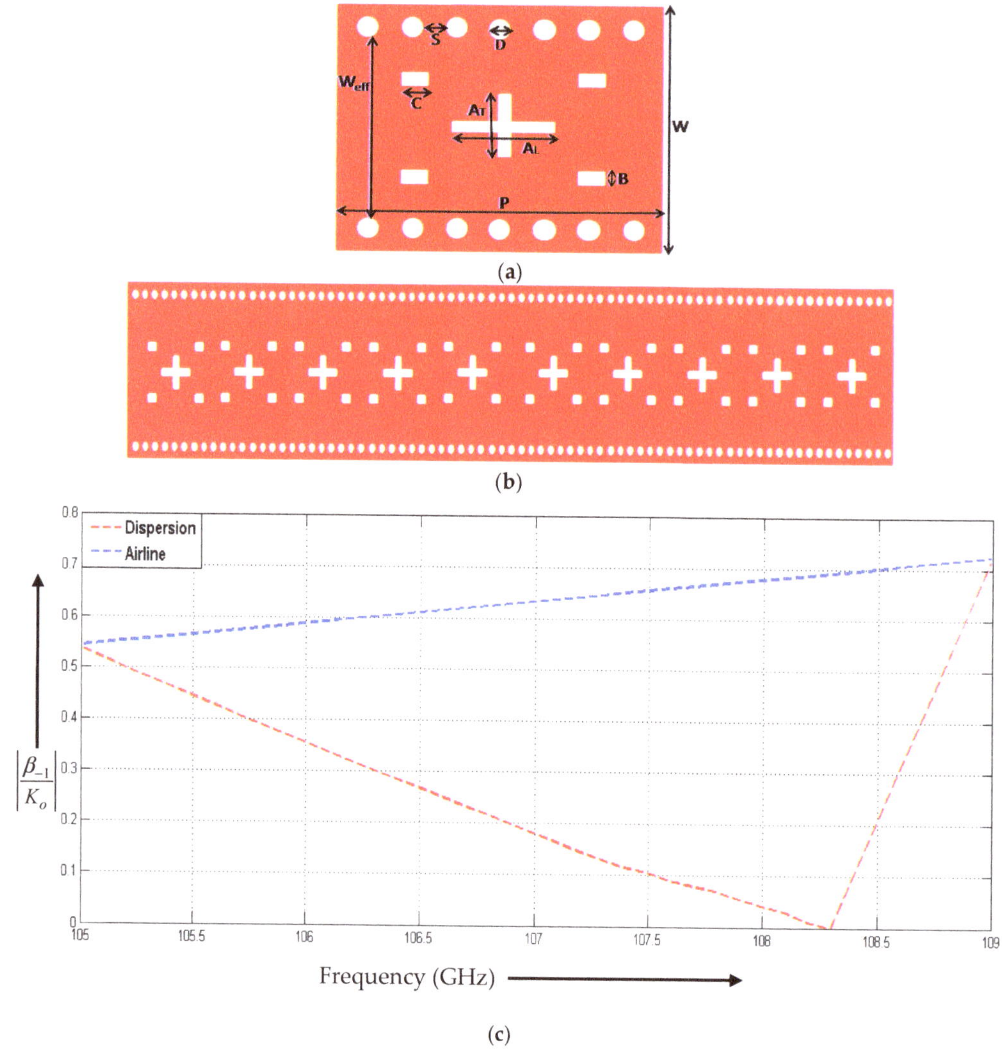

Figure 2. (**a**) SIW LWA unit cell, (**b**) SIW LWA structure, (**c**) Dispersion plot for a unit cell.

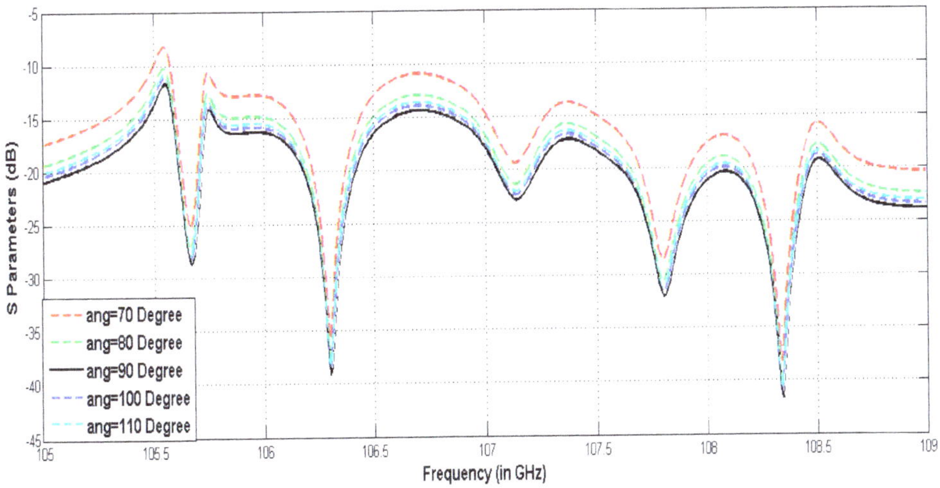

Figure 3. Parametric measurements for the angles between transversal and longitudinal slots at 70°, 80°, 90°, 100°, and 110°.

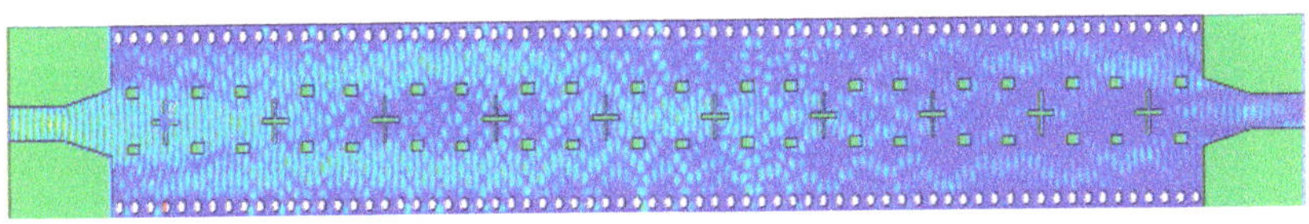

Figure 4. SIW electric field distribution.

3. Experiment Results

3.1. Geometry

An LWA for THz applications based on the SIW technology that gives CBS was designed. The SIW LWA structure for THz applications is depicted in Figure 2. The antenna was fed via a tapered microstrip line, which was terminated with a load matching the characteristic impedance. Using ANSYS HFSS software, the tapered part dimensions were optimized. The tapering line was 6.25 and 7.6 mm in length and breadth, respectively. The antenna's substrate had a tangent loss of 0.009, a height of 0.787 mm, and 2.2 dielectric permittivity.

3.2. S Parameters

Figure 5 depicts the reflection coefficient (S_{11}) for the designed SIWLWA. According to the simulated curve of the reflection coefficient (S_{11}) of the proposed antenna, the broadside (about 108.3 GHz) impedance matching was nearly perfect. As the open-stop-band issue is the biggest challenge for the researchers, the open-stop-band was removed perfectly during the simulation process without using any other components; as seen from Figure 5, there was no open-stop-band in the frequency range of 105 GHz to 109 GHz. The simulated results of the reflection coefficient (S_{11}) showed that this proposed leaky wave antenna worked well in the frequency band range of 105 GHz to 109 GHz.

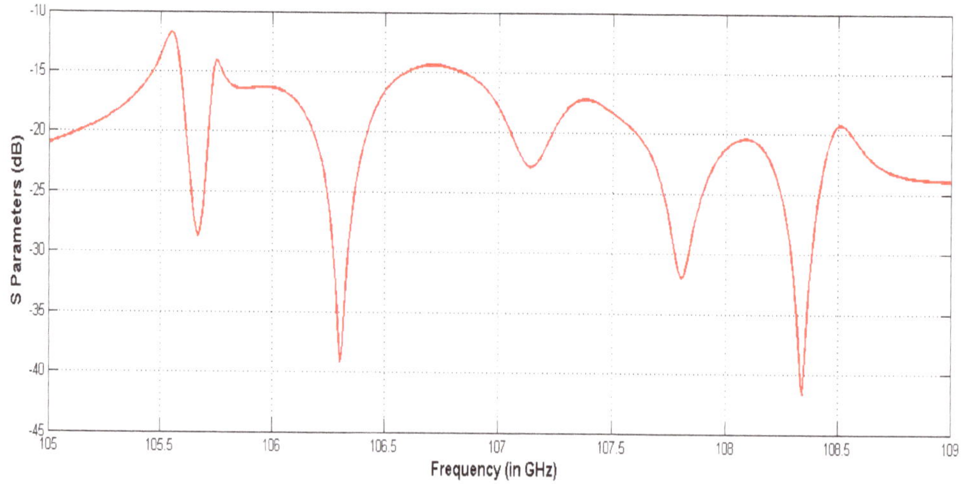

Figure 5. Simulated measured S_{11} parameter for SIW-LWA.

3.3. Radiation Features

For the simulation of the proposed antenna, the ANSYS HFSS software was utilized. First, the SIW and slots dimensions were optimized as per the frequency range, and after its simulation work was carried out. The simulated normalized radiation patterns of the suggested THz-based SIW LWA in y-z plane are depicted in Figure 6. The simulation for all the frequencies from 105 GHz to 109 GHz was performed by considering the difference of 0.5 GHz. When the frequency was swept in the 105~109 GHz range, the beam scanned from +78° to −6° for the SIW leaky wave antenna. The simulation yielded an entire beam scanning range of 84°. This beam went from +78° at 105 GHz via the broadside (108.3 GHz) to the rearward direction as well as pointed at −6° at 109 GHz with a rise in frequency. For the designed SIWLWA, this showed continuous beam scanning in the THz applications.

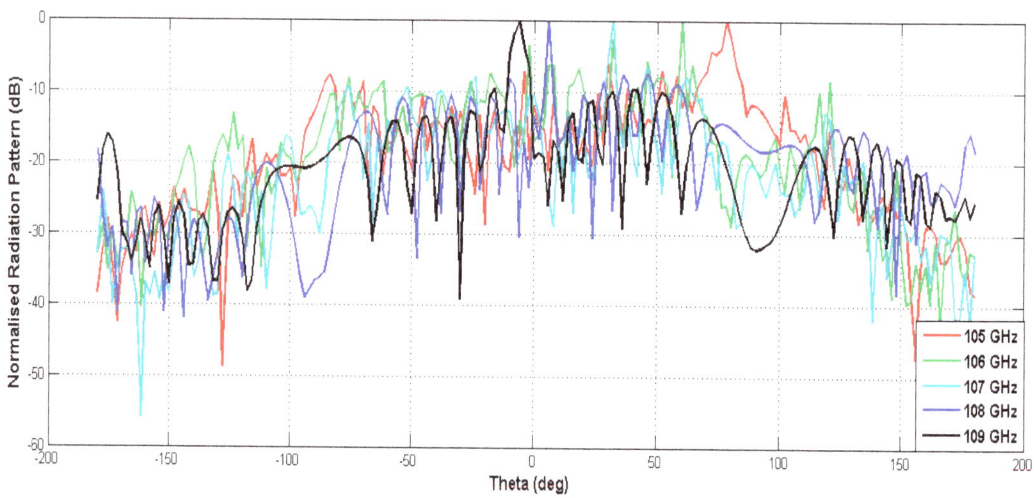

(a)

Figure 6. *Cont.*

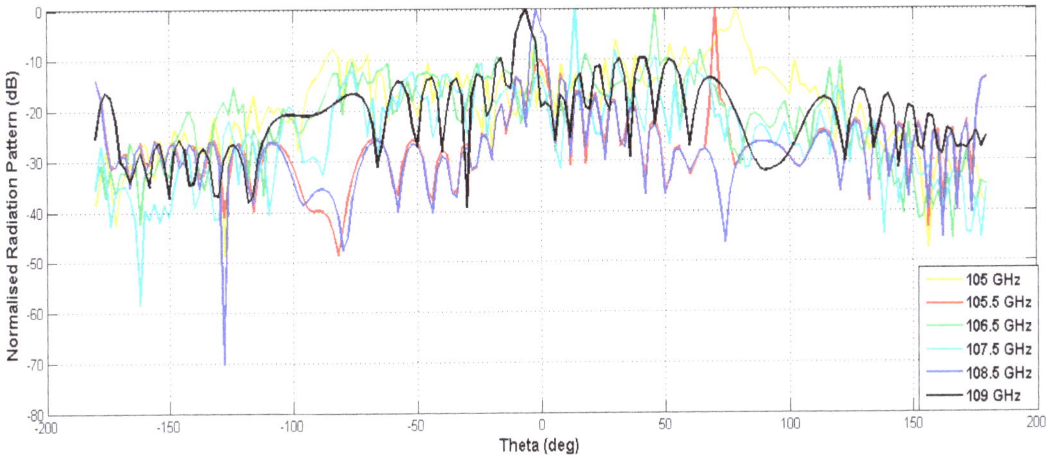

(**b**)

Figure 6. (**a**,**b**) Simulated 2D radiation patterns of SIW LWA in y-z plane.

The primary lobe directions for the proposed antenna are represented in Figure 7. The designed SIWLWA scanning range was higher than that of the current SIW-based THz leaky wave antennas, as the radiation from every longitudinal slot in the proposed SIW LWA disappeared at end-fire. As shown in Figure 7, at 108.3 GHz, scanning took place at 0 degrees (at broadside), after 108.3 GHz, scanning took place on the backward side, and before 108.3 GHz, scanning only occurred on the forward side. Figure 8 shows the simulated gain patterns for the SIWLWA and the maximum gain obtained was 16.02 dBi at 107 GHz; Table 2 shows that the recommended antenna was superior and more compact than other SIW-based leaky wave antennas for THz applications.

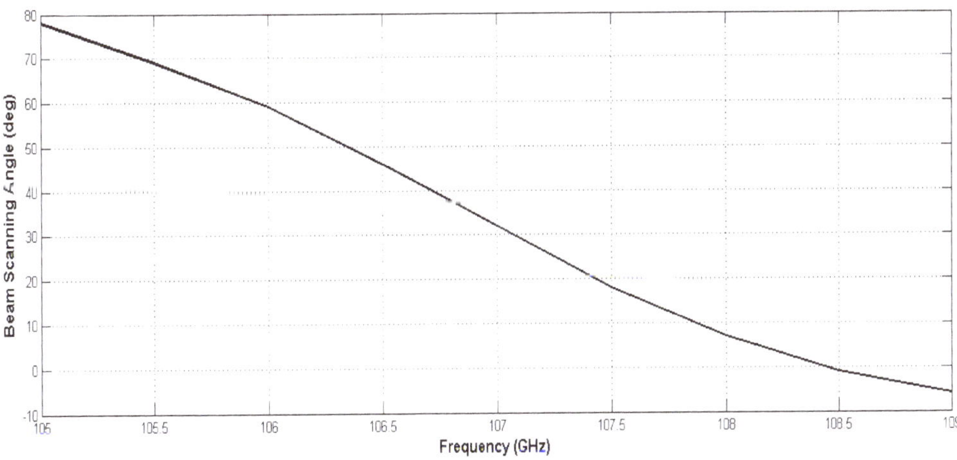

Figure 7. Various radiation angles for the proposed SIW LWA.

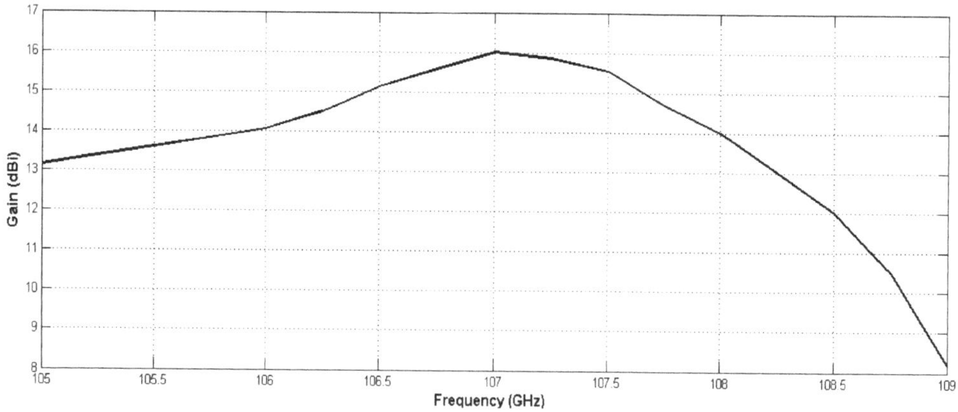

Figure 8. Simulated gain patterns for SIW LWA.

Table 2. Comparisons with other reported designs.

References	Broadside Radiation	Radiator Length	Range of Scanning Frequency (GHz)	Range of Beam Scanning (Degree)	Max. Gain
[15]	No	~$10\lambda_0$	220 to 300	−75° to −30° (Backward Only)	~28.5 dBi
[38]	Yes	~$11.1\lambda_0$	75 to 85	−10° to −8°	~12.7 dBi
[39]	No	~$2.6\lambda_0$	58 to 67	+7° to +38° (Forward Only)	~11.7 dBi
[40]	Yes	~$6\lambda_0$	86 to 106	−31° to +10°	~11 dBi
[53]	No	~$12\lambda_0$	55 to 67	4° to 18° (Forward Only)	~6 dBi
[54]	Yes	~$8\lambda_0$	230 to 245	−25° to 25°	~29 dBi
[44]	Yes	~$6.9\lambda_0$	157.5 to 206	−23° to +38°	~15 dBi
This work	Yes	~$6.84\lambda_0$	105 to 109	+78° to −6°	~16.02 dBi

3.4. Fabrication as Well as Measurement Issues

The recommended antenna design had a good structure operating as the LWA with outstanding characteristics in THz frequencies; nevertheless, its fabrication could be a little difficult. For instance, the expected air gap between the ground plane and the dielectric layer during manufacturing might result in minor differences between the simulation and measurement findings. Additionally, it might be difficult to achieve fewer reflection losses in THz frequencies since even a small variation in the feature sizes and the value of the dielectric constant can have a direct impact on antenna performance. Furthermore, because of technological restrictions, it could be difficult to measure the radiation patterns of the antenna throughout the whole 180° range. Indeed, the radiation pattern at such high frequencies can be altered at angles close to the backfire by a variety of sources of diffraction linked to antenna-feeding impacts (i.e., plastic screw, waveguide flange, etc.). Additionally, as most testing equipment is waveguide-based and operates at THz frequencies without connections, the transitions are crucial to the suggested design. A straightforward inline transition between the microstrip line and a rectangle waveguide may have been needed for the measurement in this study as the microstrip line was employed for feeding. The simulation did not include any representation of the transitions to rectangular waveguide.

4. Conclusions

In this work, a new LWA-based SIW technology was presented for CBS that provided a wider range of beam scanning with high gain. The proposed antenna system employed a SIW and a combination of longitudinal as well as transverse slots to address the OSB issue supplied by this waveguide. A 10-element linear array of radiating elements was constructed. Specific features of the suggested design were reviewed, including its feeding network, radiation characteristics, and difficulties in its production and measurement. From the broadside direction, a gain of 12.33 dBi was attained. This antenna provided continuous beam scanning in the range of +78° to −6° via the broadside while exhibiting good radiation performance over the working frequency band of 105 GHz to 109 GHz, with a peak gain of 16.02 dBi. The THz leaky wave antenna put forth in this paper was easily adaptable for realization at higher frequencies.

Author Contributions: Conceptualization, V.A., T.G. and H.M.R.A.-K.; methodology, V.A. and H.M.R.A.-K.; software, V.A. and H.M.R.A.-K.; validation, V.A. and H.M.R.A.-K.; formal analysis, V.A. and T.G.; investigation, V.A. and T.G.; resources, V.A. and H.M.R.A.-K.; writing—original draft preparation, V.A., T.G. and H.M.R.A.-K.; writing—review and editing, V.A., T.G. and H.M.R.A.-K.; visualization, V.A. and T.G.; supervision, T.G.; project administration, T.G. All authors have read and agreed to the published version of the manuscript.

Funding: This research received no external funding.

Data Availability Statement: Not applicable.

Conflicts of Interest: The authors have no conflict of interest to declare.

References

1. Hansen, W.W. Radiating Electromagnetic Waveguide. U.S. Patent No. 2,402,622,1940, 26 November 1940.
2. Marcuvitz, N. On field representation in terms of leaky modes or eigenmodes. *IRE Trans.* **1956**, *4*, 192–194.
3. Nishida, S. Leaky wave antennas. *Electron. Commun. Jpn.* **1965**, *48*, 42–48.
4. Goldstone, L.O.; Oliner, A.A. Leaky-wave antennas—Part I: Rectangular waveguides. *IRE Trans. Antennas Propag.* **1959**, *7*, 307–319. [CrossRef]
5. Cheng, Y.J.; Hong, W.; Wu, K.; Fan, Y. Millimeter-wave substrate integrated waveguide long slot leaky-wave antennas and two-dimensional multibeam applications. *IEEE Trans. Antennas Propag.* **2011**, *59*, 40–47. [CrossRef]
6. Lyu, Y.; Member, S.; Liu, X. Leaky-wave antennas based on non-cutoff substrate integrated waveguide supporting beam scanning from backward to forward. *IEEE Trans. Antennas Propag.* **2016**, *64*, 2155–2164. [CrossRef]
7. Zhou, W.; Liu, J.; Long, Y. Investigation of shorting vias for suppressing the open stopband in an SIW periodic leaky-wave structure. *IEEE Trans. Microw. Theory Tech.* **2018**, *66*, 2936–2945.
8. Agrawal, R.; Belwal, P.; Gupta, S. Asymmetric substrate integrated waveguide leaky wave antenna with open stop band suppression and radiation efficiency equalization through broadside. *Radioengineering* **2018**, *27*, 409–416. [CrossRef]
9. Liu, J.; Jackson, D.R.; Long, Y. Substrate integrated waveguide (SIW) leaky-wave antenna with transverse slots. *IEEE Trans. Antennas Propag.* **2012**, *60*, 20–29. [CrossRef]
10. Liu, J.; Tang, X.; Li, Y.; Long, Y. Substrate integrated waveguide leaky-wave antenna with H-shaped slots. *IEEE Trans. Antennas Propag.* **2012**, *60*, 3962–3967. [CrossRef]
11. Liu, L.; Caloz, C.; Itoh, T. Dominant mode (DM) leaky wave antenna with backfire-to-endfire scanning capability. *Electronics Lett.* **2003**, *38*, 1414–1416. [CrossRef]
12. Caloz, C.; Itoh, T. *Electromagnetic Metamaterials: Transmission Line Theory and Microwave Applications: The Engineering Approach*; Wiley: Hoboken, NJ, USA, 2006.
13. Guglielmi, M.; Jackson, D. Broadside radiation from periodic leaky-wave antennas. *IEEE Trans. Antennas Propag.* **1993**, *41*, 31–37. [CrossRef]
14. Paulotto, S.; Baccarelli, P.; Frezza, F.; Jackson, D.R. A novel technique for open-stopband suppression in 1-D periodic printed leaky-wave antennas. *IEEE Trans. Antennas Propag.* **2009**, *57*, 1894–1906. [CrossRef]
15. Gomez-Torrent, A.; Garcia-Vigueras, M.; Le Coq, L.; Mahmoud, A.; Ettorre, M.; Sauleau, R.; Oberhammer, J. A Low-Profile and High-Gain Frequency Beam Steering Subterahertz Antenna Enabled by Silicon Micromachining. *IEEE Trans. Antennas Propag.* **2020**, *68*, 672–682. [CrossRef]
16. Monnai, Y. Terahertz Radar Based on Leaky-Wave Coherence Tomography. In Proceedings of the 2020 Conference on Lasers and Electro-Optics Pacific Rim, CLEO-PR 2020—Proceedings, Sydney, Australia, 2–6 August 2020; pp. 1–2.
17. Rikkinen, K.; Kyosti, P.; Leinonen, M.E.; Berg, M.; Parssinen, A. THz Radio Communication: Link Budget Analysis toward 6G. *IEEE Commun. Mag.* **2020**, *58*, 22–27. [CrossRef]
18. Kwon, H.; Kim, Y.; Yoon, H.; Choi, D. Selective Audio Adversarial Example in Evasion Attack on Speech Recognition System. *IEEE Trans. Inf. Forensics Secur.* **2020**, *15*, 526–538. [CrossRef]
19. Brown, E.R. Fundamentals of Terrestrial Millimeter-Wave and THz Remote Sensing. *Int. J. High Speed Electron. Syst.* **2003**, *13*, 995–1097. [CrossRef]

20. Golubiatnikov, G.Y.; Koshelev, M.A.; Tsvetkov, A.I.; Fokin, A.P.; Glyavin, M.Y.; Tretyakov, M.Y. Sub-Terahertz High-Sensitivity High-Resolution Molecular Spectroscopy with a Gyrotron. *IEEE Trans. Terahertz Sci. Technol.* **2020**, *10*, 502–512. [CrossRef]
21. Guerboukha, H.; Shrestha, R.; Neronha, J.; Ryan, O.; Hornbuckle, M.; Fang, Z.; Mittleman, D.M. Efficient Leaky-Wave Antennas at Terahertz Frequencies Generating Highly Directional Beams. *Appl. Phys. Lett.* **2020**, *117*, 261103. [CrossRef]
22. Agarwal, R.; Agarwal, A.; Dwivedi, A.; Sharma, A. Leaky Wave Antenna for Millimeter Wave Utilization. *J. Phys. Conf. Ser.* **2021**, *1921*, 012026. [CrossRef]
23. Ghalibafan, J.; Hashemi, S.M. Leaky-Wave Centerline Longitudinal Slot Antenna Fed by Transversely Magnetized Ferrite. *IEEE Trans. Magn.* **2016**, *52*, 4000104. [CrossRef]
24. Zheng, D.; Lyu, Y.-L.; Wu, K. Transversely Slotted SIW Leaky-Wave Antenna Applications. *IEEE Trans. Antennas Propag.* **2020**, *68*, 4172–4185. [CrossRef]
25. Ranjan, R.; Ghosh, J. SIW-Based Leaky-Wave Antenna Supporting Wide Range of Beam Scanning Through Broadside. *IEEE Antennas Wirel. Propag. Lett.* **2019**, *18*, 606–610. [CrossRef]
26. Saghati, A.P.; Mirsalehi, M.M.; Neshati, M.H. A HMSIW Circularly Polarized Leaky-Wave Antenna With. *IEEE Trans. Antennas Propag.* **2014**, *13*, 451–454. [CrossRef]
27. Chen, Y.; Wang, K.; Li, Y.; Long, Y. Periodic Microstrip Leaky Wave Antenna with Double-Sided Shorting Pins and Pairs of Slots. *Int. J. Antennas Propag.* **2020**, *2020*, 7101752. [CrossRef]
28. Martinez-Ros, A.J.; Gómez-Tornero, J.L.; Goussetis, G. Planar Leaky-Wave Antenna with Flexible Control of the Complex Propagation Constant. *IEEE Trans. Antennas Propag.* **2012**, *60*, 1625–1630. [CrossRef]
29. Grbic, A.; Eleftheriades, G.V. Leaky CPW-Based Slot Antenna Arrays for Millimeter-Wave Applications. *IEEE Trans. Antennas Propag.* **2002**, *50*, 494–1504. [CrossRef]
30. Alibakhshikenari, M.; Virdee, B.S.; Khalily, M.; Shukla, P.; See, C.H.; Abd-Alhameed, R.; Falcone, F.; Limiti, E. Beam-Scanning Leaky-Wave Antenna Based on CRLH-Metamaterial for Millimetre-Wave Applications Mohammad. *IET Microw. Antennas Propag.* **1941**, *74*, 535–546.
31. Pozar, D.M. *Microwave Engineering*, 4th ed.; John Wiley & Sons: Hoboken, NJ, USA, 2012; ISBN 978-0-470-63155-3.
32. Xu, S.; Gao, H. Double Dielectric Grating Leaky-Wave Antenna-Improved Perturbation Analysis. *Int. J. Infrared Millim. Waves* **1989**, *10*, 1103–1119. [CrossRef]
33. Hammad, H.F.; Antar, Y.M.; Freundorfer, A.P.; Sayer, M. Frequency. *IEEE Trans. Antennas Propag.* **2004**, *52*, 36–44. [CrossRef]
34. Bai, Y.; Liu, S. A Novel Dual-Beam Terahertz Leaky-Wave Antenna Based On Spoof Surface Plasmon Waveguide. *Optoelectron. Lett.* **2022**, *18*, 404–407. [CrossRef]
35. Choi, J.H.; Itoh, T. Beam-Scanning Leaky-Wave Antennas. In *Handbook of Antenna Technologies*; Chen, Z., Ed.; Springer: Singapore, 2015. [CrossRef]
36. Salman, A.O. On the antenna efficiencies for the dielectric leaky-wave antennas with a sinusoidal metallic diffraction grating coupled from the broad and the narrow face of the dielectric. *Microw. Opt. Technol. Lett.* **2011**, *53*, 2030–2034.
37. Aysu, B.; Filiz, G.; Merih, P.; Ozlem, T.; Peyman, M. 3D EM Data Driven Surrogate Based Design Optimization of Traveling Wave Antennas for Beam Scanning In X-Band: An Application Example. *Wirel. Netw.* **2022**, *28*, 1827–1834. [CrossRef]
38. Zandamela, A.; Al-Bassam, A.; Heberling, D. Circularly Polarized Periodic Leaky-Wave Antenna Based on Dielectric Image Line for Millimeter-Wave Radar Applications. *IEEE Antennas Wirel. Propag. Letter.* **2021**, *20*, 938–942. [CrossRef]
39. Cheng, Y.J.; Guo, Y.X.; Bao, X.Y.; Ng, K.B. Millimeter-Wave Low Temperature Co-Fired Ceramic Leaky-Wave Antenna and Array Based on the Substrate Integrated Image Guide Technology. *IEEE Trans. Antennas Propag.* **2014**, *62*, 669–676. [CrossRef]
40. Patrovsky, A.; Wu, K. Substrate Integrated Image Guide Array Antenna for the Upper Millimeter-Wave Spectrum. *IEEE Trans. Antennas Propag.* **2007**, *55*, 2994–3001. [CrossRef]
41. Liu, L.; Wang, J.; Yin, X.; Chen, Z.N. Wide-Angle Beam Scanning Leaky-Wave Antenna Using Spoof Surface Plasmon Polaritons Structure. *Electronics* **2018**, *7*, 348. [CrossRef]
42. Fuscaldo, W.; Zografopoulos, D.C.; Imperato, F.; Burghignoli, P.; Beccherelli, R.; Galli, A. Analysis and Design of Tunable THz 1-D Leaky-Wave Antennas Based on Nematic Liquid Crystals. *Appl. Sci.* **2022**, *12*, 11770. [CrossRef]
43. Sharma, J.; De, A. Full-Wave Analysis of Dielectric Rectangular Waveguides. *Prog. Electromagn. Res. M* **2010**, *13*, 121–131.
44. Torabi, Y.; Dadashzadeh, G.; Hadeie, M.; Oraizi, H.; Lalbakhsh, A. A Wide-Angle Scanning Sub-Terahertz Leaky-Wave Antenna Based on a Multilayer Dielectric Image Waveguide. *Electronics* **2021**, *10*, 2172.
45. Tesmer, H.; Razzouk, R.; Polat, E.; Wang, D.; Jakoby, R.; Maune, H. Temperature Characterization of Liquid Crystal Dielectric Image Line Phase Shifter for Millimeter-Wave Applications. *Crystals* **2021**, *11*, 63. [CrossRef]
46. Lalbakhsh, A.; Afzal, M.U.; Hayat, T.; Esselle, K.P.; Mandal, K. All-Metal Wideband Metasurface for near-Field Transformation of Medium-to-High Gain Electromagnetic Sources. *Sci. Rep.* **2021**, *11*, 9421. [CrossRef] [PubMed]
47. Mirmozafari, M.; Zhang, Z.; Gao, M.; Zhao, J.; Honari, M.M.; Booske, J.H.; Behdad, N. Mechanically Reconfigurable, Beam-Scanning Reflectarray and Transmitarray Antennas: A Review. *Appl. Sci.* **2021**, *11*, 6890, ISBN 0000000211512. [CrossRef]
48. Goudarzi, A.; Honari, M.M.; Gharaati; Mirzavand, R. A Cavity-Backed Antenna with a Tilted Directive Beam for 5G Applications. In Proceedings of the 2021 IEEE International Symposium on Antennas and Propagation and USNC-URSI Radio Science Meeting (APS/URSI), Singapore, 4–10 December 2021; pp. 1737–1738.
49. Afzal, M.U.; Matekovits, L.; Esselle, K.P.; Lalbakhsh, A. Beam-Scanning Antenna Based on near-Electric Field Phase Transformation and Refraction of Electromagnetic Wave through Dielectric Structures. *IEEE Access* **2020**, *8*, 199242–199253. [CrossRef]

50. Ibrahim, I.M.; Ahmed, M.I.; Abdelkader, H.M. A Novel Compact High Gain Wide-Band Log Periodic Dipole Array Antenna for Wireless Communication Systems. *J. Infrared Millim. Terahertz Waves* **2022**, *43*, 872–894. [CrossRef]
51. Kishihara, M.; Ohta, I.; Okubo, K.; Yamakita, J. Analysis of Post-Wall Waveguide Based on H-Plane Planar Circuit Approach. *IEICE Trans. Electron.* **2009**, *E92-C*, 63–71. [CrossRef]
52. Cassivi, Y.; Perregrini, L.; Arcioni, P.; Bressan, M.; Wu, K.; Conciauro, G. Dispersion Characteristics of Substrate Integrated Rectangular Waveguide. *IEEE Microw. Wirel. Compon. Lett.* **2002**, *12*, 333–335. [CrossRef]
53. Ghasemi, A.; Laurin, J.-J. A Continuous Beam Steering Slotted Waveguide Antenna Using Rotating Dielectric Slabs. *IEEE Trans. Antennas Propag.* **2021**, *67*, 6362–6370. [CrossRef]
54. Sarabandi, K.; Jam, A.; Vahidpour, M.; East, J. A Novel Frequency Beam-Steering Antenna Array for Submillimeter-Wave Applications. *IEEE Trans. Terahertz Sci. Technol.* **2018**, *8*, 654–665. [CrossRef]
55. Arya, V.; Garg, T.; Al-Khafaji, H.M.R. High Gain and Wide-Angle Continuous Beam Scanning SIW Leaky-Wave Antenna. *Electronics* **2023**, *12*, 370.

Disclaimer/Publisher's Note: The statements, opinions and data contained in all publications are solely those of the individual author(s) and contributor(s) and not of MDPI and/or the editor(s). MDPI and/or the editor(s) disclaim responsibility for any injury to people or property resulting from any ideas, methods, instructions or products referred to in the content.

electronics

MDPI

Article

Exploring the Validity of Plane and Spherical Millimeter-Wave Incidences for Multiple-Diffraction Calculations in Wireless Communication Systems

Alba López-Segovia [1], Ignacio Rodríguez-Rodríguez [2], José-Víctor Rodríguez [1,*], Leandro Juan-Llácer [1], María Campo-Valera [1] and Wai Lok Woo [3]

[1] Departamento de Tecnologías de la Información y las Comunicaciones, Universidad Politécnica de Cartagena, Antiguo Cuartel de Antigones, Plaza del Hospital, 1, Cartagena, 30202 Murcia, Spain
[2] Departamento de Ingeniería de Comunicaciones, Universidad de Málaga, Avda. Cervantes, 2, 29071 Málaga, Spain
[3] Department of Computer and Information Sciences, Northumbria University, Newcastle upon Tyne NE1 8ST, UK; wailok.woo@northumbria.ac.uk
* Correspondence: jvictor.rodriguez@upct.es; Tel.: +34-968-326-548

Abstract: The focus of this work is to determine at which threshold can the results for both plane and spherical wave incidence assumptions either converge or deviate when performing multiple diffraction attenuation calculations. The analysis has been carried out—for various millimeter-wave frequencies, inter-obstacle spacings, and angles of incidence—by employing a pair of two-dimensional (2D) hybrid formulations based on both the uniform theory of diffraction and physical optics (UTD-PO). This way, we seek to demonstrate under which circumstances each wave incidence assumption can be valid in environments that entail millimeter-wave bands. Based on this, we may ensure the minimum necessary distance from the transmitter to the first diffracting obstacle for the convergence of the spherical wave incidence solution onto that of the plane wave with a relative error below 0.1%. Our results demonstrate that for less than four diffracting elements, the minimum necessary distance engages in quasi-linear behavior under variations in both the angle of incidence and obstacle spacing. Notably, the considered frequencies (60–100 GHz) have almost no bearing on the results. Our findings will facilitate the simplified, more accurate and realistic planning of millimeter-wave radio communication systems, with multiple diffractions across various obstacles.

Keywords: radio communication systems; multiple diffraction; millimeter-wave frequency band; uniform theory of diffraction

Citation: López-Segovia, A.; Rodríguez-Rodríguez, I.; Rodríguez, J.-V.; Juan-Llácer, L.; Campo-Valera, M.; Woo, W.L. Exploring the Validity of Plane and Spherical Millimeter-Wave Incidences for Multiple-Diffraction Calculations in Wireless Communication Systems. *Electronics* **2023**, *12*, 2020. https://doi.org/10.3390/electronics12092020

Academic Editors: Syed Muzahir Abbas, Yang Yang and Adão Silva

Received: 6 March 2023
Revised: 25 April 2023
Accepted: 26 April 2023
Published: 27 April 2023

Copyright: © 2023 by the authors. Licensee MDPI, Basel, Switzerland. This article is an open access article distributed under the terms and conditions of the Creative Commons Attribution (CC BY) license (https://creativecommons.org/licenses/by/4.0/).

1. Introduction

The analysis of the multiple diffraction experienced by radio waves due to the presence of obstacles in their propagation path has been extensively studied through numerous formulations. These solutions are usually based on the uniform theory of diffraction (UTD) [1,2], physical optics (PO) [3,4], or both theories [5,6], and aim to predict the losses caused by this phenomenon by modeling the mentioned obstacles as knife-edges and assuming a plane wave incidence over them in most cases. In this case, when the number of these obstacles is large, the multiply diffracted field is approximated by the so-called Q factor, as proposed in [3].

However, in micro/picocellular environments where the transmitting antenna may be located at a short distance from the series of diffracting elements, the assumption of a plane wave incidence on them may not yield realistic results. Therefore, in such a case, considering a spherical wave incidence could be more appropriate in terms of obtaining predictions of losses due to multiple diffraction that are more in line with reality.

On the other hand, future wireless telecommunication systems are required to offer higher data rates and capacity to cope with the next generation of high-bandwidth multimedia services. In this context, frequency bands located between 60 and 100 GHz have captured the interest of researchers due to the large available bandwidth [7] and the small frequency reuse distance it offers [8], which allows for the development of low-range indoor systems with low-power transmitters, but with transmission rates of up to several gigabits per second [9].

Regarding the above, numerous papers have been published on radio wave propagation models at millimeter frequencies [10–14], and specifically, several works have addressed the analysis of multiple diffraction at such frequencies [15–17]. However, none of the previous work addresses the fact of checking beforehand, whether the distance between the transmitter and the first obstacle is large enough for a plane-wave incidence assumption to be valid, which could lead to inaccurate or unrealistic results. On the other hand, it is also not verified if when a spherical-wave incidence is being considered, a plane-wave assumption could be perfectly used instead, with the improvement in terms of computational efficiency that this would entail.

In this work, in order to clarify the distance range between the transmitter and the first diffracting element in which each type of incidence would be valid, a comparison of the losses due to multiple diffractions caused by a series of obstacles for both plane and spherical-wave assumptions is presented, while considering a range of frequencies between 60 and 100 GHz, as well as several angles of incidence and spacing between diffracting elements. The analysis is carried out using two hybrid UTD-PO formulations (for plane and spherical-wave incidences, respectively) developed by the authors [5,6], which are more computationally efficient than other existing solutions. Moreover, the study considers an environment in which millimeter-wave communication systems are being used, and attenuation due to multiple diffraction needs to be calculated over a series of objects. In other words, the main novelty of the paper is the calculation of the limits at which both UTD-PO multiple-diffraction formulations—based on two different wave-incidence assumptions (plane and spherical)—converge or differ, at millimeter-wave frequencies. In this sense, the results will establish the validity of both types of incidence depending on the distance between the transmitter and the first diffracting obstacle, thereby alerting about an improper use of one or the other formulation, which may lead to inaccurate or computationally demanding results.

2. Propagation Environment

Figure 1 shows a diagram of the propagation environment considered, where n obstacles are modeled as parallel and absorbent edges, separated by a constant distance w and at the same height H relative to the transmitter Tx. Additionally, the transmitter is assumed to be located above the height of the obstacles and at a distance d from the first obstacle. Furthermore, the reference point where the received field will be obtained is assumed to be located at a distance w from the last obstacle considered.

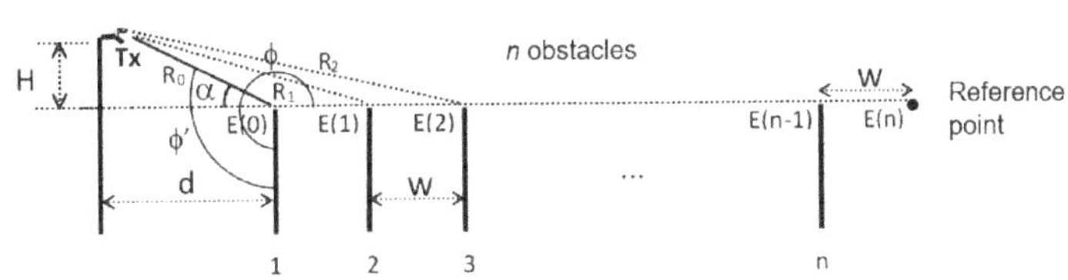

Figure 1. Scheme of the considered propagation environment.

3. Theoretical Models

The following theoretical models are described to evaluate multiple diffraction in both the case of a plane wave incident on the obstacles and a spherical wave.

3.1. Plane Wave Incidence

For $n = 1$, considering UTD, the field that reaches the reference point in Figure 1 can be calculated as:

$$E(1) = E(0)\left[\exp(-jkw\cos\alpha) + \frac{1}{\sqrt{w}}D\left(\varphi = \frac{3\pi}{2}, \varphi' = \frac{\pi}{2} + \alpha, L = w\right)\exp(-jkw)\right] \tag{1}$$

where $E(0) = E_0$ (being that E_0 is the amplitude of the transmitted plane wave, which is assumed to be 1), k is the wave number, and $D(\phi, \phi', L)$ is the diffraction coefficient for an absorbent knife-edge proposed in [18] (ϕ' is the angle between the diffracting obstacle and the incident ray, ϕ is the angle between the diffracting obstacle and the diffracted ray, and L is a distance parameter, as can be observed in Figure 1).

For $n = 2$, multiple diffraction arises, and following the same PO recursive procedure as that presented by Saunders and Bonar in [4], the field at the reference point in Figure 1 can be obtained as the average of two contributions expressed in terms of UTD single diffractions:

$$E(2) = \frac{1}{2}\left[\begin{array}{l} E(0)\left[\exp(-jk2w\cos\alpha) + \frac{1}{\sqrt{2w}}D\left(\varphi = \frac{3\pi}{2}, \varphi' = \frac{\pi}{2} + \alpha, L = 2w\right)\exp(-jk2w)\right] \\ +E(1)\left[\exp(-jkw\cos\alpha) + \frac{1}{\sqrt{w}}D\left(\varphi = \frac{3\pi}{2}, \varphi' = \frac{\pi}{2} + \alpha, L = w\right)\exp(-jkw)\right] \end{array}\right] \tag{2}$$

Therefore, if the previous process is generalized for n knife-edges—by considering the hybrid UTD-PO methodology presented in [5] for the analysis of multiple diffraction of plane waves—the total field that reaches the reference point in Figure 1 can be expressed, assuming $d >> nw$ and for $n \geq 1$ as:

$$\begin{array}{c} E(n) = \frac{1}{n}\sum\limits_{m=0}^{n-1} E(m)\{\exp[-jk(n-m)w\cos\alpha] \\ +\frac{1}{\sqrt{(n-m)w}}D\left(\varphi = \frac{3\pi}{2}, \varphi' = \frac{\pi}{2} + \alpha, L = (n-m)w\right)\exp[-jk(n-m)w]\} \end{array} \tag{3}$$

The main advantage of this formulation is that, due to its recursion, the calculations of each iteration are expressed in terms of single diffractions, thus avoiding the consideration of higher order terms in the diffraction coefficients (*slope* diffraction). In this way, a simpler solution is obtained from the mathematical point of view, and therefore becomes computationally more efficient, without this fact entailing a loss of precision.

3.2. Spherical Wave Incidence

In this case, for $n = 1$ and by applying UTD, the field that reaches the reference point in Figure 1 can be calculated as:

$$\begin{array}{l} E(1) = \frac{E_0}{R_1}\exp(-jkR_1) + \frac{E_0}{R_0}\exp(-jkR_0)\sqrt{\frac{R_0}{w(R_0+w)}}D_1\left(\varphi = \frac{3\pi}{2}, \varphi' = \frac{\pi}{2} + \alpha, L = \frac{R_0 w}{R_0+w}\right)\exp(-jkw) \\ = E(0)\left(\frac{R_0}{R_1}\exp(-jk(R_1 - R_0)) + \sqrt{\frac{R_0}{w(R_0+w)}}D_1\left(\varphi = \frac{3\pi}{2}, \varphi' = \frac{\pi}{2} + \alpha, L = \frac{R_0 w}{R_0+w}\right)\exp(-jkw)\right) \end{array} \tag{4}$$

where

$$E(0) = \frac{E_0}{R_0}\exp(-jkR_0) \tag{5}$$

with E_0 being the relative amplitude of the spherical source, which is assumed to be 1, k is the wave number, $D(\phi, \phi', L)$ is again the diffraction coefficient for an absorbent knife-edge proposed in [18], and R_0, R_1 are the distances that can be observed in Figure 1.

For $n = 2$, multiple diffraction emerges, and following the same methodology—based on virtual spherical sources—as presented in [6] (which, in turn, is based on the PO recursivity proposed in [4,5]), in the plane-wave case, the field at the reference point in Figure 1 can be calculated as the average of two contributions expressed in terms of UTD single diffractions:

$$E(2) = \frac{1}{2}\left[\begin{array}{l} E(0)\left(\frac{R_0}{R_2}\exp(-jk(R_2 - R_0)) + \sqrt{\frac{R_0}{2w(R_0+2w)}}D_1\left(\varphi = \frac{3\pi}{2}, \varphi' = \frac{\pi}{2} + \alpha, L = \frac{R_0 2w}{R_0+2w}\right)\exp(-jk2w)\right) \\ +E(1)\left(\frac{R_0}{R_1}\exp(-jk(R_2 - R_1)) + \sqrt{\frac{R_0}{w(R_0+w)}}D_2\left(\varphi = \frac{3\pi}{2}, \varphi' = \frac{\pi}{2} + \alpha, L = \frac{R_0 w}{R_0+w}\right)\exp(-jkw)\right) \end{array}\right] \tag{6}$$

Regarding the above, if we generalize the previous process for the case of n knife-edges, the total field existing at the reference point indicated in Figure 1, for $n \geq 1$, can be evaluated—by considering the UTD-PO recursive procedure presented in [6] for the analysis of losses due to multiple diffraction when the incident wavefront is spherical—as:

$$
\begin{aligned}
E(n) = \frac{1}{n}\sum_{m=0}^{n-1} E(m)\cdot\Big[& \frac{R_0}{R_{n-m}}\exp(-jk(R_n - R_m)) + \sqrt{\frac{R_0}{(n-m)w[R_0+(n-m)w]}} \\
& \cdot D\left(\varphi = \frac{3\pi}{2}, \varphi' = \frac{\pi}{2} + \alpha, L = \frac{R_0(n-m)w}{R_0+(n-m)w}\right)\cdot \exp(-jk(n-m)w)\Big]
\end{aligned} \tag{7}
$$

where

$$
R_x = \sqrt{H^2 + (d + x\cdot w)^2} \tag{8}
$$

being that, as can be observed in Figure 1, x represents the number of the considered incident ray and it can take values from 0 to $n-1$.

Again, we obtain a recursive solution, which is also being expressed in terms of UTD single diffractions and does not need higher order terms in the diffraction coefficients, thus proving to be computationally more efficient.

4. Results

4.1. Comparison between Plane and Sherical-Wave Incidences

In Figures 2 and 3, the variation of the total field at the reference point with respect to the free-space field (attenuation in dB), is presented for the two types of wave incidence as a function of the distance of the transmitting point Tx to the first obstacle (d). A frequency of f = 80 GHz, w = 0.5 m, and two values of n (5 and 50, respectively) have been considered, as well as three values of α (1.0°, 1.75°, and 2.5°, where H varies accordingly with d in the case of spherical wave incidence to maintain these angles). It should be noted that in the case of plane wave incidence, the attenuation is independent of d, since it has been assumed that $d >> n\cdot w$. Therefore, in that case, the results appear constant with d.

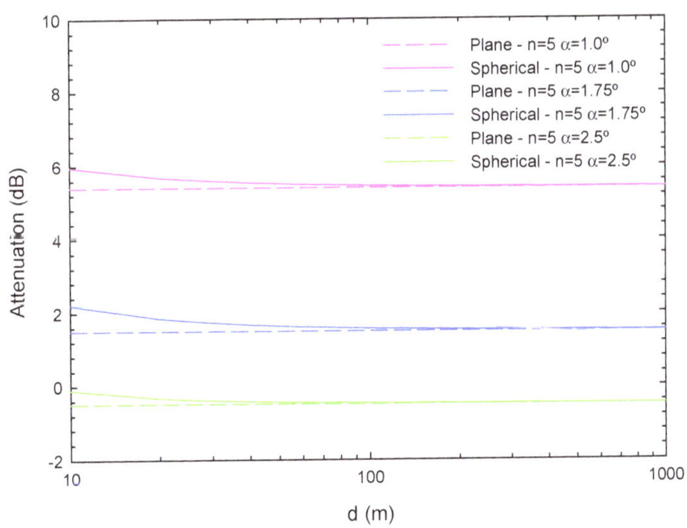

Figure 2. Variation of the total field at the reference point relative to the field in free space (attenuation), for the two types of wave incidence, as a function of the distance from the transmitting antenna to the first obstacle (d). A frequency of f = 80 GHz, w = 0.5 m, n = 5 and various values of α (1, 1.75, and 2.5°) have been considered.

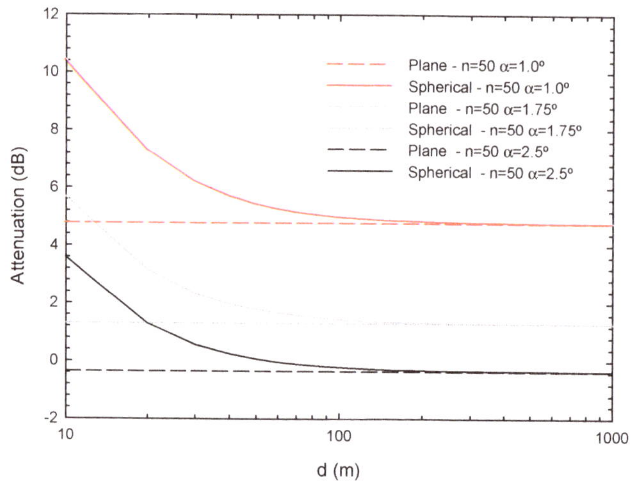

Figure 3. Variation of the total field at the reference point relative to the field in free space (attenuation), for the two types of wave incidence, as a function of the distance from the transmitting antenna to the first obstacle (d). A frequency of f = 80 GHz, w = 0.5 m, n = 50 and various values of α (1, 1.75, and 2.5°) have been considered.

It can be observed how the difference between the results of both solutions becomes more significant as d takes a lower value, and on the contrary, the two formulations converge for high values of d, as expected. Moreover, it is worth noting that the attenuation values for f = 60 GHz are the highest of the three frequencies considered—for both n = 5 and n = 50—due to the strong absorption of radiation that occurs in the atmosphere at that frequency. On the other hand, in Figure 4, the relative error (in %) between the results of the two types of wave incidence has been represented and understood as:

$$Relative\ error = \frac{Atten.Spherical\ Wave\ -\ Atten.Plane\ Wave}{Atten.Plane\ Wave} \quad (9)$$

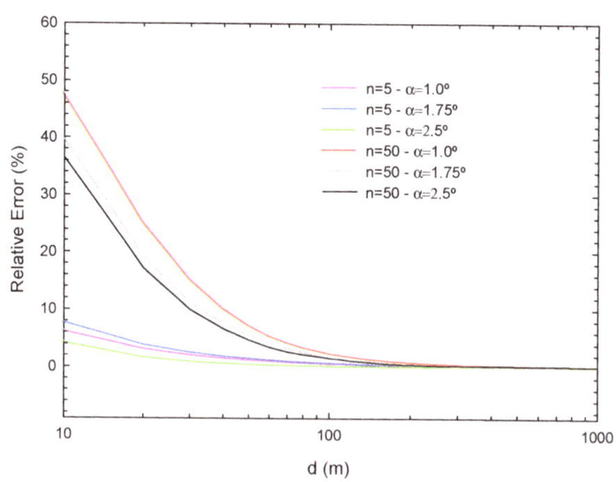

Figure 4. Relative error between the results of Figures 2 and 3 for both types of wave incidence.

From Figure 4, we can conclude that if values of d are considered increasingly, a point is reached where the relative error between both types of wave incidence becomes lower than a certain percentage (attenuation results for plane and spherical waves converge). Therefore, for distances shorter than these values of d, considering the parameters of the scenario under study, the spherical wave solution should be assumed in order to obtain more realistic predictions of multiple diffraction attenuation, since otherwise (using the plane wave formulation) errors of up to more than 5.6 dB (for $d = 10$ m, $n = 50$, and $\alpha = 1.0°$) could be reached.

In Figures 5–7, the same analysis as before is carried out (attenuation and relative error, respectively) but in this case, the parameters of $f = 80$ GHz and $\alpha = 1.5°$ are fixed, and w is varied taking values of 0.1, 0.5, and 1.0 m (also for $n = 5$ and $n = 50$).

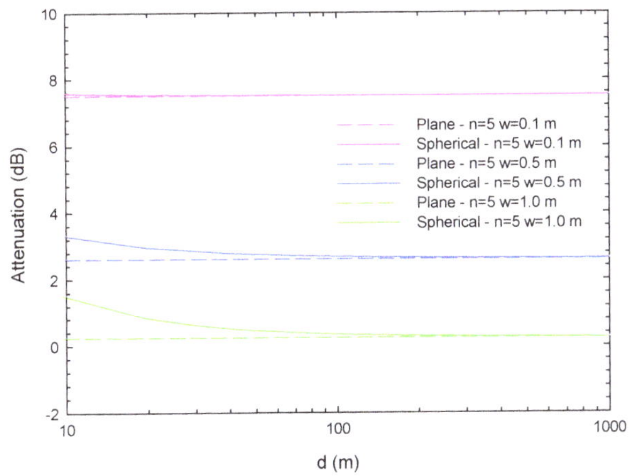

Figure 5. Variation of the total field at the reference point with respect to the free-space field (attenuation), for the two types of wave incidence, is shown as a function of the distance between the transmitting antenna and the first obstacle (d). The frequency considered is $f = 80$ GHz, the angle of incidence is $\alpha = 1.5°$. $n = 5$, and several values of w (0.1, 0.5 and 1.0 m) have been taken into account.

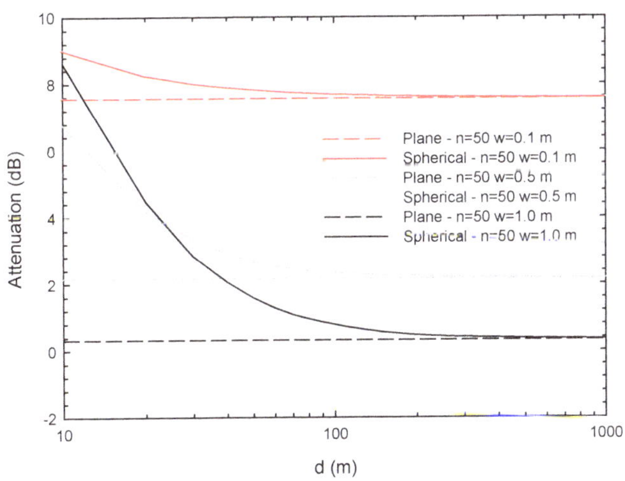

Figure 6. Variation of the total field at the reference point with respect to the free-space field (attenuation), for the two types of wave incidence, is shown as a function of the distance between the transmitting antenna and the first obstacle (d). The frequency considered is $f = 80$ GHz, the angle of incidence is $\alpha = 1.5°$. $n = 50$, and several values of w (0.1, 0.5 and 1.0 m) have been taken into account.

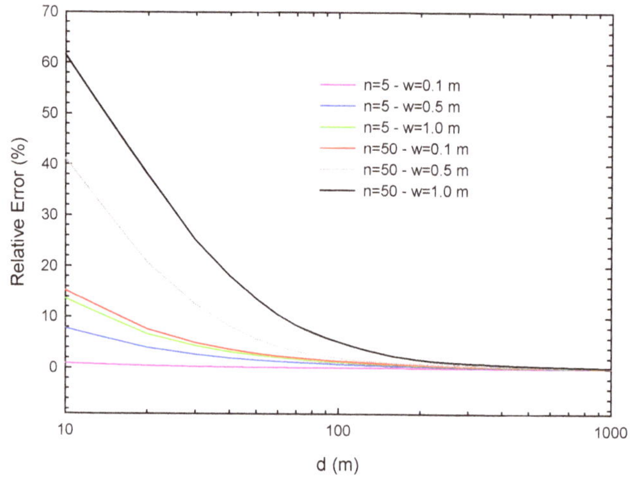

Figure 7. Relative error between the results of Figures 5 and 6 for the two types of wave incidence.

Again, the convergence of the spherical wave solution to the plane wave solution is observed, and values of d are considered larger, thus reducing the relative error between both types of incidence. On the other hand, the maximum difference between both solutions is obtained for $d = 10$, $n = 50$, and $w = 1.0$ m with 8.3 dB.

Finally, in Figures 8–10, the results of attenuation and relative error, assuming $\alpha = 1.5°$, $w = 0.5$ m, $n = 5$, and 50, are shown with three different millimeter-wave frequencies (60, 80, and 100 GHz), yielding identical considerations. In this case, the maximum difference between the two formulations is obtained for $d = 10$, $n = 50$, and f = 60 GHz with 5.0 dB.

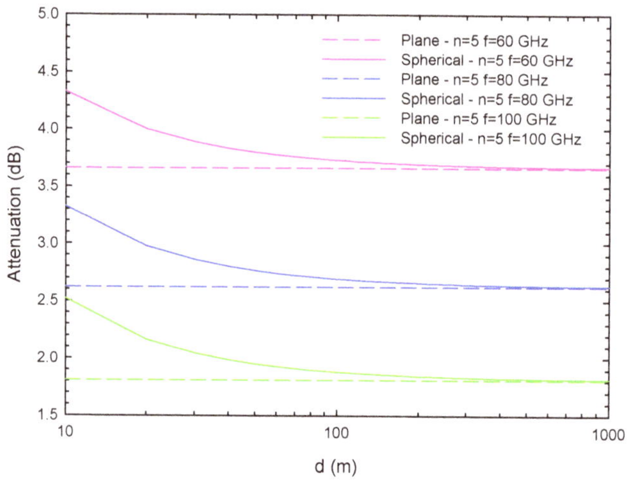

Figure 8. Variation of the total field at the reference point with respect to the free space field (attenuation) for the two types of wave incidence is shown as a function of the distance between the transmitting antenna and the first obstacle (d). The values of $n = 5$ and f (60, 80 and 100 GHz) have been considered, with $w = 0.5$ m and $\alpha = 1.5°$.

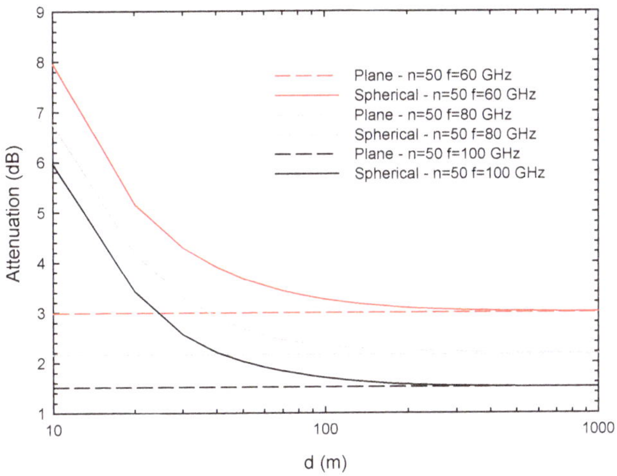

Figure 9. Variation of the total field at the reference point with respect to the free space field (attenuation), for the two types of wave incidence, is shown as a function of the distance between the transmitting antenna and the first obstacle (d). The values of $n = 50$ and f (60, 80 and 100 GHz) have been considered, with $w = 0.5$ m and $\alpha = 1.5°$.

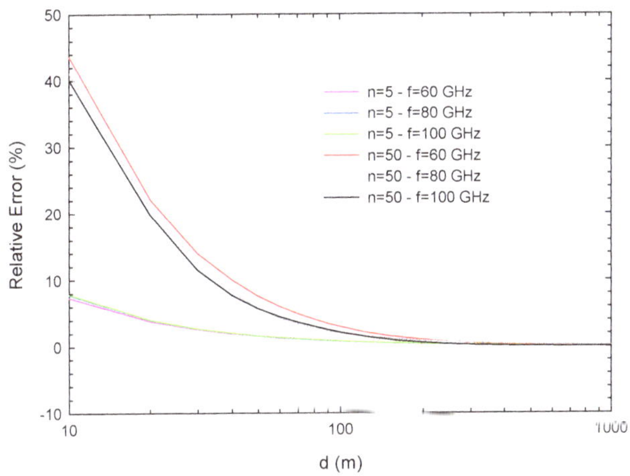

Figure 10. Relative error between the results of Figures 8 and 9 for the two types of wave incidence.

4.2. *Minimum Values of d for the Two Types of Incidence to Converge*

In order to delve deeper into the previous analysis, a study is presented below on the minimum values of d required for the two solutions (plane wave and spherical wave) to converge with a relative error less than 0.1%. These results are shown as a function of the incidence angle α (Figure 11, for f = 80 GHz and w = 0.5 m) and the spacing between obstacles w (Figure 12, for f = 80 GHz and α = 1.5°). In this sense, considering the context of wireless communication systems operating at millimeter-wave frequencies, the aforementioned study has been conducted for a small number of obstacles (n = 1, 2, 3, and 4), since it is expected that in the propagation environments where these systems operate, the signal passes through few diffracting elements on its path from the transmitter to the receiver (due to the small cell size or the consideration of indoor contexts [15]).

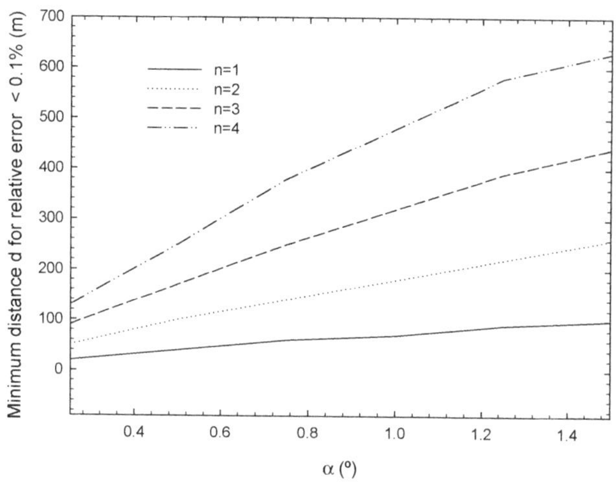

Figure 11. Minimum distance d for which the relative error between the two types of wave incidence (plane and spherical wave) is less than 0.1%, as a function of the incidence angle α and for $n = 1, 2, 3$, and 4 ($f = 80$ GHz and $w = 0.5$ m).

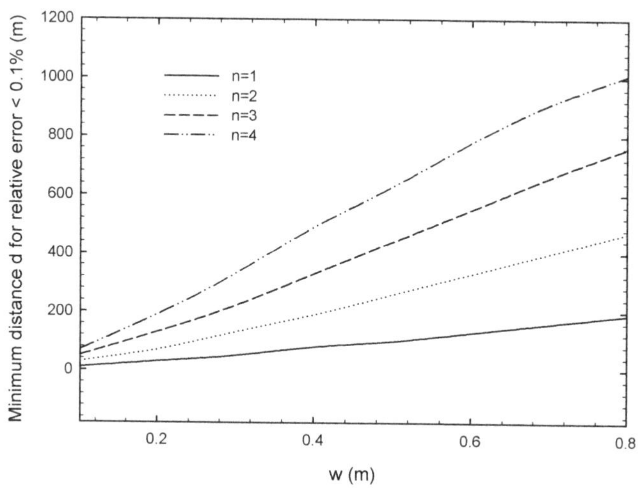

Figure 12. Minimum distance d at which the relative error between the two types of wave incidence (plane and spherical wave) is less than 0.1%, as a function of the spacing between obstacles w, for $n = 1, 2, 3$ and 4 ($f = 80$ GHz and $\alpha = 1.5°$).

As can be observed, a quasi-linear behavior of the curves is apparent in both figures, with minimum values of d for relative errors < 0.1% ranging from 20 to 630 m in the case of Figure 11 (for [$n = 1$, $\alpha = 0.25°$] and [$n = 4$, $\alpha = 1.5°$], respectively), and between 10 and 1010 m in Figure 12 (for [$n = 1$, $w = 0.1$ m] and [$n = 4$, $w = 0.8$ m], respectively).

On the other hand, in order to perform an analysis of the minimum d for relative error < 0.1% as a function of frequency, Figures 13 and 14 are presented, where several incidence angles α (0.5, 1.0, and 1.5°, with $w = 0.5$) and several spacing values w (0.3, 0.5, and 0.7 m, with $\alpha = 1.5°$) have been considered, respectively.

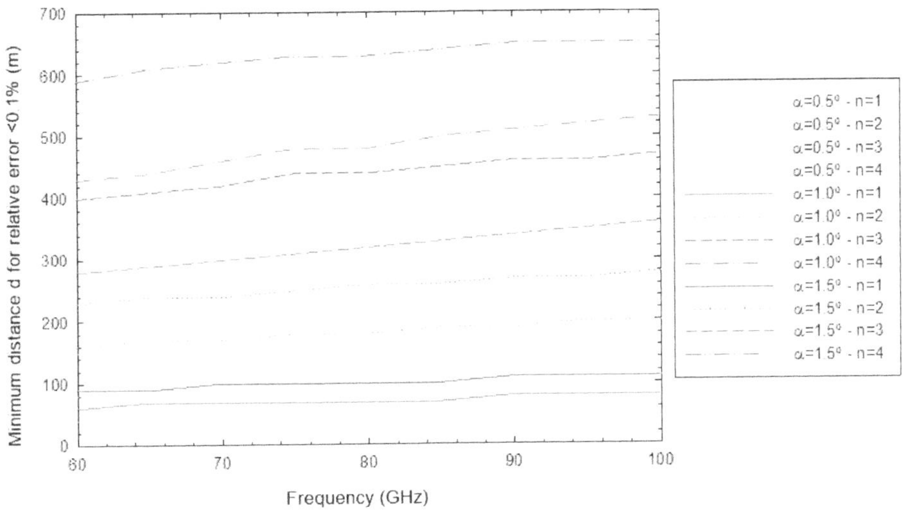

Figure 13. Minimum distance d for which the relative error between the two types of wave incidence (plane and spherical) is less than 0.1%, as a function of frequency for several values of α (0.5, 1.0, and 1.5°) and n = 1, 2, 3, and 4 (w = 0.5 m).

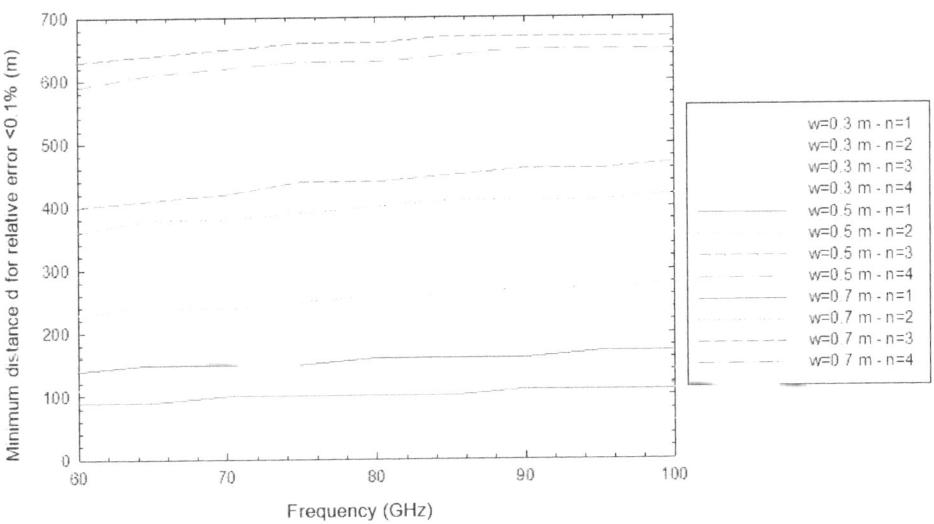

Figure 14. Minimum distance d for which the relative error between the two types of wave incidence (plane and spherical wave) is less than 0.1%, is studied as a function of frequency for several values of w (0.3, 0.5 and 0.7 m) and n = 1, 2, 3 and 4, considering an angle of incidence of α = 1.5°.

These two figures show a very interesting fact, since, as can be observed in both graphs, the curves practically show a constant behavior, so it can be inferred that, for a small number of obstacles (up to n = 4) and the considered parameters, the minimum distance d for a relative error between the two types of incidence is less than 0.1%, and can be assumed to be independent of frequency (in a range of 60 to 100 GHz).

Therefore, Figures 15 and 16 represent the minimum d for relative error <0.1% considering different values of α (0.5, 1.0, and 1.5°, with w = 0.5 m) and w (0.3, 0.5, and 0.7 m, with

α = 1.5°), respectively, as a function of the number of obstacles n (from 1 to 4). Notably considering the previous result, both figures can be perfectly valid for any millimeter-wave frequency between 60 and 100 GHz (for each given n and α/w, the average of the minimum d for relative error < 0.1% throughout the evaluated frequency range has been considered since the variation with this last parameter has been found to be minimal). As can be observed, the behavior of the curves in both figures is quasi-linear.

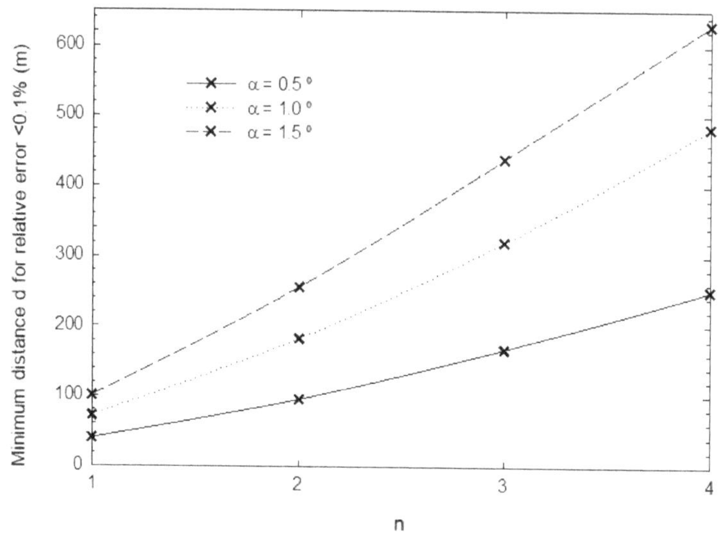

Figure 15. Minimum distance d for which the relative error between the two types of wave incidence (plane and spherical wave) is less than 0.1%, as a function of n for various values of α (0.5, 1 and 1.5°) is considered (w = 0.5 m).

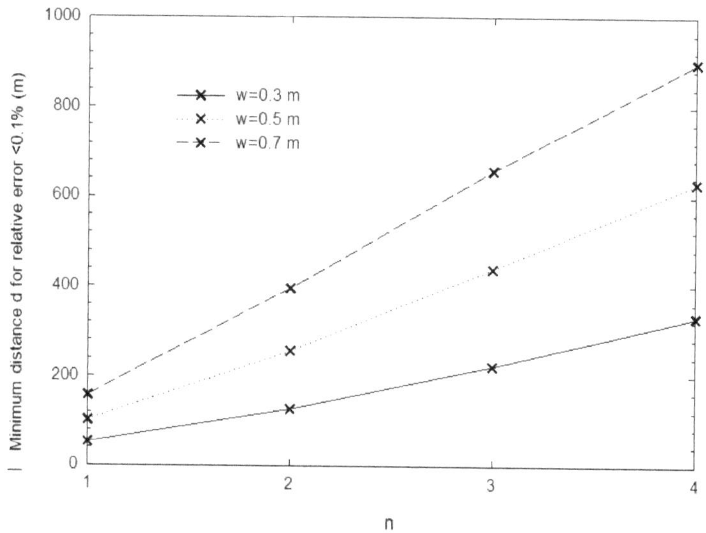

Figure 16. Minimum distance d for which the relative error between the two types of wave incidence (plane and spherical wave) is less than 0.1%, as a function of n for various values of w (0.3, 0.5 and 0.7 m) with α = 1.5°.

5. Conclusions

This work has determined at which threshold will the results for both plane and spherical wave incidence assumptions converge or deviate when performing multiple diffraction attenuation calculations. The analysis has been carried out—for various millimeter-wave frequencies, inter-obstacle spacings, and angles of incidence—by employing a pair of two-dimensional (2D) hybrid formulations based on both the uniform theory of diffraction and physical optics (UTD-PO). This way, we have calculated the minimum necessary distance from the transmitter to the first diffracting obstacle for the convergence of the spherical wave incidence solution onto that of the plane wave with a relative error below 0.1%. Our results have demonstrated that using a plane wave solution in environments where the distance between the transmitter and the first obstacle is not large enough for the spherical wave results to converge to those of the plane wave, can lead to errors in the estimation of multiple diffraction losses of more than 8 dB. Moreover, it has been demonstrated that for a few (less than four) diffracting elements, the minimum necessary distance engages in quasi-linear behavior, under variations in both the angle of incidence and obstacle spacing. Furthermore, it is interesting to note that these distance values have turned out to be practically independent of the considered millimeter-wave frequencies (60–100 GHz). The results presented in this study may contribute to a more precise and realistic planning of wireless communication systems employing millimeter wave frequencies and needing to account for the losses caused by multiple diffraction due to a series of obstacles.

Author Contributions: Conceptualization, A.L.-S., I.R.-R. and J.-V.R.; methodology, A.L.-S., I.R.-R. and J.-V.R.; software, A.L.-S., I.R.-R. and J.-V.R.; validation, A.L.-S., I.R.-R., J.-V.R., L.J.-L., M.C.-V. and W.L.W.; formal analysis, A.L.-S., I.R.-R. and J.-V.R.; investigation, A.L.-S., I.R.-R. and J.-V.R.; resources, A.L.-S., I.R.-R., J.-V.R., L.J.-L., M.C.-V. and W.L.W.; data curation, A.L.-S., I.R.-R., J.-V.R., L.J.-L., M.C.-V. and W.L.W.; writing—original draft preparation, A.L.-S., I.R.-R. and J.-V.R.; writing—review and editing, A.L.-S., I.R.-R., J.-V.R., L.J.-L., M.C.-V. and W.L.W.; visualization, A.L.-S., I.R.-R., J.-V.R., L.J.-L., M.C.-V. and W.L.W.; supervision, A.L.-S., I.R.-R., J.-V.R., L.J.-L., M.C.-V. and W.L.W.; project administration, I.R.-R., J.-V.R. and L.J.-L.; funding acquisition, I.R.-R., J.-V.R. and L.J.-L. All authors have read and agreed to the published version of the manuscript.

Funding: This work was supported by the Ministerio de Ciencia e Innovación, Spain, under Grant PID2019-107885GB-C33.

Data Availability Statement: Not applicable.

Acknowledgments: Ignacio Rodríguez-Rodríguez would like to thank Plan Andaluz de Investigación, Desarrollo e Innovación (PAIDI), Junta de Andalucía, Spain. María Campo-Valera is grateful for postdoctoral program Margarita Salas—Spanish Ministry of Universities (financed by European Union—NextGenerationEU).

Conflicts of Interest: The authors declare no conflict of interest.

References

1. Zhang, W.; Lähteenmäki, J.; Vainikainen, P. A practical aspect of over-rooftop multiple-building forward diffraction from a low source. *IEEE Trans. Electromagn. Compat.* **1999**, *41*, 115–119. [CrossRef]
2. Neve, M.J.; Rowe, G.B. Contributions towards the development of a UTD-based model for cellular radio propagation prediction. *IEE Proc. Microw. Antennas Propag.* **1994**, *141*, 407–414. [CrossRef]
3. Walfisch, J.; Bertoni, H. A theoretical model of UHF propagation in urban environments. *IEEE Trans. Antennas Propag.* **1988**, *36*, 1788–1796. [CrossRef]
4. Saunders, S.; Bonar, F. Prediction of mobile radio wave propagation over buildings of irregular heights and spacings. *IEEE Trans. Antennas Propag.* **1994**, *42*, 137–144. [CrossRef]
5. Juan-Llácer, L.; Cardona, N. UTD solution for the multiple building diffraction attenuation function for mobile radiowave propagation. *Electron. Lett.* **1997**, *33*, 92–93. [CrossRef]
6. Rodríguez, J.-V.; Molina-García-Pardo, J.-M.; Juan-Llácer, L. A new solution expressed in terms of UTD coefficients for the multiple diffraction of spherical waves by a series of buildings. *Radio Sci.* **2007**, *42*, 1–15. [CrossRef]
7. Jay, Y.; Liu, D.; Bird, N.C. Guest Editorial for the Special Issue on Antennas and Propagation Aspects of 60–90 GHz Wireless Communications. *IEEE Trans. Antennas Propag.* **2009**, *57*, 2817–2819.

8. Smulders, P. Exploiting the 60 GHz band for local wireless multimedia access: Prospects and future directions. *IEEE Commun. Mag.* **2002**, *40*, 140–147. [CrossRef]
9. Smulders, P.; Haibing, Y.; Akkermans, I. On the Design of Low-Cost 60-GHz Radios for Multigigabit-per-Second Trans-mission over Short Distances [Topics in Radio Communications]. *IEEE Commun. Mag.* **2007**, *45*, 44–51. [CrossRef]
10. Shay, W.T.; Shen, K.J.; Chen, Y.R.; Lin, Y.Y.; Tsai, Z.M.; Tarng, J.H. Millimeter-wave channel modelling and ray-tracing method validation. In Proceedings of the 2019 Photonics & Electromagnetics Research Symposium-Fall (PIERS-Fall), Xiamen, China, 17–20 December 2019; pp. 2612–2620.
11. Zhang, P.; Wang, H.; Hong, W. Radio propagation measurement and cluster-based analysis for millimeter-wave cellular systems in dense urban environments. *Front. Inf. Technol. Electron. Eng.* **2021**, *22*, 471–487. [CrossRef]
12. Wang, H.; Zhang, P.; Li, J.; You, X. Radio propagation and wireless coverage of LSAA-based 5G millimeter-wave mobile communication systems. *China Commun.* **2019**, *16*, 1–18. [CrossRef]
13. Al-Samman, A.M.; Azmi, M.H.; Al-Gumaei, Y.A.; Al-Hadhrami, T.; Rahman, T.A.; Fazea, Y.; Al-Mqdashi, A. Millimeter Wave Propagation Measurements and Characteristics for 5G System. *Appl. Sci.* **2020**, *10*, 335. [CrossRef]
14. Uwaechia, A.N.; Mahyuddin, N.M. A Comprehensive Survey on Millimeter Wave Communications for Fifth-Generation Wireless Networks: Feasibility and Challenges. *IEEE Access* **2020**, *8*, 62367–62414. [CrossRef]
15. Jacob, M.; Priebe, S.; Dickhoff, R.; Kleine-Ostmann, T.; Schrader, T.; Kurner, T. Diffraction in mm and Sub-mm Wave Indoor Propagation Channels. *IEEE Trans. Microw. Theory Tech.* **2012**, *60*, 833–844. [CrossRef]
16. Kyosti, P.; Carton, I.; Karstensen, A.; Fan, W.; Pedersen, G.F. Frequency dependency of channel parameters in urban LOS scenario for mmwave communications. In Proceedings of the 2016 10th European Conference on Antennas and Propagation (EuCAP), Davos, Switzerland, 10–15 April 2016; pp. 1–5. [CrossRef]
17. Yamada, W.; Sasaki, M.; Kita, N. Extended Walfisch-Bertoni propagation model to cover short range and millime-ter-wave bands. *Radio Sci.* **2021**, *56*, 1–8. [CrossRef]
18. Kouyoumjian, R.; Pathak, P. A uniform geometrical theory of diffraction for an edge in a perfectly conducting surface. *Proc. IEEE* **1974**, *62*, 1448–1461. [CrossRef]

Disclaimer/Publisher's Note: The statements, opinions and data contained in all publications are solely those of the individual author(s) and contributor(s) and not of MDPI and/or the editor(s). MDPI and/or the editor(s) disclaim responsibility for any injury to people or property resulting from any ideas, methods, instructions or products referred to in the content.

MDPI
St. Alban-Anlage 66
4052 Basel
Switzerland
www.mdpi.com

Electronics Editorial Office
E-mail: electronics@mdpi.com
www.mdpi.com/journal/electronics

Disclaimer/Publisher's Note: The statements, opinions and data contained in all publications are solely those of the individual author(s) and contributor(s) and not of MDPI and/or the editor(s). MDPI and/or the editor(s) disclaim responsibility for any injury to people or property resulting from any ideas, methods, instructions or products referred to in the content.

www.ingramcontent.com/pod-product-compliance
Lightning Source LLC
Chambersburg PA
CBHW061552210926
51850CB00009B/2077